T0275720

# The Social Dog

# The Social Dog
## Behaviour and Cognition

Edited by

## Juliane Kaminski
Psychology Department, University of Portsmouth, Portsmouth, UK

## Sarah Marshall-Pescini
Dipartimento di Fisiopatologia Medico-Chirurgica e dei Trapianti, Sezione di Neuroscienze, Università degli Studi di Milano, Milan, Italy

Comparative Cognition, Messerli Research Institute, University of Veterinary Medicine, University of Vienna, Vienna, Austria

Wolf Science Centre, Ernstbrunn, Austria

AMSTERDAM • BOSTON • HEIDELBERG • LONDON
NEW YORK • OXFORD • PARIS • SAN DIEGO
SAN FRANCISCO • SINGAPORE • SYDNEY • TOKYO

Academic Press is an imprint of Elsevier

Academic Press is an imprint of Elsevier
525 B Street, Suite 1800, San Diego, CA 92101-4495, USA
32 Jamestown Road, London NW1 7BY, UK
225 Wyman Street, Waltham, MA 02451, USA

**Notice**
No responsibility is assumed by the publisher for any injury and/or damage to
persons or property as a matter of products liability, negligence or otherwise,
or from any use or operation of any methods, products, instructions or ideas
contained in the material herein. Because of rapid advances in the medical
sciences, in particular, independent verification of diagnoses and drug dosages
should be made

**British Library Cataloguing in Publication Data**
A catalogue record for this book is available from the British Library

**Library of Congress Cataloging-in-Publication Data**
A catalog record for this book is available from the Library of Congress

ISBN: 978-0-12-407818-5

For information on all Academic Press publications
visit our web site at store.elsevier.com

Printed and bound in the US

14 15 16 17   10 9 8 7 6 5 4 3 2 1

# Contents

## Section II
## Social Behaviour
*Dog–Dog*

### 3. The Social Organisation of a Population of Free-Ranging Dogs in a Suburban Area of Rome: A Reassessment of the Effects of Domestication on Dogs' Behaviour

*Roberto Bonanni and Simona Cafazzo*

### 4. Social Behaviour among Companion Dogs with an Emphasis on Play

*Barbara Smuts*

## 5. Auditory Communication in Domestic Dogs: Vocal Signalling in the Extended Social Environment of a Companion Animal

*Anna Magdalena Taylor, Victoria Frances Ratcliffe, Karen McComb, and David Reby*

### Dog–Human

## 6. The Immaterial Cord: The Dog–Human Attachment Bond

*Emanuela Prato Previde and Paola Valsecchi*

## 12. Do Dogs Show an Optimistic or Pessimistic Attitude to Life?
*Oliver Burman*

## 13. Wagging to the Right or to the Left: Lateralisation and What It Tells of the Dog's Social Brain
*Marcello Siniscalchi and Angelo Quaranta*

The domestic dog seems to be the new star in behavioural research. Forgotten for some time, it has been rising again in the past two decades. Research on dog behaviour and cognition has exploded, with so many papers being published that it is hard to keep track. There are probably many reasons why research with dogs has increased so substantially, one being that several researchers have raised the hypothesis that dogs might have evolved specialized social skills as a result of living with humans and adapting to the human environment. This hypothesis has sparked the interest of both the scientific community and a wider audience, reflected in many of the contributions to this volume.

The main purpose of this book is to summarize the growing body of research on dogs' social behaviour and cognition, as well as to highlight some of the controversial issues and open questions still needing answers. The intention is to make this exciting new field more accessible to scholars from both similar and different fields, to students, and to dog specialists whose practical work with dogs is informed by the research highlighted throughout the book. Our emphasis, reflected in the organization of this volume, has been to draw the reader's attention both to studies focusing on dog-to-dog social interactions as well as those directed at a deeper understanding of the dog-to-human relationship, since we think both are necessary to gain a more complete understanding of dogs.

This project was endorsed enthusiastically by both the publishers and the many scientists who have contributed to the book. Indeed, we combine 13 original contributions, from 27 authors and co-authors, from many disciplines including biology, veterinary science, and psychology. The diversity of the authors' scientific backgrounds allows the reader to gain 'a feel' for how dogs are being studied, from many different perspectives. Each contribution is a review of a particular topic pertaining to dogs' social behaviour and/or cognition and is written to stand alone. However, each chapter is also a 'slice of the cake', and as a whole the result is an interdisciplinary book that provides an overview of progress in understanding many aspects of dogs' social behaviour and cognition.

**Juliane Kaminski**
**Sarah Marshall-Pescini**

# Contributors

**Roberto Bonanni**  Department of Neuroscience, University of Parma, Italy

**Juliane Bräuer**  Max Planck Institute for Evolutionary Anthropology, Leipzig, Germany

**Oliver Burman**  University of Lincoln, Riseholme Park, Lincoln, UK

**Simona Cafazzo**  Wolf Science Center, Ernstbrunn, Austria

**Juliane Kaminski**  Psychology Department, University of Portsmouth, Portsmouth, UK

**Anna Kis**  Institute of Cognitive Neuroscience and Psychology, Research Centre for Natural Sciences, Hungarian Academy of Sciences, Budapest, Hungary; Department of Ethology, Eötvös University, Budapest, Hungary

**Enikő Kubinyi**  Department of Ethology, Eötvös University, Budapest, Hungary

**Sarah Marshall-Pescini**  Comparative Cognition, Messerli Research Institute, University of Veterinary Medicine, University of Vienna, Vienna, Austria; Wolf Science Centre, Ernstbrunn, Austria; Dipartimento di Fisiopatologia Medico-Chirurgica e dei Trapianti, Sezione di Neuroscienze, Università degli Studi di Milano

**Karen McComb**  School of Psychology, University of Sussex, Falmer, UK

**Ádám Miklósi**  Department of Ethology, Eötvös University, Budapest, Hungary

**Daniel Mills**  Animal Behaviour Cognition and Welfare Group, School of Life Sciences, University of Lincoln, Riseholme Park, Lincoln, UK

**Katalin Oláh**  Institute of Cognitive Neuroscience and Psychology, Research Centre for Natural Sciences, Hungarian Academy of Sciences, Budapest, Hungary; Department of Cognitive Psychology, Eötvös University, Budapest, Hungary

**Péter Pongrácz**  Department of Ethology, Biological Institute, Eötvös Loránd University, Budapest, Hungary

**Emanuela Prato Previde**  Dipartimento di Fisiopatologia Medico-Chirurgica e dei Trapianti, Sezione di Neuroscienze, Università degli Studi di Milano, Segrate, Italy

**Angelo Quaranta**  Department of Veterinary Medicine, Section of Behavioural Sciences and Animal Bioethics, University of Bari 'Aldo Moro', Italy

**Friederike Range**  Comparative Cognition, Messerli Research Institute, University of Veterinary Medicine, Vienna, Austria; Medical University of Vienna, Vienna, Austria; University of Vienna, Vienna, Austria; Wolf Science Centre, Ernstbrunn, Austria

**Victoria Frances Ratcliffe**  School of Psychology, University of Sussex, Falmer, UK

**David Reby**  School of Psychology, University of Sussex, Falmer, UK

**Marcello Siniscalchi**  Department of Veterinary Medicine, Section of Behavioural Sciences and Animal Bioethics, University of Bari 'Aldo Moro', Italy

**Barbara Smuts**  Department of Psychology, University of Michigan, Ann Arbor, MI, USA

**Anna Magdalena Taylor**  School of Psychology, University of Sussex, Falmer, UK

**József Topál**  Institute of Cognitive Neuroscience and Psychology, Research Centre for Natural Sciences, Hungarian Academy of Sciences, Budapest, Hungary

**Borbála Turcsán**  Department of Ethology, Eötvös University, Budapest, Hungary

**Paola Valsecchi**  Dipartimento di Neuroscienze, Università degli Studi di Parma, Parma, Italy

**Emile van der Zee**  School of Psychology, University of Lincoln, Brayford Pool, Lincoln, UK

**Zsófia Virányi**  Comparative Cognition, Messerli Research Institute, University of Veterinary Medicine, Vienna, Austria; Medical University of Vienna, Vienna, Austria; University of Vienna, Vienna, Austria; Wolf Science Centre, Ernstbrunn, Austria

**Helen Zulch**  Animal Behaviour Cognition and Welfare Group, School of Life Sciences, University of Lincoln, Riseholme Park, Lincoln, UK

# Theoretical Aspects

# Chapter 1

# The Social Dog: History and Evolution

Sarah Marshall-Pescini[1,2,3] and Juliane Kaminski[4]

[1]*Dipartimento di Fisiopatologia Medico-Chirurgica e dei Trapianti, Sezione di Neuroscienze, Università degli Studi di Milano, Milan, Italy,* [2]*Comparative Cognition, Messerli Research Institute, University of Veterinary Medicine, University of Vienna, Vienna, Austria,* [3]*Wolf Science Centre, Ernstbrunn, Austria,* [4]*Psychology Department, University of Portsmouth, Portsmouth, UK*

## 1.1 WHERE DO DOGS' SOCIALITY AND SOCIO-COGNITIVE ABILITIES COME FROM? THE CANID STORY

The explosion of studies on dogs' social behaviour and cognitive abilities since the turn of the twenty-first century has been impressive (see Bensky et al., 2013, for a comprehensive review), and the many hypotheses as to the causes behind dog's remarkable socio-cognitive abilities have engendered lively debates in journals and at conferences. However, most debates revolve around the wolf–dog comparison (the wolf being dog's closest living relative), neglecting the fact that the dog's canine family is much larger and shows some unique and intriguing features that may well have played a role in allowing dogs' emergence as our favoured social companions. Hence, in the first part of this chapter, we introduce dogs' canine family, presenting some of these intriguing social features and highlighting some of the characteristics that may have played a fundamental role in allowing the emergence of one species' unique history with humans.

### 1.1.1 Introducing Dogs' 'Canine' Family

The domestic dog belongs to the Canidae family, consisting of 35 related species that diverged within the last ten million years (Wayne et al., 1997; Ostrander & Wayne, 2005). In recent years, there has been considerable interest in the evolutionary relationships between canids that has resulted in analyses based on both morphological (Berta, 1987; Tedford et al., 1995; Lyras & Van Der Geer, 2003; Zrzavý & Řičánková, 2004) and molecular data (Wayne et al., 1987; Wayne et al., 1989), including, more recently, DNA sequencing (Wayne et al., 1997; Bardeleben et al., 2005; Linblad-Toh et al., 2005; Wong et al., 2010). The development of methodologies for the sequencing of DNA has allowed researchers to

The Social Dog. http://dx.doi.org/10.1016/B978-0-12-407818-5.00001-2

reconstruct the dog's family tree, with a certain amount of accuracy (although a few grey areas still exist).

Taken together, current results converge in showing three major groupings within the dog's family: (1) the red fox–like canids, (2) South American canids, and (3) wolf-like canids. Together, these three clades contain 93% of all living canids. A separate lineage comprising the grey fox seems to be the most primitive and suggests a North American origin of the living canids about ten million years ago (Ostrander & Wayne, 2005; Bardeleben et al., 2005; Lindblad-Toh et al., 2005; Graphodatsky et al., 2008) (see Figure 1-1).

When one looks more closely at the wolf-like canids, results place grey wolves as the closest living 'cousins' of domestic dogs, followed by a close affiliation with coyotes, golden jackals, and Ethiopian wolves. These phylogenetic relationships imply that the dog has several close relatives within its genus, confirmed by results showing that all members of *Canis* can produce fertile hybrids, and several species may have genomes that reflect hybridisation in the wild (Wayne & Jenks, 1991; Gottelli et al., 1994; Roy et al., 1996; Adams et al., 2003). Closest to the *Canis* group are the dhole and African wild dog (thus completing the members of the wolf-like canids). Dhole and African wild dogs do not, however, form a monophyletic group, and their exact relationship to the *Canis* genus is still somewhat unclear (Bardeleben et al., 2005; Zhang & Chen, 2011). Finally, results from genetic analyses also appear to support an African origin for the wolf-like canids because the two African jackals are the most basal members of this clade (Lindblad-Toh et al., 2005).

## 1.1.2 Evolution of the Canid Brain: A Socially Driven Phenomenon?

Studies on the evolution of canids show that this family separated from the other mammals around 40 million years ago. Interestingly, a shift in canid encephalisation and architectural reorganisation of the brain (i.e., expansion of the prorean gyrus at the anterior end of the neocortex, general increase in the amount of infolding of the frontal lobe, and expansion of the prefrontal cortex; Radinsky, 1969, 1973; Lyras & Van der Geer, 2003) appears to have occurred sometime in the late Miocene or early Pliocene period, roughly coinciding also with a sudden taxonomic diversification (Van Valkenburgh, 1991) and expansion of global grasslands (Cerling et al., 1997). Based on these data, authors have put forward a number of suggestions as to the possible causes driving these changes in the brain.

According to some authors, they may simply have been a by-product of a rapid taxonomic diversification in the new environment (Andersson, 2005); however, considering the energetic expenditure of big brains, it would seem more probable that such an expensive adaptation would be driven by some major adaptive advantage. Work by Van Valkenburgh and colleagues puts forward the possibility that, in fact, the onset of cooperative pack hunting (Van Valkenburgh

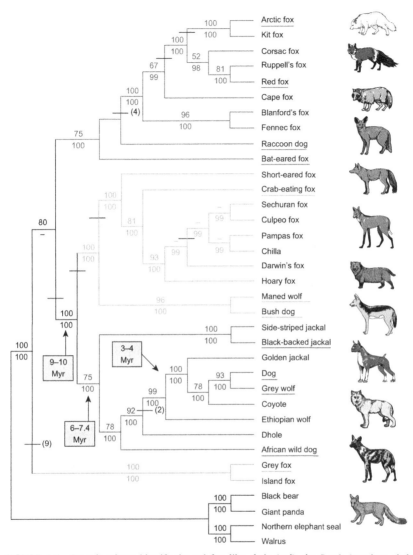

**FIGURE 1-1**   Branch colours identify the red fox–like clade (red), the South American clade (green), the wolf-like clade (blue), and the grey and island fox clade (orange). *(From Lindblad-Toh et al., 2005.)*

et al., 2003; Van Valkenburgh et al., 2004) may have driven this change. However, Finarelli's (2008) analysis, taking into account a larger sample of both extinct and living canids, suggests that encephalisation increased in the three major living clades (wolf-like, fox-like, and South American canids) at the same time, yet most of the smaller-bodied canids (except for the South American bush dog) are not cooperative hunters. The trait that most canids share, however, is 'monogamy' (defined as a single male and single female mating exclusively with each other over multiple reproductive cycles) and (to differing extents) the cooperative rearing of the young. Hence, these authors suggest that it may have been these traits driving encephalisation and that pack hunting would then have emerged as a second-order adaptation, facilitated by, but not directly causally related to, the reorganisation of the canid brain. This hypothesis seems to be plausible given another study showing that, among mammals and birds in general, it is not the size of the social group that correlates with brain size (as in primates), but instead it is species that live in pair-bonded social systems that have the largest brains (when phylogeny and a range of life history and ecological variables are partialled out; Shultz & Dunbar, 2007; Dunbar, 2009).

### 1.1.3 The Sociality of Canids' Mating System: Pair Bonding and Cooperative Pup Rearing

Taken as a whole, the dog's canine family is one of the most fascinating amongst carnivores. As a start, they are the most widespread, with at least one species inhabiting every continent except Antarctica (although dogs, of course, live there as well), and some spread over entire continents (Sillero-Zubiri et al., 2004). Perhaps more interestingly for the purpose of this book, canids have a number of interesting social adaptations, starting from their mating system.

Most species in this family are monogamous, a rare trait amongst mammals, but a potentially characteristic trait for canids (Kleiman, 1977, 2011). However, there is considerable interspecific variation in mating systems among canids, and polygamy, polyandry, and monogamy have been documented (Bekoff et al., 1981; Moehlman, 1989; Geffen et al., 1996; Carmicheal et al., 2007): for example, red foxes, Arctic foxes, and coyotes appear to adopt a largely monogamous strategy (Sillero-Zubiri et al., 2004; Hennessy et al., 2012; but see Carmicheal et al., 2007), whereas a recent study using genetic analyses to determine parenthood across multiple populations of African wild dogs found that there was, in fact, a greater than previously reported incidence of reproductive sharing in that both beta females and males (and not just the dominant pair as previously thought) played a significant role in producing young (Spiering et al., 2010), suggesting a more promiscuous strategy in this species.

Canids are also unique in that *intra*specific variation in mating systems may be as great as *inter*specific variation (Moehlman, 1989); for example, swift foxes, Arctic foxes, and urban red foxes may all live either in monogamous pairs or polygamous groups depending on their ecological context (Baker et al., 2004; Kamler et al., 2004; Carmicheal et al., 2007; and see Geffen et al., 1996 for a comprehensive review of canid social flexibility). Indeed, it appears that the choice between a monogamous or polygamous strategy may even change within the same population, most probably in relation to food availability (red foxes: von Shantz, 1984; Zabel & Taggart, 1989). Such flexibility is not confined to the fox-like canids; indeed, Ethiopian wolves show a similar level of flexibility depending on ecological factors affecting food availability (Sillero-Zubiri et al., 1996; Marino et al., 2012).

Alongside pair bonding, canids also show a variety of parental care strategies. In some species, mostly the female appears to care for the young—the most common strategy in mammals (e.g., bat-eared foxes and swift foxes: Kamler et al., 2004; Poessel & Gese, 2013); however, in others, male provisioning of both pups and lactating females occurs (e.g., black-backed jackals, red foxes, coyotes: Moehlman, 1989; Zabel & Taggart, 1989; Gese, 1998); and in yet others, both breeding individuals and their older offspring (that delay dispersal: Emlen, 1991) are involved in pup rearing (e.g., grey wolves, African wild dogs, dhole: Fritts & Mech, 1981; Malcolm & Marten, 1982; Creel & Creel, 2002; Sillero-Zubiri et al., 2004; Venkataraman & Johnsingh, 2004). In fact, cooperative breeding, involving also the non-breeding pack members and including 'helping behaviour' such as den-site attendance; provisioning (including regurgitation); play; and grooming are perhaps among the defining characteristics of a number of canid species (Macdonald, 1979; Moehlman, 1986; Mech et al., 1999; Packard, 2003).

Such helping behaviour has been shown to have an adaptive value in a number of cooperative breeding species (banded mongoose: Hodge, 2005; meerkats: Russell et al., 2007), although evidence in the canid family has been harder to obtain (Gusset & MacDonald, 2010; but see Wright, 2006, for paternal investment on pup rearing success in bat-eared foxes and helper effect in red wolves; Sparkman et al., 2011). However, sociality in many canid species is not just apparent in pup rearing but emerges strongly also in relation to hunting and territorial defence.

## 1.1.4 The Sociality of Canids' Hunting and Defence Strategies

Canid hunting strategies vary widely, ranging from species that feed largely solitarily on fish and insects, like the short-eared dog (Sillero-Zubiri et al., 2004), to hypercarnivore species that hunt in packs, such as the African hunting dog, bush dog, wolf, and dhole (Venkataraman et al., 1995; Creel & Creel, 2002; Mech & Boitani, 2003; Sillero-Zubiri et al., 2004). Hunting techniques are intimately linked with sociality levels in canids, because species that hunt cooperatively are also described as the most social. Indeed, a number of authors have suggested that pack hunting may be the major force behind canid sociality (Alexander, 1974; Pulliam & Caraco, 1984; Clark & Mangel, 1986). Group hunting, in theory, may allow the capture of larger prey. However, the cause and effect of such an argument may be difficult to tease apart: do some canid species live in larger groups because in so doing they can hunt larger prey, or does the availability of predominantly larger prey 'oblige them' to live in groups? In a chapter reviewing the topic, MacDonald et al. (2004) suggest that at present the evidence cannot tease these two alternatives apart, although novel methods of modelling hunting success seem to suggest that, at least in African wild dogs, considering multiple factors such as energetic expenditure, frequency of hunts, and pup guarding (so reduced participation due to helper role), the optimum hunting party is a group of 10–15 animals, but when groups become larger, costs to hunting increase, no longer providing a viable alternative (Creel & Creel, 1995; Courchamp et al., 2002; Creel & Creel, 2002; MacDonald et al., 2004). In wolves, based on recent data looking at multiple packs differing in size and composition, groups larger than 4 individuals appear to have a significantly reduced hunting success due to an increase in free-riders (individuals that are present during the hunt but, in fact, avoid the riskier parts of the process, such as attacking and killing of the prey; MacNulty et al., 2012). Interestingly, it is the non-breeders of the pack (that have no offspring) that tend to become free-riders, which appears to be consistent with reports of the breeding pair being responsible for leading group hunts.

Within the hunting routine, however, the capture of the prey is only one of a suite of actions required for a successful outcome (see Bailey et al., 2013, for an in-depth review of group hunting strategies). Indeed, another important aspect that has emerged in the evaluation of the benefits of numbers in foraging strategies is the presence of scavengers. In wolves, it appears that in the absence of scavengers, hunting in pairs is more productive than in large groups; however, when ravens are put in the equation, the optimal group size increases (Vucetich et al., 2004). Indeed, protection of prey from both intra- and interspecific competitors appears to be a major factor that may tip the balance in favour of group hunting, not just in wolves, but also for dholes driving tigers away from kills and African wild dogs defending prey from hyaenas (Malcolm & Marten, 1982; Creel & Creel, 1995; Carbone et al., 1997; Carbone et al., 2005). Taken together, although the jury is out as to whether the effect of group size on hunting success *per se* is enough to promote the formation and maintenance of large groups in canids, it is undoubtedly the case that in many species, hunting is likely to be a social affair, whether it is in dyads or larger parties.

Similarly, territorial defence is also in many cases a group activity, where numbers seem to count. Thus, for example, in coyotes, domestic dogs, red and Arctic foxes, as well as Ethiopian wolves and African wild dogs, clashes between packs always result in the larger chasing off the smaller (coyotes: Wells & Bekoff, 1982; domestic dogs: Bonanni et al., 2011; red foxes: Voigt & MacDonald, 1984; Arctic foxes: Frommolt et al., 2003; Ethiopian wolf: Sillero-Zubiri et al., 2004; African wild dog: Creel & Creel, 1998).

### 1.1.5 Why Wolves?

Based on genetic, morphological, and behavioural data, it is clear that the domestic dog's closest living relative is the grey wolf (*Canis lupus*); hence, the comparative approach between wolves and dogs has naturally been the focus of most studies (see following sections). However, one of the most intriguing questions (yet to be addressed) is why, amongst canids, it was indeed the wolf (and not, say, the dhole) that underwent the profound changes required for the domestication process to occur. The Siberian farm fox experiment explained in more detail in Section 1.2.2 clearly shows the potential for a process somewhat similar in another canid species.

Indeed, archaeological evidence as well as a number of reports from various sources suggest that in South America, where the arrival of domestic dogs occurred relatively late, a number of wild canids were at least 'tamed' and kept in close contact with humans (see Stahl, 2012, 2013, for exhaustive reviews of the evidence). Hamilton-Smith (1839), for example, reported maned wolves and other canids being kept by indigenous people and accompanying them in hunting expeditions, and reported on interbreeding between endemic canids and domestic dogs, which eventually replaced them as the preferred human companion. The species most often reported are the crab-eating fox, the bush dog, and, to a lesser degree, the maned wolf (Stahl, 2013). More recently, in an exhaustive report of all the status and known facts of all canid species, Sillero-Zubiri et al. (2004) report of a number of South American canids, showing very docile behaviour towards humans (e.g., short-eared dog) or that have historically been kept as pets by indigenous populations (e.g., crab-eating foxes and bush dogs).

Taming is not the same process as domestication, in that whereas the former involves the raising of a wild animal with human contact, the latter involves a process of genetic selection, which ultimately results in an animal that is substantially different from its ancestor. Nevertheless, taken together, these different lines of research suggest that there may be specific canid (and not just wolfish) characteristics that rendered this family particularly well suited (pliable) for one of its species to have undergone the domestication process. Given the review regarding their sociality, both in terms of their reproductive strategy and in relation to hunting and territorial defence; potentially their 'monogamy' in terms of their ability to establish a long-term relationship with one individual; and their *inter-* and, perhaps even more importantly, their *intraspecific flexibility* in social organisation, these are perhaps the traits that allowed one particular species to start associating with human beings, over time becoming our most cherished companion. However, what characteristics may further set wolves apart, or if wolves were simply in the right place at the right time, remains an open question.

Unfortunately, at present, the social behaviour of many species is still relatively poorly understood, with some of even the more basic information missing for many species (see Macdonald & Sillero-Zubiri, 2004a, and Sillero-Zubiri et al., 2004, for the most recent comprehensive and complementary reviews). As has been pointed out by other authors in this book, most studies on canids have focused on the socio-behavioural ecology of the species (e.g., prey–predator relationship, ranging activities, territorial defense), whereas relatively little data are available regarding their social behaviour in terms of, for example, social organisation, affiliative relationships, coalition/alliance formation, group/ fission–fusion dynamics, intergroup dynamics, etc. This area of research remains largely unexplored, both with wild and captive populations of canids; hence, it is not possible to identify whether, compared to other canids, wolves may have specific characteristics particularly suited for the establishment of a relationship with humans. Yet, whether by lucky coincidence or because of some special trait in this species, it was indeed wolves that met humans, so we now turn to when, where, and how this meeting took place.

## 1.2  WHERE DO DOGS' SOCIALITY AND SOCIO-COGNITIVE ABILITIES COME FROM? THE DOG–HUMAN STORY

There is no doubt that dogs are a very successful species. Wherever there are men, there are dogs, be it the Inuits' dogs in the Arctic regions or the Saluki in the Arabic deserts. Dogs are everywhere. It is the only carnivore species that, even though it has the potential to significantly harm (even kill) a human being, we allow to exist in close proximity. Humans allow dogs in their households, share their lives with dogs, and, in some cultures, accept dogs as part of their families. One reason humans can do so is that domestication has changed dogs so drastically that they can not only *cope* with the human lifestyle, but may in fact have specifically adapted to it (see Chapters 2 and 11 of this book). As is outlined further in the following sections, and in other chapters in this book, selection pressure during domestication may have shaped dogs' behaviour and even their cognitive skills such that the latter are in some respects functionally equivalent to human behaviour and human social skills (see Chapter 11). However, one of the intriguing questions that fascinates researchers and is spurring a growing amount of research is when, where, and how humans and wolves/dogs met. In the following sections, we outline the various answers to the when, where, and how questions, drawing both from the archaeological evidence and genetic studies conducted in the field, showing that although we may never have an answer to the how question, data emerging from these fields can help tease apart the plausibility of what would otherwise remain 'just so' stories.

We then move on to briefly outline the different hypotheses that have been formulated as to what aspects of dogs' social behaviour and cognition (both towards conspecifics and humans) may have changed in the course of domestication. We conclude the chapter by highlighting where further research is needed to allow us to tease apart the competing hypotheses on how dogs became man's best friend.

### 1.2.1  When and Where Did Dogs and Humans Meet?

Looking at the history of the relationship between dogs and men, we know that dogs are the first species humans domesticated. While canid remains were found together with human remains dated at over 100,000 years BP (e.g., Olsen, 1985), these remains were still wolf-like and showed no clear signs of domestication. Morphological features used in archaeology to identify signs of domestication in a prehistoric canid skull are the shortening of the facial region and reduction in tooth size, which leads to crowding of teeth (Olsen, 1985; Dayan, 1994; Clutton-Brock, 1995). Morey (1994) states that the best indication of domestication taking place is a morphological pattern of mandibular and maxillary (snout) reduction accompanied by a lesser reduction of tooth size.

The fact that wolf and human remains were found together over quite a time span and at several locations is seen as an indicator that humans and wolves overlapped and shared a similar habitat for quite some time during the Pleistocene age (e.g., Nowak, 2003; Mech & Boitani, 2003). So even though humans and wolves must have met on many occasions for thousands of years, archaeological evidence suggests clear signs of domestication on canid skulls found near humans only quite late in the Pleistocene. One of the most important of these findings was that of the Natufian site of Mallaha, Israel (Davis & Valla, 1978; Clutton-Brock, 1995). The site is dated as ≈12,000 BP and therefore falls into the late Pleistocene (see also Tchernov & Valla, 1997, for evidence that other canid remains found at Natufian sites show clear signs of domestication). The Natufian people were thought to have been hunter-gatherers but with stable settlements, foraging for a wide variety of foods to then settle down to be what people believe were the earliest farmers in history (Bar-Yosef, 1998). The importance of dogs to the Natufian people not only becomes apparent in the joint burials of dogs and men (Morey & Wiant, 1992; Tchernov & Valla, 1997), which could be seen as evidence that dogs had a similar status as people (Losey et al., 2011), but also becomes evident from the fact that jewellery excavated at Natufian sites included one figurine that had an owl's head at one and a dog's head at the other end (Bar-Yosef, 1998). So this supports the idea that dogs and people were not just sharing the same habitat, but that there was a change in humans' perception and ideology towards dogs (maybe animals in general).

In the late Pleistocene and the beginning of the Holocene age, humans started to settle. In many areas, agriculture now started to replace the more hunter-gatherer–like culture. The near east is seen as the region in which agriculture first developed ≈10,000 BP, but from there agriculture spread around the world and to most corners of the globe (Richerson et al., 2001). There are several reasons agriculture started then, climate change being one of them, as agriculture would have been impossible in the glacial times of the Pleistocene (Richerson et al., 2001). The beginning of agriculture not only marks an important change in how humans organised their lives and the start of a rapid growth of the human population, but also is clearly the starting point for the domestication of different plant and animal species. So apart from dogs, we see goats being the first ungulate to be domesticated ≈10,000 years ago (e.g., Zeder & Hesse, 2000) and the domestication of pigs and cows (Loftus et al., 1994; Giuffra et al., 2000).

Because human settlement and agriculture during the end of the Pleistocene and beginning of the Holocene is seemingly so closely connected to the first findings of dog specimens, it is not surprising that earlier hypotheses about the domestication of dogs developed a scenario in which human settlement played a major role (e.g., Zeuner et al., 1967; Zimen, 1992; Clutton-Brock, 1995). Researchers speculated that wolves probably approached human settlements, maybe even continuously lived around human settlements, living from human waste and leftovers. The hypothesis is that during this process humans either actively selected nice and friendly wolf puppies to be their companions (Zimen, 1992) or that this was more like a process of self-domestication in the absence of any intentional selection (Zeuner et al., 1967; Clutton-Brock, 1995; for a review, see Hare et al., 2012).

## 1.2.2 How Did Wolves Become Dogs? The Self-Domestication Hypothesis

The central argument of the self-domestication hypothesis is that the first stage of the domestication of wolves was based on a selection against aggression or fear. This was not intentional, meaning that humans did not actively pick certain individuals based on their perception of them being less aggressive, but rather wolf individuals that were less aggressive or less fearful had a selection advantage (Zeuner, 1967; Hare et al., 2012). The reason that they might have had a selection advantage could be that being less aggressive and less fearful towards humans gave them the opportunity to live in closer proximity to them, hence the opportunity to exploit new and potentially more reliable food sources (Zeuner, 1967). One series of studies that shows how strongly a selection against aggression and fear can affect a species' behaviour and morphology is the famous silver fox farm experiments carried out in Siberia.

This experimental study was started by Russian evolutionary biologist Dmitri Belyaev. In this long-term study, researchers simulated selection pressures during domestication by selecting a group of silver foxes for tameness and against aggression towards humans. Foxes were divided in different groups based on how they reacted towards the hand of a human reaching into their cage. While the ones that behaved aggressively (or fearfully) against the hand were selected to be in the non-domesticated group (Class III), the ones that stayed calm and showed no signs of aggression towards the hand were moved to the 'intermediate' (Class II) group, and the ones that showed clear signs of friendliness towards the humans (like approaching the hand, licking the hand, etc.) were moved to the 'domesticated' (Class I) group (Trut et al., 2009). Interestingly, after several generations of selection based on levels of tameness being the only selection criterion, the Class I foxes showed clear signs of paedomorphic features and 'dog-like' behaviour, such as whining, tail wagging, exhibiting signs of clear submission towards humans, approaching humans, etc. (Trut et al., 2004; Trut et al., 2009). There were also changes in morphology including, for example, floppy ears and curly tails compared to the non-selected foxes (Trut et al., 2009). These experiments therefore support the hypothesis that selection against aggression or fear towards humans can lead to behavioural and morphological changes in a canid species, and create something like a 'proto-dog', which researchers have suggested to be the type of dog that was created by this first wave of domestication (Zeuner, 1967; Coppinger & Coppinger, 2001; Hare & Tomasello, 2005).

## 1.2.3 The Recent Genetic Revolution

Recently, the view that dog domestication started in the late Pleistocene and early Holocene period has been seriously challenged by both archaeological evidence and studies tracing the genetic origins of the domestic dog. Two main findings suggest the story of domestication may change completely. First, a number of skulls, showing clear signs of differentiation through domestication, were found in the Goyet Cave in Belgium and the Altai Mountains in Russia (Ovodov et al., 2011). A recent detailed analysis of these skulls dates them as >30,000 years old (Germonpré et al., 2009; Druzhkova et al., 2013), making them the oldest remains currently available. The dating of the skulls to 30,000 years is also supported by the second major line of research, which is based on the comparison of DNA collected from these specimens (the Goyet skull and the Altai skull) to that of modern dog breeds, modern wolves, and prehistoric wolf specimens. Thalmann et al. (2013) collected partial mitochondrial genomes from prehistoric canids, modern wolves of Eurasian and American origin, and modern dogs. The data suggest three things. First, it shows that the 30,000-year-old skulls are indeed dogs, not wolves (see also Druzhkova et al., 2013; Ovodov et al., 2013). Second, it suggests a European origin of the dog, as there was a strong association of the sequences from modern dogs with ancient European specimens and with European wolves, but no association of modern wolf sequences from the Middle East or East Asia with modern dogs (Thalman et al., 2013); and last, it suggests that the wolf species, which was dogs' ancestor, was more likely from a now-extinct branch than from a modern wolf species still in existence today. Recent evidence presented by Freedmann et al. (2014) suggests a very similar story. They compared the DNA from three different grey wolves, chosen from Europe, the Middle East, and East/South Asia, which represent the regions where domestication is hypothesised to have taken place (see Savolainen et al., 2002, and Wang et al., 2013, for support of the hypothesis that dogs originated in the Middle East and East Asia; and Thalmann et al., 2013, for support of a European origin) to basenji DNA and dingo DNA and used a golden jackal as an outgroup. Their data suggest that after dogs separated from wolves, wolves went through a sharp bottleneck during which the number of wolf lineages was significantly reduced. This might support the view also brought up by Thalmann et al. (2013) that dogs' ancestor is now extinct and not represented by any of the modern wolf species.

If, as suggested by recent evidence, dogs were domesticated 30,000 years ago, the story behind how wolves became dogs changes entirely. At that time, humans were still hunter-gatherers, long before humans settled down and long before agriculture started (as explained previously, agriculture together with human settlement started ≈12,000 years ago [Larsen, 1995]). So a likely scenario, then, would be that wolves probably followed humans during hunting and were scavenging their share of the meat or simply fed on the leftovers (Thalmann et al., 2013). In this context, it is interesting that there is evidence of signs of gnawing damage to ancient bones from larger mammals of the Pleistocene period, which were found near human remains (Haynes, 1983). Haynes (1983) compared these traces on the ancient bones to traces that modern wolves would leave on bones from larger mammals and found that they were highly comparable. This suggests that the marks on the ancient bones were also left from wolves or at least some larger canid species (potentially proto-dogs). This therefore would support the hypothesis that wolves (or proto-dogs) basically were around humans while they were hunting for larger mammals. There is even some speculation that dogs played a role in the massive extinction of Megafauna (e.g., mammoths) roaming earth during that period, for example, by transmitting diseases to the mammals of areas that did not previously have dogs (see, e.g., Fiedel, 2005) or because they increased humans' hunting success so significantly during the Pleistocene period. This, however, is purely speculative.

Yet, undoubtedly, for this type of relationship to have lasted over time, and developed to the extent that it did, it must have brought substantial advantages to both species, and likely these advantages were related to food intake. An interesting study was conducted on the hunting success of today's humans on moose, in relation to the group size of the hunting party and the presence/absence of dogs (Ruusila & Pesonen, 2004). Results clearly showed that the presence of dogs increased hunting success, especially in human parties of fewer than ten individuals, and where moose density was lower. Indeed, the role of dogs was mostly to halt the prey, allowing one hunter to shoot down the animal down, and the other hunters to be strategically placed to cut off the animal's escape route.

It is therefore quite likely that a similar pattern occurred also in the past. In the northern hemisphere, large ungulates such as moose were likely one of the main food sources for both humans and wolves; however, rather than enter in competition, what may have happened is that wolves and humans were able to take advantage of each other's specific skills. The hypothesis is that while humans in those times already had the necessary tools (e.g., spears) to successfully kill large prey (Grayson & Meltzer, 2002), wolves would have been much faster and more skilled at tracking and immobilising the prey, as indeed dogs are today. The possibility is therefore that humans may have learned to follow wolves to their selected prey and, with their more advanced technology, would have succeeded in more rapidly finishing off the large animals after the wolves had immobilised them. Assuming that then, as now, the presence of wolves/proto-dogs increased hunting success, it would explain why our ancestors were willing to share their spoils with the wolves/proto-dogs they hunted next to.

So while the timing of events changes the picture of dog domestication quite a lot, the self-domestication hypothesis could still explain this first wave of domestication. Both species—humans and wolves—were hunting for the same prey. Both species possibly benefitted from hunting 'together' (i.e., alongside each other). Maybe both species picked up on certain signs coming from the other, indicating when prey was around and a hunt was imminent. Both species probably followed each other's routes in the expectation to find prey, though here it is more likely that humans followed wolves, given that wolves' ability to detect prey were most likely much more advanced back then, as they are today. Once prey was detected and the hunting started, the much faster wolves were likely with the prey long before humans arrived and maybe started to attack, or surround and immobilise, the prey while humans arrived later to finish the kill with tools. The question is what may have happened next. Most likely, wolves were not eager to let go of the prey after they had started the attack. So it could be that in order to obtain it, the humans would have had to chase the wolves away (not unlike how hyaenas react to African wild dogs at prey sites) or vice versa. It is interesting to note that data so far suggest that hunting in packs for a number of canid species is advantageous because it allows them to keep scavengers away, rather than because it allows larger prey to be taken down (see Section 1.1.4 earlier). Hence, success at holding on to the prey may have depended on numerical factors in both species. However, humans may at some point have recognised the benefit of these hunting 'allies' and chose to share the spoils by cutting off pieces of the prey for the wolves (perhaps those parts which humans have more trouble with, e.g., bones), not unlike a modern human–dog hunting scenario. Following this scenario, it may be that the initial relationship was one based on a mixture of reciprocal scavenging/scrounging of each other's prey, where on occasions, both species contributed to the actual killing, slowly consolidating the advantages of following each other on the same hunts.

In this scenario, the self-domestication hypothesis can still easily explain the first wave of domestication. A change of temperament due to self-domestication would still mean a selection advantage in that wolf individuals that were less fearful and aggressive around humans were probably the ones that were more likely to join these hunting scenarios and probably also to benefit from scavenging and leftovers. Most likely another wave of domestication occurred later, when humans had settled and established agriculture. That changes during times of agriculture actually put additional selection pressures on dogs is supported by recent work, which suggests that dogs have adapted specifically to the starch-rich diet typical for modern human agricultural societies (Axelsson et al., 2013). One can only speculate as to what extent and how these additional selection pressures to cope with life in close proximity to humans in times of agriculture also affected dogs' behaviour.

Some authors, however, go one step further and argue for a co-evolution of dogs and humans (e.g., Schleidt & Shalter, 2003). A co-evolution of two species would mean that not only did both species co-exist (and evolve independently while living next to each other), but that one species affected the other's evolutionary path and vice versa, e.g., the arms race of two species that are in a predator–prey relationship. So if we consider the scenario of a co-evolution of dogs and humans, we would have to show evidence that being with dogs imposed a selection pressure on humans, which resulted in a new trait emerging, and show that dogs acquired novel adaptions as a consequence to life with humans. Miklósi and Szabo (2012) use a nice example to explain what kind of evidence we would be looking for, using human communication. We know that in modern times humans mostly guide dogs through gestures. So one could speculate, very hypothetically of course, that humans who used this kind of communication had a selective advantage because they were more successful in guiding their dogs during hunting and hence were more successful hunters, received more meat, and had more offspring. The 'gestural communication' trait therefore would have established quickly in the human population, leading to humans regularly and frequently using pointing gestures in their communication—a trait that is seen to be unique to our species (Tomasello, 2008). This scenario, however, is pure fiction, just an example meant to illustrate the kind of process that would have had to occur to justify the use of the term 'co-evolution' to refer to the dog–human evolutionary relationship. As of yet, we, however, see no evidence of any trait being established in the human population as a direct consequence of being with dogs. We therefore think, similarly to Miklósi and Szabo (2012), that using the term 'co-evolution' is rather misleading. In fact, at present, what is certain is that the two species, wolves and humans, co-existed and evolved in parallel for a long time, but at a certain point the co-existence led to advantages that were significant enough for both species to continue seeking/maintaining such proximity. Indeed, viewed in this light, the term 'symbiotic mutualism', which indicates both species' benefitting from living in this kind of relationship, would be more appropriate (Coppinger & Coppinger, 2001; Reid, 2009).

## 1.2.4  Effects of Domestication on Social Behaviour and Cognition

One crucial question, especially for a book like this one, is to what extent domestication affected dogs' behaviour and more specifically dogs' social behaviour. Clearly, domestication affected dogs' morphology in ways paralleled by other domesticated species (e.g., goats). Compared to wolves, dogs, for example, have a reduced cranial capacity, have smaller canines, and also show a great variety of fur pigmentation. Dogs also show a variety of paedomorphic features in their morphology (e.g., large eyes, short nose; see Wayne, 1986), and some speculate that they may also show paedomorphic features in their behaviour (e.g., more play behaviour in dogs than wolves; Topál et al., 2005; see also Bekoff, 1974, and Chapter 4 of this book). One selection pressure that has been suggested to be crucial for the evolution of paedomorphic features is the selection against aggression (Trut et al., 2009), although, in fact, whether dogs are indeed less aggressive than wolves, particularly in their intraspecific interaction, has been questioned (Fedderson-Peterson, 2007; and see Chapter 2 of this book for an in-depth discussion of this issue). Nonetheless, regardless of how specific paedomorphic features emerged in dogs, a recent study by Waller et al. (2013) suggests that they may indeed give dogs a selection advantage, because of the preference humans show for these characteristics. In this study, the researchers used dog selection from a shelter as a proxy to study which features most affected humans' choices. Interestingly, dogs that produced a facial movement (inner eyebrow raise) that made the dogs' eyes look larger were selected from the shelter quicker than dogs that did not produce that same facial movement as often. Other behaviours such as tail wagging had no effect on selection. Results from this study therefore suggest that humans (most likely unintentionally) select along their preferences for paedomorphic features in dogs' behaviour (Waller et al., 2013).

Undoubtedly, domestication has had various repercussions on dogs' sociality and cognition both in terms of their behaviour towards humans (see Chapter 2 and preceding section), and, equally interestingly, also in terms of dogs' intraspecific social organisation and behaviour (Chapters 2 and 3).

In terms of how domestication may have affected dogs' social behaviour, there is evidence that dogs, if living in conditions under which they are able to form social groups with conspecifics, form stable, hierarchical groups, with a social structure most likely resembling that of free-living wolves (see Chapter 3). Nonetheless, a number of differences in the intraspecific social behaviour/organisation of wolves and dogs have been suggested by various authors, the most prominent being dogs' lack of pair bonding as the primary mating strategy (Lord et al., 2013; Cafazzo et al., submitted); differences in parental care investment, with a reduced involvement of fathers and other pack members in litter raising in dogs (Pal, 2005; Lord et al., 2013); a decreased reliance on cooperative hunting strategies in dogs (Boitani & Ciucci, 1995; Boitani et al., 2007) and less stability or reliance on the social group (Boitani & Ciucci, 1995; Coppinger & Coppinger, 2001); a reduced/altered presence of a hierarchical organisation within dog social groups (Boitani & Ciucci, 1995; Boitani et al., 2007); and an increased tolerance (reduced aggressiveness) towards conspecifics in dogs (Frank & Frank, 1982; Miklósi, 2007; Hare et al., 2012), but also the opposite, that is, a decreased tolerance (or heightened aggression) towards conspecifics in dogs and consequent decrease in cooperation (Fedderson-Peterson, 2007) (see Chapters 2 and 3 for a discussion of most of these issues).

As can be deduced from the conflicting hypotheses being presented, to date, firm conclusions on the effect of domestication on intraspecific social behaviour are impossible to draw. Indeed, this area of research has noticeably lagged behind, probably due to the fact that although an estimated 83% of dogs worldwide are thought to live largely free of direct human influence (Lord et al., 2013), studying these populations is substantially harder than analysing the behaviour of owned (pet) dogs, and perhaps because being human, we are more interested in how dogs may have adapted to life with us, rather than the potential effects of these adaptations to their behaviour towards each other.

Indeed, more has emerged as regards the potential effect of domestication on dogs' social behaviour towards humans. When dogs live with humans, they do show behaviours that we do not seem to find in wolves raised in similar conditions. Evidence that the bond between dogs and humans might be 'special' comes from research showing that dogs develop levels of attachment to humans that might be comparable to the attachment of children to their parents (see Chapter 6; see also Chapter 8) and which have not been found in wolves raised in a similar way to dogs (Topál et al., 2005). Furthermore, in a number of studies, similarly raised wolves were slower or more reluctant to establish eye contact with a human partner (Gácsi et al., 2005; Virányi et al., 2008; Gácsi et al., 2009), and appeared to be less likely to adapt their behaviour in accordance with a human's change in attitude from threatening to friendly (Gácsi et al., 2013). However, the latter study was conducted with pet dogs, rather than dogs raised and living in the same manner as the tested wolves; hence, it is not clear whether dogs being better attuned to the human partners' behavioural changes was due to experience or indeed an effect of the domestication process. Indeed, studies comparing identically raised wolves and dogs show that the former accept the humans they were raised with as social partners, inasmuch as they readily use them as a source of information (e.g., following their gaze; Range & Virányi, 2011) and learn from them in various tasks (Range & Virányi, 2013, 2014). Although, given the relative shyness wolves exhibit towards human strangers compared to the ready friendliness of dogs, it may be that extending (or generalising) the relationship of social partnerships to humans in general may be less likely to occur in wolves. A recent study has nicely shown how oxytocin may be involved in the establishment of the human–dog partnership both towards the owner and towards strangers (Kis et al., 2014); hence, this may also provide important implications as to the potential genetic underpinnings of the selective changes from wolves to dogs in their sociality towards humans.

The strongest evidence to date, however, as regards to how the selection pressures during domestication may have affected dogs' social cognition comes from research showing that dogs seem to have extraordinary skills in understanding human forms of communication (for a review, see Kaminski & Nitzschner, 2013; see also Chapter 11). Dogs seem to stand out in the animal kingdom in how sensitive they are to human communication. One hypothesis that has been put forward is the domestication hypothesis, which suggests that through selection pressures during domestication, dogs evolved special social cognitive skills that are in some domains functionally equivalent to that of humans (Hare et al., 2002; Miklósi et al., 2003; see also Chapter 2 in this book for a discussion).

The hypothesis that selection pressures during domestication might have affected dogs' skills when it comes to the sensitivity to human communication is supported by several facts. First, dogs outperform other species when it comes to their understanding of cooperative communicative signals (e.g., the human pointing gesture). Dogs even outperform humans' closest living relatives, the chimpanzees, in how well they respond to these gestures (Hare et al.,

2002; Bräuer et al., 2006; Kirchhoffer et al., 2012). Dogs also outperform their own closest living relative, the wolf, when it comes to reading subtle communicative cues and without receiving any special training (Hare et al., 2002; Miklósi et al., 2003; Virányi et al., 2008; but see Udell et al., 2008, and Hare et al., 2010, for a recent discussion). Finally, dog puppies, from an early age on and again without receiving any major training, already follow human gestures, suggesting that major learning during ontogeny cannot account for dogs' behaviour in this domain (Riedel et al., 2008; Virányi et al., 2008). So while the evidence is accumulating that selection during domestication seems to have affected dogs' skills in this domain, several hypotheses exist for how and to what extent this may have occurred. As this is very much the topic of the chapter by Viranyi and Range (Chapter 2), we do not go into much detail here and rather simply summarise the hypotheses as falling into four main categories: (1) the category that would predict generally advanced social cognitive skills in dogs compared to other species (including wolves) (Hare & Tomasello, 2005); (2) the category that would predict different (maybe more advanced) social cognitive skills in some areas of dog social cognition as a special adaptation to life with humans (Gasci et al., 2009; Kaminski & Nitzschner, 2013; Miklósi & Topál, 2013); (3) the category which predicts that dogs' social cognitive skills are shared with wolves (and potentially other canids), and dogs' uniqueness lies in their capacity/willingness to direct these more easily towards human social partners (see Chapter 2); and (4) the last category, which predicts no true differences between dogs' cognitive skills and that of their closest living relative the wolf, even when directed towards humans (e.g., Udell et al., 2008). It should be noted, however, that researchers in this last category also do not categorically rule out the effects of domestication on dogs, but rather they attribute possible differences between the species on environmental influences and differences in the development of dogs and wolves. Despite differing predictions, what is interesting to note is that all hypotheses recognise some impact of domestication on dog's behaviour and cognition; hence, what is rather being debated is more specifically which behaviour and/or mechanisms may have been affected and how such effects may have been brought about (e.g., specific changes in genetic, epigenetic aspects of the two species). Unfortunately, currently, a number of problems do not allow us to draw firm conclusions as regards the validity of the suggested hypotheses.

One important point emerging from various studies is the need for more comparable or standardised methodologies. Thus, for example, the hypothesis put forward by Udell et al. (2010) is mainly based on one study in which Udell et al. (2008) compared the behaviour of hand-raised (and specially trained) wolves with that of normal pet dogs and former street dogs. The study showed that, in fact, the hand-raised wolves outperformed the former street dogs, suggesting that ontogeny might have affected the individuals more than any selection during domestication. However, there is a major difference between this study compared to all other former studies looking at dogs' (and wolves') understanding of the human pointing gesture. The paradigm that most researchers use is the so-called objects choice paradigm (Anderson et al., 1995). In this paradigm, the animal is presented with two (ore more) containers, one of which contains a piece of food that was hidden outside the animal's view. So, in the absence of any direct information about the location of the food, the animal is then presented with some social information, e.g., a pointing gesture indicating where the food is hidden. After the animal receives that information, it is encouraged to make a choice between the presented containers. A correct choice would be counted as a choice in which the animal follows the gesture and therefore finds the food, and consequently, an incorrect choice would be any choice for the container the human did not point to and hence did not contain the reward. Udell et al. (2008), however, also counted trials in which subjects refused to make any choice as incorrect choices. This means that any trial in which a subject was maybe too shy or not motivated at all to participate was treated identically to trials during which subjects chose the incorrect location, hence the location the human had *not* pointed towards. A re-analysis of the data excluding trials during which subjects did not make any choice changed the picture completely, suggesting that the influence of ontogeny was not at all as strong as the authors had suggested (Hare et al., 2010).

A second fundamental issue is the need for comparable subject populations when testing wolves and dogs (see also Chapter 2) and the complementary need for more diverse populations of pet dogs and captive/wild wolves to be included in the subject pool.

The first aspect is crucial because a number of studies advocating dogs' superior understanding of human communicative skills were based on results comparing wolves that, although raised by humans during their first few months of life, were then allowed to live in a captive-like setting with frequent but not intensive human contact. The dogs in these studies, however, came from a pet dog population, living in human homes and therefore with a far more intensive everyday experience of human communication (e.g., Topál et al., 2009; Gácsi et al., 2013). Similarly problematic, although from the opposite perspective, the wolves that participated in the study by Udell et al. (2008) were not only hand-raised by humans, but also received specific and rigorous training in order to interact with humans. One training method used was the so-called clicker training. The wolves were rewarded with the sound of a clicker upon performing correctly. That this might have influenced the wolves' behaviour to quite an extent is illustrated by a study by Pongrácz et al. (2013), which showed that clicker training (as performed on the wolves tested by Udell et al., 2008) strongly and positively influenced dogs' response to human gestures. Considering the wolves' performance on these tasks was compared to dogs with no such training experience, it is perhaps not surprising that they performed on average better than dogs. Hence, in both of these cases, the differing effects of experience on the two species cannot be ruled out, making any claims to results referring to 'domestication effects' highly problematic. Indeed, where identically raised and kept wolves and dogs have been tested, results have shown substantially smaller differences than expected (Range & Virányi, 2013).

Widening the subject pool for both species is equally important, for a number of reasons. First, today's wolves are many generations away from the shared wolf–dog ancestor; hence, a certain amount of caution appears to be necessary when concluding that domestication is directly responsible for *all* the potential differences observed between these two species. Indeed, considering the history of heavy persecution of wolves, it is quite probable that their current shyness towards human beings may have been indirectly selected for by the extermination of the 'bolder' wolves that did not keep well away from human establishments. This inadvertent selection for shyer wolves may have resulted in a suite of other by-products being passed on across generations as 'covariants' to shyness. Although it is reasonable to assume that the change from common ancestor to today's wolves may have been less rapid or dramatic than that of dogs, we cannot assume that wolves as a species have remained unchanged since their divergence from the wolf–dog ancestor. Indeed, this seems to be supported by evidence of a serious genetic bottleneck and consequent loss of genetic variability occurring in grey wolf populations both in Europe and North America (Leonard et al., 2005; Sastre et al., 2011). In fact, if the extermination of wolf populations by humans has resulted in the survival of the shyer animals, it may be that differences between today's wolves and dogs are the result of opposite selection pressures, one for increased shyness and the other for an increased tolerance/affiliation towards humans. Furthermore, there is growing evidence from genetic studies that dogs may have, in fact, derived from a now-extinct subspecies (Thalman et al., 2013). Considering the above, it seems all the more important to broaden the focus of studies to include as many subspecies of wolves as possible. Although this may be difficult as regards experimental studies, a comparison between subspecies on their intraspecific social behaviour may already shed light on the potential variability within this species.

The second aspect worth considering in support of widening the subject pool is that although a lot more is known about the social behaviour of grey wolves than of many other canid species, the number of populations studied (both in the wild and in captivity) is, in fact, relatively small, and given the high *intra*specific variability displayed by many canids depending on ecological factors (Macdonald & Sillero-Zubiri, 2004b; Macdonald et al., 2004), it is quite probable that with more populations being studied, a greater social flexibility will emerge amongst wolf populations. The exact same argument is even more applicable to free-ranging and feral dogs, where even fewer studies are available (see Chapter 3 for a more in-depth discussion of this point). Hence, for example, current conclusions on the potential differences between the intraspecific social behaviour of these two species will need to be re-assessed once a larger dataset on both species (in the wild and captivity) becomes available.

A similar argument applies also to cognition studies, where to be able to get at questions relating to the relative importance of experience versus 'domestication', it would be overly important to include a variety of study populations, particularly those with a reduced experience of humans (e.g., free-ranging dogs with more or less human experience).

A third issue needing attention is that in the context of many of the socio-cognitive abilities under examination, a comparison needs to be made not just with wolves but also with other canids and other social mammals. A nice example, in historical terms, is that of the study of dogs' understanding of human attentional states. As mentioned previously, dogs' potentially extraordinary abilities first emerged as regards their understanding of human communicative gestures, where dogs appear to be outstanding in the animal kingdom (Hare et al., 2002); however, it later emerged that dogs are also very sensitive to humans' attentional states (e.g., Call et al., 2003) in that they seem to understand when a human is watching them and will avoid stealing food when the human has her eyes open versus closed or her back turned towards them (Call et al., 2003; Schwab & Huber, 2006; for a review, see Chapter 10 of this book). Based on this evidence, claims were made that this seemingly sophisticated understanding of a human's attentional state might also be the result of processes occurring during domestication, leading to hypothesise that not only have dogs evolved specialised skills in one area (sensitivity to gestures) but that they have evolved generally more sophisticated skills in all areas of social cognition (Hare & Tomasello, 2005).

However, a later study found that dogs' understanding of attentional states was also displayed in conspecific social interaction, since in a detailed analysis of dogs at play, Horowitz (2009) showed that not only did the signallers take into account the receiver's attentional state prior to emitting a particular signal, but they also flexibility-adjusted the combination and strength of the signals in accordance with their audiences' state of attention. Hence, based on this evidence, it would seem that either dogs' ability to read human attentional states was an extension to their ability to read their conspecifics' state of attention or vice versa.

Studies on a wider range of species, however, eventually highlighted the fact that many different social living mammals understand when others are or are not attentive and, in many cases, not just conspecifcs but humans as well (see Rosati & Hare, 2009; and Kaminski & Nitzschner, 2013, for a review). So the current evidence suggests that, in fact, this ability is widespread in the animal kingdom and must be underpinned by an urgent evolutionary function (see Emery, 2000). Hence, studies including a wider range of species show that at least in the case of dogs' comprehension of attentional states, their performance is far from unique, calling into question claims that domestication has somehow allowed dogs to evolve a more 'human-like set of socio-cognitive skills'. Indeed, later evidence also showed that wolves (like many other social mammals tested) behave very similarly in tests looking at their comprehension of human attentional states (see Udell et al., 2011, for evidence of this).

A final issue is that there appears to be an urgent need for better integration between studies focusing on intraspecific and interspecific social skills to allow questions regarding the effects of domestication to be addressed. As an example, it has been suggested that the unique adaptation of dogs within the human social environment is their understanding of communicative gestures as 'referential' (see Chapter 11 of this book). Indeed, as summarised previously, this idea is supported by data showing dogs' superior skills, as compared to wolves and also various primate species, at understanding human communicative cues such as pointing; however, only one study (based on a very small sample) has also addressed whether dogs may see other dogs' behaviour as referential (Hare et al., 1999). In this study, the authors compared whether dogs use cues from conspecifics indicating the location of the hidden food equally well as they use human cues. Results showed that, indeed, dogs were able to find food in an object choice setting equally successfully regardless of the species communicating with them (i.e., human pointing, dog head turning, etc.; Hare et al., 1999). If wolves are also found to understand their conspecifics' gestures (e.g., the head turn) as referential, the uniqueness of dogs' ability may lie in their capacity to easily generalise their conspecific skills to humans, rather than any new ability having emerged following the domestication process. A nice example of a more integrated approach to this specific topic is given in Chapter 9, where dogs' social learning abilities have been assessed both with conspecifics and humans, allowing a more comprehensive picture to emerge. However, until more integration between conspecific and interspecific studies is carried out in a more systematic way, the relative merits of the different hypotheses are difficult to assess.

## 1.3 CONCLUSIONS AND BOOK OVERVIEW

As we hope we have been able to suggest in the current chapter, dogs' sociality and their potentially 'special' socio-cognitive skills likely emerge both from the specific characteristics of their canid ancestry (e.g., a reliance on 'the pack' in many contexts, their willingness to form life-long bonds, and a remarkable social flexibility) and the unique event of having encountered and started living alongside humans.

Because we think that both dogs' intraspecific canid skills and their meeting with humans played a crucial role in shaping the companion we have today, the chapters of this book focus as much as possible on dogs' social behaviour and cognitive skills towards other dogs as well as humans. In the case of cognition, because so little has been done within an 'intraspecific' framework, the emphasis cannot but be more on humans.

Hence, following this introductory chapter, and given the importance held by the different domestication hypotheses in spurring further research, Virányi and Range (in Chapter 2) outline the major theories, highlighting the importance of 'older' studies comparing wolves and dogs, and emphasising the need for these to be taken into account when developing new hypotheses.

Together, we hope these first two chapters set the theoretical framework in which many of the following issues addressed by each author can be read. The remaining chapters of the book are organised in two major areas: (1) studies focusing on dogs' social behaviour directed towards other dogs and humans; and (2) studies focusing on dogs' social cognition.

The dogs' social behaviour in their conspecific setting is first addressed in Chapter 3 by Bonanni and Cafazzo, who review the theories regarding the effects of domestication on dogs' social structure and organisation and give a comprehensive overview of their studies on a population of free-ranging dogs living in the outskirts of Rome, integrating their own results with those of other researchers. In the subsequent chapter (Chapter 4), Smuts presents a number of results from studies on the social behaviour and social organisation of pet dogs frequenting a 'dog day care centre' and hones in on play as one of the only affiliative behaviours currently being studied in dogs. Taken together, these chapters appear to highlight the variability and flexibility of dogs' social behaviours and clearly show that many more studies are needed in this field of research to allow us to better comprehend the variables affecting dogs' intraspecific social behaviour.

Taylor and colleagues, in Chapter 5, nicely bridge the gap between conspecific and interspecific studies, by presenting an overview of studies on dogs' acoustic communication, both in relation to other dogs and to humans. Indeed, a comparison with the wolf communication system leads the authors to suggest that a number of vocal adaptations in dogs may indeed be in response to their increased need to communicate with us.

In Chapter 7, Miklósi and colleagues give an in-depth and critical overview of studies on dog personality, highlighting exciting new research on the genetic underpinnings of personality traits and presenting the wide range of implications of dog personality amongst others for the establishment of the dog–human bond and dog welfare issues. Following on from this chapter, the dog–human bond is focused on more specifically by both Prato-Previde and Valsecchi (Chapter 6) and Mills and colleagues (Chapter 8). In the former chapter, the authors give a comprehensive overview of studies on 'attachment', starting from a historical introduction of this concept and following on with studies looking at different dog populations and discussing what the origins of the dog–human attachment bond may be. The latter chapter shows how behaviours considered problematic by owners may indeed arise from expectations established in the social relationship between our two species, or misrepresentation of species-specific behaviours. As in any social relationship, a number of elements may 'go wrong', and the current chapter has important implications both from the theoretical and applied perspective.

The final five chapters of the book (Chapters 9 to 13) focus on major areas being investigated under the umbrella term 'social cognition'. In the first chapter of this section (Chapter 9), Pongrácz, gives a critical overview of studies conducted investigating social learning in dogs, setting these studies in the context of the wider research being carried out on social learning in animals. As mentioned previously, this area of research nicely integrates both conspecific and interspecific studies on the same question, setting a potential blueprint for the future.

In the subsequent chapter (Chapter 10), Bräuer tackles this difficult question: Do dogs understand humans as mental agents? Drawing on numerous studies looking at how dogs relate to humans in different contexts, Bräuer presents interesting results in the field of dog social cognition research. By comparing results to those with other species, the author sets dogs in a wider context, concluding that although dogs may have 'special talents' in understanding humans' communicative intent, they might not 'read' humans as fully intentional agents, as humans do one another.

In Chapter 11, Topál and colleagues expand on dogs' 'special talent', presenting an in-depth overview of the studies carried out on dog–human communication, contrasting these with results from wolf studies and drawing the parallel but also highlighting the differences between infant–adult and dog–adult communication. Together, these chapters cover some of the major topics being discussed in the field.

In the final two chapters (Chapters 12 and 13), some of the more innovative aspects of the dog cognition research are presented. In Chapter 12, Burman gives a critical overview of studies on dogs' 'cognitive bias', setting these in the context of similar studies in other species and highlighting the potential implications for animal welfare. Whereas in the final chapter, Siniscalchi and Quaranta present results from their studies on dog brain lateralisation, showing how this may link to dogs' emotions, cognition, and communication, and be used as a tool in assessing dog welfare.

The study of dog social behaviour and cognition is relatively young; hence, it not surprising that, in many areas, authors mention the need for more studies. Because it is our hope that this book will spur the next generation of dog researchers to continue applying their talents to this field, we explicitly asked all authors to include further directions in the field, highlighting which, according to their expertise, are the major questions still needing an answer. We hope this will inspire readers to 'go find' these answers.

## ACKNOWLEDGEMENTS

The writing of this chapter and editing of the entire volume were done whilst being supported by grants from the University of Milan and funding from the European Research Council under the European Union's Seventh Framework Programme (FP/ 2007-2013)/ERC Grant Agreement n. [311870] to Sarah Marshall-Pescini. We wish to thank Simona Cafazzo for constructive discussions and useful comments on parts of this manuscript.

# REFERENCES

Adams, J.R., Leonard, J.A., Waits, L.P., 2003. Widespread occurrence of a domestic dog mitochondrial DNA haplotype in southeastern US coyotes. Mol. Ecol. 12 (2), 541–546.

Alexander, R.D., 1974. The evolution of social behavior. Annu. Rev. Ecol. Syst. 5, 325–383.

Anderson, J.R., Sallaberry, P., Barbier, H., 1995. Use of experimenter-given cues during object-choice tasks by capuchin monkeys. Anim. Behav. 49 (1), 201–208.

Andersson, K., 2005. Were there pack-hunting canids in the Tertiary, and how can we know? Paleobiology 31 (1), 56–72.

Axelsson, E., Ratnakumar, A., Arendt, M.L., Maqbool, K., Webster, M.T., Perloski, M., et al., 2013. The genomic signature of dog domestication reveals adaptation to a starch-rich diet. Nature 495 (7441), 360–364.

Bailey, I., Myatt, J.P., Wilson, A.M., 2013. Group hunting within the Carnivora: physiological, cognitive and environmental influences on strategy and cooperation. Behav. Ecol. Sociobiol. 67 (1), 1–17.

Baker, P.J., Funk, S.M., Bruford, M.W., Harris, S., 2004. Polygynandry in a red fox population: implications for the evolution of group living in canids? Behav. Ecol. 15 (5), 766–778.

Bardeleben, C., Moore, R.L., Wayne, R.K., 2005. A molecular phylogeny of the Canidae based on six nuclear loci. Mol. Phylogenet. Evol. 37 (3), 815–831.

Bar–Yosef, O., 1998. The Natufian culture in the Levant, threshold to the origins of agriculture. Evolutionary Anthropol. 6 (5), 159–177.

Bekoff, M., 1974. Social play in coyotes, wolves, and dogs. Bioscience 24 (4), 225–230.

Bekoff, M., Diamond, J., Mitton, J.B., 1981. Life-history patterns and sociality in canids: body size, reproduction, and behavior. Oecologia 50, 386–390.

Bensky, M.K., Gosling, S.D., Sinn, D.L., 2013. The world from a dog's point of view: a review and synthesis of dog cognition research. Adv. Study Behav. 45, 209–406.

Berta, A., 1987. Origin, diversification, and zoogeography of the South American Canidae. Fieldiana Zool. 39, 455–471.

Boitani, L., Ciucci, P., 1995. Comparative social ecology of feral dogs and wolves. Ethol. Ecol. Evol. 7, 49–72.

Boitani, L., Ciucci, P., Ortolani, A., 2007. Behaviour and social ecology of free-ranging dogs. In: Jensen, P. (Ed.), The behavioural biology of dogs. CAB International, Wallingford, UK, pp. 147–165.

Bonanni, R., Natoli, E., Cafazzo, S., Valsecchi, P., 2011. Free-ranging dogs assess the quantity of opponents in intergroup conflicts. Anim. Cogn. 14 (1), 103–115.

Bräuer, J., Kaminski, J., Riedel, J., Call, J., Tomasello, M., 2006. Making inferences about the location of hidden food: social dog, causal ape. J. Comp. Psychol. 120 (1), 38.

Cafazzo, S., Bonanni, R., Valsecchi, P., & Natoli, E. (submitted). Social variables affecting mate preferences, copulation and reproductive outcome in a pack of free-ranging dogs.

Call, J., Bräuer, J., Kaminski, J., Tomasello, M., 2003. Domestic dogs (*Canis familiaris*) are sensitive to the attentional state of humans. J. Comp. Psychol. 117 (3), 257.

Carbone, C., Du Toit, J.T., Gordon, I.J., 1997. Feeding success in African wild dogs: does kleptoparasitism by spotted hyenas influence hunting group size? J. Anim. Ecol. 66, 318–326.

Carbone, C., Frame, L., Frame, G., Malcolm, J., Fanshawe, J., FitzGibbon, C., et al., 2005. Feeding success of African wild dogs (*Lycaon pictus*) in the Serengeti: the effects of group size and kleptoparasitism. J. Zool. 266 (02), 153–161.

Carmichael, L.E., Szor, G., Berteaux, D., Giroux, M.A., Cameron, C., Strobeck, C., 2007. Free love in the far north: plural breeding and polyandry of Arctic foxes (*Alopex lagopus*) on Bylot Island, Nunavut. Can. J. Zool. 85 (3), 338–343.

Cerling, T.E., Harris, J.M., MacFadden, B.J., Leakey, M.G., Quade, J., Eisenmann, V., et al., 1997. Global vegetation change through the Miocene/Pliocene boundary. Nature 389 (6647), 153–158.

Clark, C.W., Mangel, M., 1986. The evolutionary advantages of group foraging. Theor. Popul. Biol. 30 (1), 45–75.

Clutton-Brock, J., 1995. Origins of the dog: domestication and early history. In: Serpell, J. (Ed.), The domestic dog: its evolution, behaviour and interactions with people. Cambridge University Press, Cambridge, UK, pp. 7–20.

Coppinger, R., Coppinger, L., 2001. Dogs: a startling new understanding of canine origin, behaviour and evolution. Scribner, New York.

Courchamp, F., Rasmussen, G.S., Macdonald, D.W., 2002. Small pack size imposes a trade-off between hunting and pup-guarding in the painted hunting dog Lycaon pictus. Behav. Ecol. 13 (1), 20–27.

Creel, S., Creel, N.M., 1995. Communal hunting and pack size in African wild dogs, Lycaon pictus. Anim. Behav. 50 (5), 1325–1339.

Creel, S., Creel, N.M., 1998. Six ecological factors that may limit African wild dogs, Lycaon pictus. Anim. Conservation 1 (1), 1–9.

Creel, S., Creel, N.M., 2002. The African wild dog: behavior, ecology and conservation. Princeton University Press, Princeton, NJ.

Davis, S.J., Valla, F.R., 1978. Evidence for domestication of the dog 12,000 years ago in the Natufian of Israel. Nature 276, 608–610.

Dayan, T., 1994. Early domesticated dogs of the Near East. J. Archaeol. Sci. 21 (5), 633–640.

Dunbar, R.I.M., 2009. The social brain hypothesis and its implications for social evolution. Ann. Hum. Biol. 36 (5), 562–572.

Druzhkova, A.S., Thalmann, O., Trifonov, V.A., Leonard, J.A., Vorobieva, N.V., Ovodov, N.D., et al., 2013. Ancient DNA analysis affirms the canid from Altai as a primitive dog. PloS One 8 (3), e57754.

Emery, N.J., 2000. The eyes have it: the neuroethology, function and evolution of social gaze. Neurosci. Biobehav. Rev. 24 (6), 581–604.

Emlen, S.T., 1991. Evolution of cooperative breeding in birds and mammals. In: Krebs, J.R., Davies, N.B. (Eds.), Behavioural ecology: an evolutionary approach. Blackwell Scientific, Oxford, UK, pp. 301–337.

Feddersen-Petersen, D.U., 2007. Social behaviour of dogs and related canids. In: Jensen, P. (Ed.), The behavioural biology of dogs. CAB International, Wallingford, UK, pp. 105–119.

Fiedel, S.J., 2005. Man's best friend—mammoth's worst enemy? A speculative essay on the role of dogs in Paleoindian colonization and megafaunal extinction. World Archaeol. 37 (1), 11–25.

Finarelli, J.A., 2008. Testing hypotheses of the evolution of brain-body size scaling in the Canidae (Carnivora, Mammalia). Paleobiology 34, 35–145.

Frank, H., Frank, M.G., 1982. On the effects of domestication on canine social development and behavior. Appl. Anim. Ethol. 8, 507–525.

Freedman, A.H., Gronau, I., Schweizer, R.M., Ortega-Del Vecchyo, D., Han, E., Silva, P.M., et al., 2014. Genome sequencing highlights the dynamic early history of dogs. PLoS Genetics 10 (1), e1004016.

Fritts, S.H., Mech, L.D., 1981. Dynamics, movements, and feeding ecology of a newly protected wolf population in northwestern Minnesota. Wildlife Monogr. 80, 3–79.

Frommolt, K.H., Goltsman, M.E., Macdonald, D.W., 2003. Barking foxes, Alopex lagopus: field experiments in individual recognition in a territorial mammal. Anim. Behav. 65 (3), 509–518.

Gácsi, M., Gyori, B., Miklósi, Á., Virányi, Z., Kubinyi, E., 2005. Species-specific differences and similarities in the behavior of hand-raised dog and wolf pups in social situations with humans. Dev. Psychobiol. 47, 111–122.

Gácsi, M., Gyoöri, B., Virányi, Z., Kubinyi, E., Range, F., Belényi, B., et al., 2009. Explaining dog wolf differences in utilizing human pointing gestures: selection for synergistic shifts in the development of some social skills. PLoS One 4 (8), e6584.

Gácsi, M., Vas, J., Topál, J., Miklósi, Á., 2013. Wolves do not join the dance: sophisticated aggression control by adjusting to human social signals in dogs. Appl. Anim. Behav. Sci. 145, 109–122.

Geffen, E., Gompper, M.E., Gittleman, J.L., Luh, H.K., MacDonald, D.W., Wayne, R.K., 1996. Size, life-history traits, and social organization in the Canidae: a reevaluation. Am. Nat. 147 (1), 140–160.

Germonpré, M., Sablin, M.V., Stevens, R.E., Hedges, R.E., Hofreiter, M., Stiller, M., et al., 2009. Fossil dogs and wolves from Palaeolithic sites in Belgium, the Ukraine and Russia: osteometry, ancient DNA and stable isotopes. J. Archaeol. Sci. 36 (2), 473–490.

Gese, E.M., 1998. Response of neighboring coyotes (Canis latrans) to social disruption in an adjacent pack. Can. J. Zool. 76 (10), 1960–1963.

Giuffra, E.J.M.H., Kijas, J.M.H., Amarger, V., Carlborg, Ö., Jeon, J.T., Andersson, L., 2000. The origin of the domestic pig: independent domestication and subsequent introgression. Genetics 154 (4), 1785–1791.

Gottelli, D., Sillero–Zubiri, C., Applebaum, G.D., Roy, M.S., Girman, D.J., Garcia–Moreno, J., et al., 1994. Molecular genetics of the most endangered canid: the Ethiopian wolf Canis simensis. Mol. Ecol. 3 (4), 301–312.

Graphodatsky, A., Perelman, P.L., Sokolovskaya, N., Beklemisheva, V.R., Serdukova, N.A., Dobigny, G., et al., 2008. Phylogenomics of the dog and fox family (Canidae, Carnivora) revealed by chromosome painting. Chromosome. Res. 16, 129–143.

Grayson, D.K., Meltzer, D.J., 2002. Clovis hunting and large mammal extinction: a critical review of the evidence. J. World Prehistory 16 (4), 313–359.

Gusset, M., Macdonald, D.W., 2010. Group size effects in cooperatively breeding African wild dogs. Anim. Behav. 79 (2), 425–428.

Hamilton-Smith, C., 1839. The natural history of dogs: Canidae or genus Canis of authors; including also the genera Hyaena and Proteles. vol. 1. W.H. Lizars, Edinburgh.

Hare, B., Brown, M., Williamson, C., Tomasello, M., 2002. The domestication of social cognition in dogs. Science 298 (5598), 1634–1636.

Hare, B., Rosati, A., Kaminski, J., Bräuer, J., Call, J., Tomasello, M., 2010. The domestication hypothesis for dogs' skills with human communication: a response to Udell et al. (2008) and Wynne, et al. (2008). Anim. Behav. 79 (2), e1–e6.

Hare, B., Tomasello, M., 1999. Domestic dogs (Canis familiaris) use human and conspecific social cues to locate hidden food. J. Comp. Psychol. 113 (2), 173.

Hare, B., Tomasello, M., 2005. Human-like social skills in dogs? Trends Cogn. Sci. 9 (9), 439–444.

Hare, B., Wobber, V., Wrangham, R., 2012. The self-domestication hypothesis: evolution of bonobo psychology is due to selection against aggression. Anim. Behav. 83 (3), 573–585.

Haynes, G., 1983. Frequencies of spiral and green-bone fractures on ungulate limb bones in modern surface assemblages. Am. Antiq. 48, 102–114.

Hennessy, C.A., Dubach, J., Gehrt, S.D., 2012. Long-term pair bonding and genetic evidence for monogamy among urban coyotes (Canis latrans). J. Mammal. 93 (3), 732–742.

Hodge, S.J., 2005. Helpers benefit offspring in both the short and long-term in the cooperatively breeding banded mongoose. Proc. Royal Soc. B. Biol. Sci. 272 (1580), 2479–2484.

Horowitz, A., 2009. Attention to attention in domestic dog (Canis familiaris) dyadic play. Anim. Cogn. 12 (1), 107–118.

Kaminski, J., Nitzschner, M., 2013. Do dogs get the point? A review of dog–human communication ability. Learn. Motiv. 44 (4), 294–302.

Kamler, J.F., Ballard, W.B., Gese, E.M., Harrison, R.L., Karki, S.M., 2004. Dispersal characteristics of swift foxes. Can. J. Zool. 82 (12), 1837–1842.

Kirchhofer, K.C., Zimmermann, F., Kaminski, J., Tomasello, M., 2012. Dogs (Canis familiaris), but not chimpanzees (Pan troglodytes), understand imperative pointing. PloS One 7 (2), e30913.

Kleiman, D., 1977. Monogamy in mammals. Q. Rev. Biol. 52, 39–69.

Kleiman, D., 2011. Canid mating systems, social behavior parental care and ontogeny: are they flexible? Behav. Genet. 41, 803–809.

Kis, A., Bence, M., Lakatos, G., Pergel, E., Turcsán, B., et al., 2014. Oxytocin receptor gene polymorphisms are associated with human directed social behavior in dogs (Canis familiaris). PLoS One 9 (1), e83993.

Larsen, C.S., 1995. Biological changes in human populations with agriculture. Annu. Rev. Anthropol. 24 (1), 185–213.

Leonard, J.A., Vila, C., Wayne, R.K., 2005. FAST TRACK: Legacy lost: genetic variability and population size of extirpated US grey wolves (Canis lupus). Mol. Ecol. 14 (1), 9–17.

Lindblad-Toh, K., Wade, C.M., Mikkelsen, T.S., Karlsson, E.K., Jaffe, D.B., Kamal, M., et al., 2005. Genome sequence, comparative analysis and haplotype structure of the domestic dog. Nature 438 (7069), 803–819.

Loftus, R.T., MacHugh, D.E., Bradley, D.G., Sharp, P.M., Cunningham, P., 1994. Evidence for two independent domestications of cattle. Proc. Natl. Acad. Sci. 91 (7), 2757–2761.

Lord, K., Feinstein, M., Smith, B., Coppinger, R., 2013. Variation in reproductive traits of members of the genus Canis with special attention to the domestic dog (Canis familiaris). Behav. Processes. 92, 131–142.

Losey, R.J., Bazaliiskii, V.I., Garvie-Lok, S., Germonpré, M., Leonard, J.A., Allen, A.L., et al., 2011. Canids as persons: early Neolithic dog and wolf burials, Cis-Baikal, Siberia. J. Anthropol. Archaeol. 30 (2), 174–189.

Lyras, G.A., Van der Geer, A.A.E., 2003. External brain anatomy in relation to the phylogeny of Caninae (Carnivora: Canidae). Zool. J. Linnean Soc. 138 (4), 505–522.

Macdonald, D.W., 1979. "Helpers" in fox society. Nature 282 (5734), 69–71.

Macdonald, D.W., Creel, S., Mills, M.G.L., 2004. Canid society. In: Macdonald, D.W., Sillero-Zubiri, C. (Eds.), Biology and conservation of wild canids. Oxford University Press, New York, pp. 85–106.

Macdonald, D.W., Sillero-Zubiri, C., 2004a. Biology and conservation of wild canids. Oxford University Press, New York, pp. 85–106.

Macdonald, D.W., Sillero-Zubiri, C., 2004b. Dramatis personae. In: Macdonald, D.W., Sillero-Zubiri, C. (Eds.), Biology and conservation of wild canids. Oxford University Press, New York, pp. 3–36.

MacNulty, D.R., Smith, D.W., Mech, L.D., Vucetich, J.A., Packer, C., 2012. Nonlinear effects of group size on the success of wolves hunting elk. Behav. Ecol. 23 (1), 75–82.

Malcolm, J.R., Marten, K., 1982. Natural selection and the communal rearing of pups in African wild dogs (Lycaon pictus). Behav. Ecol. Sociobiol. 10 (1), 1–13.

Marino, J., Sillero-Zubiri, C., Johnson, P.J., Macdonald, D.W., 2012. Ecological bases of philopatry and cooperation in Ethiopian wolves. Behav. Ecol. Sociobiol. 66 (7), 1005–1015.

Mech, L.D., Boitani, L., 2003. Wolves: behaviour, ecology, and conservation. University of Chicago Press, Chicago.

Mech, L.D., Wolf, P.C., Packard, J.M., 1999. Regurgitative food transfer among wild wolves. Can. J. Zool. 77 (8), 1192–1195.

Miklósi, Á., 2007. Dog behaviour, evolution and cognition. Oxford University Press, Oxford.

Miklósi, Á., Kubinyi, E., Topál, J., Gácsi, M., Virányi, Z., Csányi, V., 2003. A simple reason for a big difference: wolves do not look back at humans, but dogs do. Curr. Biol. 13 (9), 763–766.

Miklósi, Á., Szabo, D., 2012. Modeling behavioural evolution and cognition in canines: some problematic issues. Jap. J. Anim. Psychol. 62 (1), 69–89.

Miklósi, Á., Topál, J., 2013. What does it take to become 'best friends'? Evolutionary changes in canine social competence. Trends Cogn. Sci. 17 (6), 287–294.

Moehlman, P.D., 1986. Ecology and cooperation in canids. In: Rubenstein, D.I., Wrangham, R.W. (Eds.), Ecological aspects of social evolution: birds and mammals. Princeton University Press, Princeton, NJ, pp. 64–86.

Moehlman, P.D., 1989. Intraspecific variation in canid social systems. In: Gittleman, J.L. (Ed.), Carnivore behavior, ecology and evolution, vol. 1. Cornell University Press, Ithaca, NY, pp. 143–163.

Morey, D.F., 1994. The early evolution of the domestic dog. Am. Sci. 82 (4), 336–347.

Morey, D.F., Wiant, M.D., 1992. Early Holocene domestic dog burials from the North American Midwest. Curr. Anthropol. 33 (2), 224–229.

Nowak, R.M., 2003. Wolf evolution and taxonomy. In: Mech, L.D., Boitani, L. (Eds.), Wolves: behavior, ecology and conservation. University of Chicago Press, Chicago, pp. 239–258.

Olsen, S.J., 1985. Origins of the domestic dog: the fossil record. University of Arizona Press, Tucson.

Ostrander, E.A., Wayne, R.K., 2005. The canine genome. Genome. Res. 15 (12), 1706–1716.

Ovodov, N.D., Crockford, S.J., Kuzmin, Y.V., Higham, T.F., Hodgins, G.W., van der Plicht, J., (2011). A 33,000-year-old incipient dog from the Altai Mountains of Siberia: evidence of the earliest domestication disrupted by the Last Glacial Maximum. Plos One, 6 (7), e22821.

Packard, J.M., 2003. Wolf behavior: reproductive, social and intelligent. In: Mech, L.D., Boitani, L. (Eds.), Wolves: behavior, ecology and conservation. University of Chicago Press, Chicago, pp. 35–65.

Pal, S.K., 2005. Parental care in free-ranging dogs, *Canis familiaris*. Appl. Anim. Behav. Sci. 90 (1), 31–47.

Poessel, S.A., Gese, E.M., 2013. Den attendance patterns in swift foxes during pup rearing: varying degrees of parental investment within the breeding pair. J. Ethol. 31 (2), 1–9.

Pongrácz, P., Gácsi, M., Hegedüs, D., Péter, A., Miklósi, Á., 2013. Test sensitivity is important for detecting variability in pointing comprehension in canines. Anim. Cogn. 16 (5), 1–15.

Pulliam, H.R., Caraco, T., 1984. Living in groups: is there an optimal group size? In: Krebs, J.R., Davies, N.B. (Eds.), Behavioral ecology: an evolutionary approach, second ed. Sinauer, Sunderland, MA, pp. 122–127.

Radinsky, L., 1969. Outlines of canid and felid brain evolution. Ann. NY. Acad. Sci. 167 (1), 277–288.

Radinsky, L., 1973. Evolution of the canid brain. Brain. Behav. Evol. 7 (3), 169–185.

Range, F., Virányi, Z., 2011. Development of gaze following abilities in wolves (*Canis lupus*). PloS One 6 (2), e16888.

Range, F., Virányi, Z., 2013. Social learning from humans or conspecifics: differences and similarities between wolves and dogs. Frontiers Psychol. 4 (December), 868.

Range, F., Virányi, Z., 2014. Wolves are better imitators of conspecifics than dogs. PLoS One 9 (1), e86559.

Reid, P.J., 2009. Adapting to the human world: dogs' responsiveness to our social cues. Behav. Processes 80 (3), 325–333.

Richerson, P.J., Boyd, R., Bettinger, R.L., 2001. Was agriculture impossible during the Pleistocene but mandatory during the Holocene? A climate change hypothesis. Am. Antiq. 66 (3), 387–411.

Riedel, J., Schumann, K., Kaminski, J., Call, J., Tomasello, M., 2008. The early ontogeny of human–dog communication. Anim. Behav. 75 (3), 1003–1014.

Rosati, A.G., Hare, B., 2009. Looking past the model species: diversity in gaze-following skills across primates. Curr. Opin. Neurobiol. 19 (1), 45–51.

Roy, M.S., Geffen, E., Smith, D., Wayne, R.K., 1996. Molecular genetics of pre–1940 red wolves. Conservation Biol. 10 (5), 1413–1424.

Russell, A.F., Young, A.J., Spong, G., Jordan, N.R., Clutton-Brock, T.H., 2007. Helpers increase the reproductive potential of offspring in cooperative meerkats. Proc. Royal Soc. B: Biol. Sci. 274 (1609), 513–520.

Ruusila, V., Pesonen, M., 2004. Interspecific cooperation in human (Homo sapiens) hunting: the benefits of a barking dog (Canis familiaris). Ann. Zool. Fennici 41, 545–549.

Sastre, N., Vilà, C., Salinas, M., Bologov, V.V., Urios, V., Sánchez, A., et al., 2011. Signatures of demographic bottlenecks in European wolf populations. Conservation Genet. 12 (3), 701–712.

Savolainen, P., Zhang, Y.P., Luo, J., Lundeberg, J., Leitner, T., 2002. Genetic evidence for an East Asian origin of domestic dogs. Science 298 (5598), 1610–1613.

von Schantz, T., 1984. 'Non-breeders' in the red fox Vulpes vulpes: a case of resource surplus. Oikos, 37, 59–65.

Schleidt, W.M., Shalter, M.D., 2003. Co-evolution of humans and canids. Evol. Cogn. 9, 57–72.

Shultz, S., Dunbar, R.I., 2007. The evolution of the social brain: anthropoid primates contrast with other vertebrates. Proc. Royal Soc. B: Biol. Sci. 274 (1624), 2429–2436.

Schwab, C., Huber, L., 2006. Obey or not obey? Dogs (Canis familiaris) behave differently in response to attentional states of their owners. J. Comp. Psychol. 120 (3), 169.

Sillero-Zubiri, C., Gottelli, D., Macdonald, D.W., 1996. Male philopatry, extra-pack copulations and inbreeding avoidance in Ethiopian wolves (Canis simensis). Behav. Ecol. Sociobiol. 38 (5), 331–340.

Sillero-Zubiri, C., Hoffman, M., Macdonald, D.W., 2004. Canids: foxes, wolves, jackals and dogs. Status Survey and Conservation Action Plan, vol. 62. IUCN, Gland, Switzerland, and Cambridge, UK.

Sparkman, A.M., Adams, J., Beyer, A., Steury, T.D., Waits, L., Murray, D.L., 2011. Helper effects on pup lifetime fitness in the cooperatively breeding red wolf (Canis rufus). Proc. Royal Soc. B: Biol. Sci. 278 (1710), 1381–1389.

Spiering, P.A., Somers, M.J., Maldonado, J.E., Wildt, D.E., Gunther, M.S., 2010. Reproductive sharing and proximate factors mediating cooperative breeding in the African wild dog (Lycaon pictus). Behav. Ecol. Sociobiol. 64 (4), 583–592.

Stahl, P.W., 2012. Interactions between humans and endemic canids in Holocene South America. J. Ethnobiol. 32 (1), 108–127.

Stahl, P.W., 2013. Early dogs and endemic South American canids of the Spanish Main. J. Anthropol. Res. 69 (4), 515–533.

Tchernov, E., Valla, F.F., 1997. Two new dogs, and other Natufian dogs, from the southern Levant. J. Archaeol. Sci. 24 (1), 65–95.

Tedford, R.H., Taylor, B.E., Wang, X., 1995. Phylogeny of the Caninae (Carnivora: Canidae): the living taxa. Am. Mus. Novit. 0, 1–37.

Thalmann, O., Shapiro, B., Cui, P., Schuenemann, V.J., Sawyer, S.K., Greenfield, D.L., et al., 2013. Complete mitochondrial genomes of ancient canids suggest a European origin of domestic dogs. Science 342 (6160), 871–874.

Tomasello, M., 2008. Origins of human communication. MIT Press, Cambridge.

Topál, J., Gácsi, M., Miklósi, Á., Virányi, Z., Kubinyi, E., Csányi, V., 2005. Attachment to humans: a comparative study on hand-reared wolves and differently socialized dog puppies. Anim. Behav. 70 (6), 1367–1375.

Topál, J., Gergely, G., Erdőhegyi, Á., Csibra, G., Miklósi, Á., 2009. Differential sensitivity to human communication in dogs, wolves, and human infants. Science 325 (5945), 1269–1272.

Trut, L., Oskina, I., Kharlamova, A., 2009. Animal evolution during domestication: the domesticated fox as a model. Bioessays 31 (3), 349–360.

Trut, L.N., Plyusnina, I.Z., Oskina, I.N., 2004. An experiment on fox domestication and debatable issues of evolution of the dog. Russ. J. Genet. 40 (6), 644–655.

Udell, M.A., Dorey, N.R., Wynne, C.D., 2008. Wolves outperform dogs in following human social cues. Anim. Behav. 76 (6), 1767–1773.

Udell, M.A., Dorey, N.R., Wynne, C.D., 2010. What did domestication do to dogs? A new account of dogs' sensitivity to human actions. Biol. Rev. 85 (2), 327–345.

Udell, M.A., Dorey, N.R., Wynne, C.D., 2011. Can your dog read your mind? Understanding the causes of canine perspective taking. Learning Behav. 39 (4), 289–302.

Van Valkenburgh, B., 1991. Iterative evolution of hypercarnivory in canids (Mammalia: Carnivora): evolutionary interactions among sympatric predators. Paleobiology 17 (4), 340–362.

Van Valkenburgh, B.L., Sacco, T., Wang, X., 2003. Pack hunting in Miocene Borophagine dogs: evidence from craniodental morphology and body size. In: L. Flynn (Ed.), Vertebrate fossils and their context: contributions in honor of Richard H. Tedford. Bull. Am. Museum Nat. Hist. 279, 147–162.

Van Valkenburgh, B., Wang, X., Damuth, J., 2004. Cope's rule, hypercarnivory, and extinction in North American canids. Science 306 (5693), 101–104.

Venkataraman, A.B., Arumugam, R., Sukumar, R., 1995. The foraging ecology of dhole (*Cuon alpinus*) in Mudumalai Sanctuary, southern India. J. Zool. 237 (4), 543–561.

Venkataraman, A.B., Johnsingh, A.J.T., 2004. The behavioural ecology of dholes in India. In: D.W. Macdonald and C. Sillero-Zubiri (Eds.), The biology and conservation of wild canids. Oxford University Press, Oxford, pp. 323–356.

Virányi, Z., Gácsi, M., Kubinyi, E., Topál, J., Belényi, B., Ujfalussy, D., et al., 2008. Comprehension of human pointing gestures in young human-reared wolves (*Canis lupus*) and dogs (*Canis familiaris*). Anim. Cogn. 11 (3), 373–387.

Voigt, D.R., Macdonald, D.W., 1984. Variation in the spatial and social behaviour of the red fox, *Vulpes vulpes*. Acta Zoologica Fennica 171, 261–265.

Vucetich, J.A., Peterson, R.O., Waite, T.A., 2004. Raven scavenging favours group foraging in wolves. Anim. Behav. 67 (6), 1117–1126.

Waller, B.M., Peirce, K., Caeiro, C.C., Scheider, L., Burrows, A.M., McCune, S., et al., 2013. Paedomorphic facial expressions give dogs a selective advantage. PloS One 8 (12), e82686.

Wang, G.D., Zhai, W., Yang, H.C., Fan, R.X., Cao, X., Zhong, L., et al., 2013. The genomics of selection in dogs and the parallel evolution between dogs and humans. Nat. Comm. 4, 1860.

Wayne, R.K., Benveniste, R.E., Janczewski, D.N., O'Brien, S.J., 1989. Molecular and biochemical evolution of the Carnivora. In: J.L. Gittleman (Ed.), Carnivore behavior, ecology, and evolution. Springer, London, pp. 465–494.

Wayne, R.K., Jenks, S.M., 1991. Mitochondrial DNA analysis implying extensive hybridization of the endangered red wolf *Canis rufus*. Nature 351, 565–568.

Wayne, R.K., Geffen, E., Girman, D.J., Koepfli, K.P., Lau, L.M., Marshall, C.R., 1997. Molecular systematics of the Canidae. Syst. Biol. 46 (4), 622–653.

Wayne, R.K., Nash, W.G., O'Brien, S.J., 1986. Chromosomal evolution of the Canidae. II. Divergence from the primitive carnivore karyotype. Cytogenet. Cell. Genet. 44 (2–3), 134–141.

Wayne, R.K., Nash, W.G., O'Brien, S.J., 1987. Chromosomal evolution of the Canidae. Cytogenet. Genome. Res. 44 (2–3), 123–133.

Wells, M.C., Bekoff, M., 1982. Predation by wild coyotes: behavioral and ecological analyses. J. Mammal. 63, 118–127.

Wright, H.W.Y., 2006. Paternal den attendance is the best predictor of offspring survival in the socially monogamous bat-eared fox. Anim. Behav. 71 (3), 503–510.

Wong, A.K., Ruhe, A.L., Dumont, B.L., Robertson, K.R., Guerrero, G., Shull, S.M., et al., 2010. A comprehensive linkage map of the dog genome. Genetics 184 (2), 595–605.

Zabel, C.J., Taggart, S.J., 1989. Shift in red fox, *Vulpes vulpes*, mating system associated with El Niño in the Bering Sea. Anim. Behav. 38 (5), 830–838.

Zeder, M.A., Hesse, B., 2000. The initial domestication of goats (*Capra hircus*) in the Zagros Mountains 10,000 years ago. Science 287 (5461), 2254–2257.

Zeuner, F.E., 1967. Geschichte der Haustiere. Bayrischer Landwirtschaftsverlag, München, Basel, Wien.

Zhang, H., Chen, L., 2011. The complete mitochondrial genome of dhole Cuon alpinus: phylogenetic analysis and dating evolutionary divergence within Canidae. Mol. Biol. Rep. 38, 1651–1660.

Zimen, E., 1992. Der Hund. Abstammung, Verhalten, Mensch und Hund. Goldmann, München.

Zrzavý, J., Řičánková, V., 2004. Phylogeny of recent Canidae (Mammalia, Carnivora): relative reliability and utility of morphological and molecular datasets. Zool. Scr. 33 (4), 311–333.

# Chapter 2

# On the Way to a Better Understanding of Dog Domestication: Aggression and Cooperativeness in Dogs and Wolves

Zsófia Virányi and Friederike Range
*Comparative Cognition, Messerli Research Institute, University of Veterinary Medicine, Vienna, Austria; Medical University of Vienna, Vienna, Austria; University of Vienna, Vienna, Austria; Wolf Science Centre, Ernstbrunn, Austria*

## 2.1 DOG DOMESTICATION AND HUMAN EVOLUTION: THE ROLE OF WOLF–DOG COMPARISONS

It has repeatedly been suggested that dogs can tell us about the evolution of human social behaviour and cognition (Miklósi et al., 2004; Hare & Tomasello, 2005; Fitch et al., 2010). More precisely, we expect dog–human similarities based on the hypothesis that dogs and humans went through convergent evolution, as has been proposed in the previous chapter. The core of this hypothesis is that during domestication, dogs have been selected to cooperate and communicate with humans, to live and work as members of human groups—just as has happened to humans as well. In response to the demands posed by the human environment, we expect that some genetic predispositions evolved in dogs allowing them to develop skills shared with humans. It is a mistake, however, to automatically attribute all dog–human similarities to domestication, that is, to assume that they rely on genetic changes that occurred since the dog separated from its closest wild-living relative, the wolf. Alternatively or, more likely, additionally, because pet dogs and humans grow up and live essentially in the same social environment, similar socialisation and learning can take place. Consequently, we can be sure that dog–human similarities have been supported by evolutionary changes only if we can demonstrate that wolves socialized with humans do not show those skills that dogs and humans share.

The Social Dog. http://dx.doi.org/10.1016/B978-0-12-407818-5.00002-4

## 2.1.1 How to Compare Dogs and Wolves for This Aim?

Consequently, we need to compare dogs and wolves that grew up and live in an identical environment to make sure that their behaviour differences do not originate purely from their different individual experiences. Then we can demonstrate the effects of domestication on dog social behaviour and cognition and, thus, reason about human evolution. Few comparisons on dog and wolf social cognition exist, and even fewer satisfy this requirement. In a number of studies, wolves kept in an enclosure with daily but limited contact with humans were compared to pet dogs (Agnetta et al., 2000; Hare et al., 2002); or wolves that had received an early and intensive socialisation with humans were compared to stray dogs living in a shelter (Udell et al., 2008); or, although both grew up in human families, wolves that lived in this way only for 2 to 4 months and were then returned to enclosures where they had only limited contact with humans were compared at a later age to dogs that remained in their human families (Topál et al., 2009a). Because raising and keeping a sufficient sample of dogs and wolves socialized with humans under identical conditions and satisfying the animals' needs at the same time require a lot of effort, it is understandable that many research groups work with more or less ad hoc study populations that do not allow correct comparisons. However, such comparisons of dogs and wolves with different experiences cannot provide conclusive evidence about evolutionary questions, though they may be useful to create hypotheses and to explore the behavioural plasticity of dogs or wolves living in different environments. For this latter aim, it is, of course, important to make sure that subjects of different origins are tested with the same methods.

Up to now, few research projects have had the potential to detect evolutionary differences between dogs and wolves. In the 1960s, at the Institut für Haustierkunde of the University of Kiel (Germany) led by Prof. Wolf Herre and later by Dorit Feddersen-Petersen, wolves and standard poodles (as well as coyotes and miniature poodles and golden jackals and toy poodles) were crossed in order to study the genetics of brain size and other morphological features, such as fur structure and colour. The poodle–wolf hybrids were called Puwos or Wopus, depending on whether the mother or the father was a wolf, respectively. The animals, including wolves and poodles together, were kept in packs in enclosures at the zoological garden of the institute (see Figure 2-1). The pups were raised by their natural mothers. Erik Zimen used this exceptional opportunity and began investigating the behaviour of these animals. He also hand-raised wolves, poodles, and their crosses, and compared the human-directed behaviour of these animals to the mother-raised animals (Zimen, 1987). In 1975, when the morphological studies ended, Dorit Feddersen-Petersen took over the behavioural observations of the animals at the institute. Since then, her group has collected observational data of social interactions on 189 canids (of whom 74 were hand-raised) including not only the institute's populations but also dogs of several different breeds (Feddersen-Petersen, 1991, 1994, 2000).

FIGURE 2-1   Wolves and poodles living in a captive pack established and observed by Dorit Feddersen-Petersen's research group at the University of Kiel, Germany. Very often poodles took the leading positions in the dominance hierarchy, and the wolves readily submitted to them, as can be seen in this picture. Note the ear positions, averted gaze, and back posture of both wolves in this photo. See color plate section. *(Courtesy of Dorit Feddersen-Petersen.)*

More or less in parallel, from 1979 to 1981, the University of Michigan canine information-processing project took place under the leadership of Martha and Harry Frank (Frank & Frank, 1987). This research program was of a smaller scale (including a total of 11 wolf and 4 malamute pups), but focused on comparing dogs and wolves from a behavioural and cognitive perspective. Albeit with a small sample size, it otherwise fully satisfied the methodological requirements of searching for genetically based differences between dogs and wolves. In two different years, the Franks raised 4 wolf and 4 malamute pups in the very same environment. From the age of 10 to 13 days on, all pups were taken care of by the same lactating wolf female and by two human hand-raisers. All animals were housed in the same facility, had contact with the same adult conspecifics, were fed the same diet, and were administered the same experimental tests following the same regime (Frank & Frank, 1982). A year later, 7 additional wolf pups raised exclusively by humans were tested in the same tests, but their performance differed from the 4 mother- and human-raised wolf pups, which was attributed to motivational differences (Frank et al., 1989; Frank, 2011). The animals were tested in experiments addressing their physical problem-solving and learning abilities as well as their trainability (Frank & Frank, 1988; Frank et al., 1989; Frank, 2011). Based on informal observations, their social interactions with conspecifics also were compared (Frank & Frank, 1982).

More recent attempts using somewhat bigger samples have been made by the Family Dog Project of the Department of Ethology, Eötvös Loránd University, Budapest, Hungary, led by Ádám Miklósi in the early 1990s and currently at the Wolf Science Centre, Ernstbrunn, Austria, founded by Kurt Kotrschal and the two authors of this chapter. In both research programs, the dogs and wolves have been raised under identical conditions by humans, but their goals and accordingly their raising regimes differ. In Hungary, 13 European wolves and 11 dogs (mongrels) were raised individually, each by a specific human raiser, and lived in a family environment from their first week on. Because this project aimed at comparing dogs and wolves in their communication, interactions, and relationship with humans, the animals went through very intensive socialisation in an urban environment, lived in family homes, accompanied their raisers to the university, travelled on public transportation, and thus met unfamiliar people and dogs on a daily basis—i.e., lived like well-socialized pet dogs (see Kubinyi et al., 2007, for a review). It was not possible, however, to maintain this form of keeping the wolves after the age of 2 to 4 months for reasons of safety and animal welfare. Accordingly, around this age, the wolves were returned to the captive facility where they had been born and were visited by their hand-raisers once or twice a week. Most of the dogs, however, stayed with their hand-raisers, and all continued living in a human family as a pet.

In contrast, at the Wolf Science Centre, both dogs (mongrels) and wolves (timber wolves) (currently 12 and 13, respectively) are kept in a way that is sustainable also for adult wolves: in packs in large enclosures in the Gamepark Ernstbrunn. The animals have been hand-raised in peer groups (wolves and dogs separately) and had close contact not only with a group of 6 to 10 human raisers but also with 4 to 6 pet dogs of the hand-raisers with whom the pups interacted as with adult conspecifics. The animals had continuous human contact till the age of 4 to 5 months when they were integrated in packs of older conspecifics. Importantly, the animals have continued to have daily interactions with their raisers: they receive basic obedience training and have a firm routine of being separated from the packs in order to participate in regular cognitive and behavioural testing. Because the animals are similarly well socialized with humans and with conspecifics, this raising regime allows for the comparison of intra- and interspecific interactions of the dogs and wolves using experiments as well as observations of spontaneous behaviour (see Range & Viranyi, 2011, 2013, 2014, as examples).

## 2.1.2  What Can Dog–Wolf Comparisons Really Tell Us?

Even if raised comparably, wolves and dogs do not necessarily gain the same experiences and go through the same learning processes, but may easily adapt their behaviour to different aspects of the same environment. Therefore, we can detect only epigenetic differences between wolves and dogs (Gácsi et al., 2009a; Miklósi & Topál, 2011): genetically based differences potentially enlarged by differential learning processes, which are likely to be more profound the older the investigated animals are. Still, given that their environment is identical, what dogs and wolves learn originates from their preferences and sensitivities, and thus will inform us about the domestication process. In this sense, we can talk about genetically based differences between dogs and wolves, although we need to keep in mind that evolutionary arguments can hardly be made about the effect size of such dog–wolf differences. Nonetheless, these comparisons provide us with evidence that domestication has had an influence on the behaviour of dogs.

Finding such dog–wolf differences does not mean, however, that domestication is either necessary or sufficient to explain human-like behaviour in dogs (Udell et al., 2011). Udell and colleagues (2008) showed that human-raised wolves can outperform shelter dogs that have limited experiences with humans in their use of human-given cues. These results demonstrate that (1) socialisation with and learning about humans can provide wolves with an alternative means to reach a certain dog-like performance, e.g., to pay attention to humans; and that (2) even in dogs, a minimal amount of socialisation and individual learning is needed to enable the use of some human-directed skills (see also Hare et al., 2010, for a discussion on this issue). At the same time, however, dogs can follow a difficult form of human pointing (momentary distal pointing) earlier than wolves, showing that they are genetically predisposed to develop this skill faster (Gácsi et al., 2009a). In sum, dog–wolf comparisons have been used for different purposes. On the one hand, they have been used to demonstrate that developmental processes can either mask existing or compensate for missing genetic features; and on the other hand, they have been used to show that domestication has changed a genetic predisposition in dogs compared to wolves. Important to realize, however, is that these are two independent questions and that the answer can be "Yes" to both.

## 2.2 HUMAN-LIKE BEHAVIOUR IN DOGS BUT NOT IN WOLVES: PART 1

The second question (has domestication changed a genetic predisposition in dogs compared to wolves?) is the relevant one when arguing for convergent evolution between humans and dogs. While numerous studies have reported human-like behaviour in dogs and pointed out the potential contribution of domestication (see Reid, 2009; Udell et al., 2010; Bensky et al., 2013, for reviews), only in very few cases wolves have been tested in the above-suggested rigorous manner and demonstrated to lack a similar ability. We summarise these dog–wolf comparisons in two sections in this chapter (Sections 2.2 and 2.6). The first section focuses on the performance of dogs and wolves in using a human-specific communicative cue, pointing to locate food in one of two hiding places. Interestingly, based on results produced with this single experimental paradigm, various evolutionary hypotheses have been put forward to theorise about the selection pressures that might have shaped dog social behaviour and cognition during the course of domestication. We review these hypotheses in the next section.

## 2.2.1  Following Human Pointing: A Simple Test and a Lot of Hypothesising

Quite intriguing, most studies investigated whether dogs and wolves can comprehend a uniquely human hand signal and compared how skilled they are in following human pointing. In this simple experiment (object-choice task), one of two containers is baited with food, and a human experimenter standing between them indicates to the subject where the food can be found with her extended arm and pointing finger (Figure 2-2). Many animals are successful in using simple versions of this human form of communication, but when the pointing is performed relatively far (>50 cm) from the indicated container and is not there when the animal is released to make a choice (i.e., momentary distal pointing), dogs outperform most non-human species (Miklósi & Soproni, 2006). Dogs can follow momentary distal pointing similarly to 2-year-old human infants (Lakatos et al., 2009), better than chimpanzees (Hare et al., 2002), and they do so at an early age (Riedel et al., 2008; Gácsi et al., 2009b). The initial studies that tested adult wolves in object-choice tasks found that they were less successful than pet dogs (Agnetta et al., 2000; Hare et al., 2002; see also Topál et al., 2009a) on the sensitivity of adult wolves to another form of human cuing in another task. These wolves, however, lived in captive facilities and had less experience with humans than dogs living in human families. Another study using a simpler form of static pointing, but using adult wolves that had more frequent contact with

FIGURE 2-2   Testing a young wolf in a two-way object-choice task with human momentary distal pointing (when the animal is released, the experimenter has already taken her arm back in front of her chest) (Virányi et al., 2008).

humans, found that they were able to follow pointing (Udell et al., 2008). In this study, apart from the simpler gesture used, the researchers also used a positive reinforcer to indicate the correct choice to the subject—a methodological change that has been shown to increase the performance of dogs in such a task (Pongrácz et al., 2013). Finally, Gácsi and her colleagues (2009a) demonstrated convincingly that adult wolves can follow momentary distal pointing, even if they are not kept as pets (they tested the Hungarian wolves that were hand-raised and moved to a captive facility at the age of 2 to 4 months). Thus, based on their adult performance, dogs and wolves are rather similar in this respect.

This similarity does not change the fact, though, that wolves need a longer time than dogs to develop this skill. So far, two independent samples of hand-raised wolves have been found to fail in using momentary distal pointing at the age of 4 months in contrast to hand-raised as well as pet dogs of the same age (Miklósi et al., 2003; Virányi et al., 2008; Gácsi et al., 2009a). Therefore, in sum, it seems that wolves need a longer development period and/or more exposure to humans to be able to follow momentary distal pointing as well as dogs do already at a few months of age.

## 2.3 EXPLAINING DOG–WOLF DIFFERENCES: DOMESTICATION HYPOTHESES

Despite the limited information this single test can provide regarding the social behaviour and cognition of dogs and wolves, a number of evolutionary hypotheses have been put forward to explain the differences presented in the preceding sections.

### 2.3.1 Selection for Human-Like Social Cognition (Hare et al., 2002)

Initially, the finding that dogs follow human pointing better than wolves and chimpanzees was explained by dogs having been selected directly for a set of social-cognitive abilities that enable them to communicate with humans in unique ways. Hare and colleagues (2002) proposed that dogs that were able to use social cues to predict the behaviour of humans flexibly were at a selective advantage, but they did not make a detailed suggestion in which contexts and/or with what kinds of cues such a selection process might have taken place.

### 2.3.2 Selection for Increased Attention (Miklósi et al., 2003)

Based on the finding that young dogs not only follow human pointing better but also look at humans more readily (see also Section 2.6 of this chapter), a more parsimonious idea of indirect selection has been proposed: by means of positive (both evolutionary and ontogenetic) feedback processes, dogs that pay increased attention to humans have a better chance to notice their behavioural cues and thus to achieve more complex forms of dog–human communication (Miklósi et al., 2003; Virányi et al., 2008).

### 2.3.3  Synergetic Hypothesis (Gácsi et al., 2009a)

In a more recent study, this idea was further pursued after finding that in a pointing task, the higher success and attentiveness of dogs was paralleled by another behavioural difference: less struggling and biting in dogs than in wolves when being held by a human experimenter at the starting position to wait for the pointing cue (Gácsi et al., 2009a). Gácsi and her colleagues (2009a) proposed that dogs are better in this form of visual cooperation with humans because they accept them more as social partners and thus pay more attention to them, and because they can better suppress their immediate drives in favour of delayed rewards compared to wolves (see also Gácsi et al., 2005).

### 2.3.4  The Emotional Reactivity Hypothesis: Version 1 (Hare & Tomasello, 2005)

A comparison with foxes experimentally selected for tameness further supports the idea of indirect selection (Hare et al., 2005). In the 1960s, Belyaev and his colleagues (1969) at the Siberian Division of Russian Academy of Sciences started a 50-year experiment by selecting a group of foxes against fear and aggression towards humans: always those animals were bred that readily approached the hands of humans standing in front of their cages. The selection pressure was strong: only 3% of the males and 8% to 10% of the females were bred. In this way, already in their sixth generation, the first animals belonging to the "domestication elite" appeared (Trut et al., 2004). These animals, similarly to dogs, showed no fear and aggression but rather showed tail-wagging, whining, and licking of the experimenters' hand and face when approached as well as following the experimenter when allowed out of the cage.

Importantly, from a cognitive and behavioural perspective, in comparison with a control group of farm foxes, the foxes selected for tameness also were more skilled in using human gestures in an object-choice task (Hare et al., 2005). This might be explained simply by a higher attraction of experimental foxes specifically to human hands due to the initial selection process when only the ones approaching the experimenters' hand were chosen for breeding. Nevertheless, Hare and colleagues (2005) proposed that the success of the experimental foxes in following human pointing was a by-product of being selected for reduced fear and aggression. Their hypothesis, the emotional reactivity hypothesis, suggests that the experimental foxes "were no longer constrained (e.g., by fear or disinterest) in applying previously existing social problem-solving skills to humans in interspecific interactions" (Hare et al., 2005, p. 227). That is, by losing their fear, the foxes have become able to interact with humans as they would with social partners and use their species-typical cognition and behavioural repertoire during these interactions. Based on these findings, Hare and Tomasello (2005) proposed that a similar selection for tamer temperament may explain the higher success of dogs in cooperating and communicating with humans.

This initial form of the emotional reactivity hypothesis says that through this selection for reduced fear and aggression dogs became able to use their skills evolved for within-species cooperation and communication when interacting with humans. That is, dogs can interact with humans as tolerantly and cooperatively as wolves interact with other wolves. This implies that, in dogs with good socialisation to both partners, cooperation and communication with humans and with conspecifics should be similar, and should also be similar to the cooperative and communicative interactions between wolves.

## 2.3.5  The Emotional Reactivity Hypothesis: Version 2 (Hare et al., 2012)

More recently, Hare and colleagues (2012) have extended the theory to chimpanzees, bonobos, and humans, and at the same time, suggested more fundamental changes regarding the overall reactivity and aggressiveness of dogs in comparison to wolves. According to this version of the hypothesis, during domestication, dogs became less aggressive and more tolerant than wolves in their interactions both with humans and conspecifics. This generalized version of the hypothesis suggests that in various species (including humans, bonobos, dogs, and potentially many others) social intelligence might have started to evolve as a by-product of selection on seemingly unrelated social-emotional systems. The fact that selection for tameness has led to a similar set of correlated traits—the so-called domestication syndrome (see also Section 2.4.1.2)—in various animal species, not only in the experimental foxes and in dogs (Trut et al., 2009), led to this idea. Between bonobos and chimpanzees, of course, differences in social behaviour cannot be explained by selection for reduced aggression towards humans; therefore, the second version of the emotional reactivity hypothesis shifts towards a more generalizable source of such selection. Hare and colleagues (2012) argue that not only domestication but also other forms of selection pressures favouring non-aggressive animals (in intraspecific contexts) may lead to correlated behavioural changes, including increased cooperativeness and related cognitive skills.

Indeed, there is a great variability across species regarding the severity and symmetry of their aggressive interactions, and importantly, this seems to co-vary with a range of other social behaviours, such as conciliatory behaviour and kin-bias in macaques (Thierry, 2000) or social attentiveness during feeding in birds (Kotrschal et al., 1993), forming multitrait behavioural syndromes. Based on these findings, it appears feasible that selection for decreased general aggressiveness leads to an increase in cooperativeness and other alterations of social cognitive skills. During the evolution of human social behaviour, this process might well have had an important role. We wonder, however, whether this second version of the emotional reactivity hypothesis is indeed relevant for dog domestication.

## 2.4 SELECTION FOR REDUCED AGGRESSION IN DOGS: BUT WHAT KIND OF AGGRESSION?

The two versions of the emotional reactivity hypothesis differ in respect to the social partners that dogs were selected to show reduced aggression towards. It is rather straightforward to assume that dogs were selected for reduced fear and aggression towards humans. But is this true also towards conspecifics? For either reason, can we expect an overall reduction of aggressiveness in dogs?

### 2.4.1 Selection for Reduced Aggression Towards Humans

#### 2.4.1.1 Less Fear of and Aggression Towards Humans

Defensive aggression towards humans, whom most animals seem to perceive as predators (Bilkó & Altbäcker, 2000), is thought to have been sharply reduced in most domestic animals (Price, 1999). In particular, regarding dog domestication, current theories suggest that in its initial period (called also proto-domestication), wolves, in search of food, were attracted by leftovers around human camps and thus progressively lost their fear of humans (Coppinger & Coppinger, 2001; Galibert et al., 2011). As the co-habitation of dogs and humans became closer and they started to work together in different tasks, the tractability of dogs and their ease of direct handling were likely targeted by further selection (Price, 1999). Behavioural differences observed between dogs and wolves strongly confirm these expectations. Dogs seem to be prepared to lose their fear of humans with minimal early socialisation. In dog pups, a couple of minutes of eye contact with humans suffice to remove fear responses towards them at a later age, while wolves need hours of direct daily contact and they need to be separated from their mother to obtain a comparable loss of fear of humans (Scott & Fuller, 1965; Klinghammer & Goodmann, 1987). The latter is needed because wolf pups show a clear preference for conspecifics over humans if given a choice, which is in clear contrast to dog pups that show high attraction and active greeting towards humans (Frank & Frank, 1982). At an adult age, wolves not socialized with humans early enough react with escape attempts and various fear-related behaviours to even a passive, nearby human, and there is a high risk of aggression up to a full-blown attack during the intermediate stage of a tediously long habituation (Woolpy & Ginsburg, 1967). Most animals, with no human socialisation, react to humans with similar fear: their aggression towards humans is a defensive response (Moyer, 1968).

## 2.4.1.2 By-Products of Selection for Reduced Human-Directed Defensive Aggression

The aim of the fox experiment (see Section 2.3.4) was precisely the reduction of this human-induced fear aggression, while another line of animals was selected for increased fear and aggression, both in comparison to control foxes that were bred randomly as regards their behaviour towards humans (Trut, 1980). Thus, they offer a great opportunity to test whether selection for reduced defensive aggression towards humans results in a correlated reduction of aggression in other contexts (e.g., towards conspecifics).

Various physiological and behavioural changes occurred as by-products of this targeted selection for tameness. The tame foxes not only showed dog-like behaviours but also a lower level of cortisol and adrenocorticotrophic hormone (ACTH) in the blood plasma, and their adrenal response to stress was reduced in comparison to the control foxes (Trut, 1999; Gulevich et al., 2004). The down-regulation of the hypothalamic-pituitary-adrenal axis responsible for stress reactions and the fight or flight response (Toates, 1995; Tsigos & Chrousos, 2002) can easily be linked to selection for less fearful behaviour, such as more exploration in a novel environment or approaching a human (Oskina, 1996; Künzl & Sachser, 1999). Furthermore, higher serotonin levels have been found in the midbrain region and hypothalamus of foxes selected for tameness, which again can be related to their decreased human-directed aggression because pharmacological data implicate that serotonin is an inhibitory factor in fear-induced defensive aggression (Popova, 2006). Because in the selected lines various physiological changes have been recorded in specific brain regions that are involved in the regulation of emotional defensive responses (Popova et al., 1991; Trut et al., 2000; Saetre et al., 2004), based on the underlying mechanisms it seems conceivable that the intraspecific aggression of the animals changed in parallel with their human-directed aggression. Unfortunately, since the foxes are housed in individual cages from their second month on, no specific information is available on this issue.

But more informative are studies conducted on Norway rats that were subjected to the same breeding program as the foxes. The physiological changes found were comparable to those of the selected foxes, including decreased glucocorticoid and increased serotonergic activity (Naumenko et al., 1989; Albert et al., 2008). Importantly, from our perspective, the aggressiveness of the different lines also was compared in different contexts (Nikulina & Popova, 1986; Naumenko et al., 1989; Popova et al., 1993; Plyusnina et al., 2011). The rats were selected either for a lack of defensive responses or for an aggressive reaction when a human hand appeared in front of their cage and moved around in there till the animals were pinned against the back wall. Besides this so-called glove test, the rats also were tested in other situations (Naumenko et al., 1989). For the study of their predatory aggression, an adult mouse was placed into their home cage, and the animals killing their mice were designated as "killers." To study irritable aggression, pairs of aggressive or domesticated rats were placed into a cage, and repeated electric shocks were delivered to the floor of the cage. The animals responded to the pain induced in this weird situation with increased aggression to each other. Finally, intermale aggression was studied by putting a pair of aggressive or domesticated rats into an unfamiliar cage. Their number of attacks and aggressive postures (upright and threatening) was registered and compared between the aggressive and tame lines.

In the 13th to 19th generations, no difference between the lines was found in their intermale (and predatory) aggression. However, if the threat could not be localized and escaped from, irritable aggression was higher in rats selected for increased defensive aggression towards a human hand in comparison to tame rats (Figure 2-3).

When one looks at the mechanisms underlying different forms of aggression, their independent evolution becomes understandable. In rodents, it is known that intermale and human-directed aggression are controlled by different neurochemical and hormonal processes and, thus, have a different genetic regulation (Popova et al., 1993; Nelson & Trainor, 2007). For instance, in mouse lines that differ in their aggressiveness towards conspecifics, an increased serotonin level is associated with higher aggression in contrast with the human-directed aggression of rats and foxes (Popova, 2004). Based on these findings, caution may be needed before suggesting generally decreased aggressiveness in a species that has been selected for reduced defensive aggression towards humans, as the second version of the emotional reactivity hypothesis seems to do.

Two issues may need further consideration and investigations, however. First, one study that compared intermale aggression in the three lines of rats in the 71st generation found lower intraspecific aggression in the tame rats compared to the other two lines (Plyusnina et al., 2011), and also in macaques, tolerance towards group members and towards human handling have been shown to correlate with each other (Clarke et al., 1988). Therefore, it is conceivable that after a longer selection process, intraspecific aggression also is reduced as an additional by-product of selection for reduced human-directed defensive aggression.

### 2.4.1.3 Better Acceptance of the Leading Role of Humans

Second, some may argue that in dogs not only defensive but also offensive aggression towards humans is reduced compared to wolves. Indeed, human-socialized wolves exert a constant danger on their caretakers because they tend to question their leading role, and if they do so, they often fight for their dominance with a full-blown attack (Klinghammer & Goodmann, 1987). Probably this and the predatory drives of wolves evoked by small children are the main reasons why people, wisely, rarely keep wolves at home. In contrast to this, in a relatively low percentage of the pet dog population, dominance-related aggression problems have been reported, and even if they occur, they manifest in less dangerous forms, such as house soiling, growling, or snapping at family members but only exceptionally as physically harmful

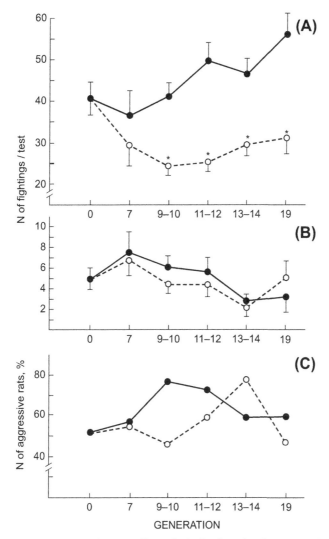

**FIGURE 2-3**   We know most about the effects of selection for reduced or increased aggression towards a human hand in regard to other forms of aggression in Norway rats. These graphs depict changes in different types of aggressive behaviour in two lines of Norway rats selected either for increased (•) or decreased aggression (○) showed in the so-called glove-test. A = irritable aggression, B = intermale aggression, C = mouse killing (for more information, see text). Vertical bars: S.E.M. * indicates $p < 0.001$. *(Reprinted from Naumenko et al., 1989.)*

behaviours (Podberscek, 2006). Therefore, the question arises whether selection for reduced human-directed offensive aggression also would decrease intraspecific aggression directed at conspecific group members. We are, however, aware of no study that investigated this question experimentally. Nevertheless, based on questionnaire data, human- and dog-directed aggression seems to be unrelated in dogs (Duffy et al., 2008; Hsu & Sun, 2010; Liinamo et al., 2007).

In sum, little evidence supports the idea that selection for reduced human-directed aggression (whether offensive or defensive) in dogs has led to an overall decrease in aggressiveness, and hence higher tolerance and cooperativeness. This is in contrast with implications of the second version of the emotional reactivity hypothesis that expects that dogs are less aggressive and more cooperative not only with humans but also with conspecifics.

## 2.4.2 Selection for Reduced Aggression Towards Conspecifics

It is still possible that during domestication, dogs also were selected directly for reduced aggression towards other dogs. It is generally assumed that domestication decreases aggressiveness towards conspecifics (Hemmer, 1990; Clutton-Brock, 1992; Price, 1998). This seems indeed necessary in species kept in crowded housing conditions such as cows, pigs, or chickens, but in this respect, dog domestication has strongly differed from that of other species kept as a source of food.

### 2.4.2.1 Reduced Between-Group Aggression in Dogs

Regarding intraspecific aggression, it has been suggested that dogs were selected for decreased between-group aggression because pet dogs, living with humans, likely encounter out-group conspecifics more often than pack-living wolves do (Miklósi, 2008). One may suppose that socialisation and human control play an important role in the tolerant behaviour of pet dogs in this context, but observations of feral dogs, in the absence of such human influence, point into the same direction. Feral dog groups, in contrast to wolf packs, rarely engage in physical aggression upon meeting (Boitani et al., 1995; Pal et al., 1999), and only one single case has been described when an out-group dog was killed after entering the territory of another group (Macdonald & Carr, 1995).

### 2.4.2.2 Reduced Within-Group Aggression in Dogs?

For the evolution of cooperation and communication, behaviours that occur between members of the same group, within-group aggression is relevant, however, and again, similarly to our previous discussion on the lack of correlation between different forms of aggression, it is questionable whether aggression towards out-group conspecifics is indicative of within-group aggression. Little is known about the neural and hormonal mechanisms underlying within-group and between-group aggression, but from a functional perspective, it is likely that these two forms of aggression evolved independently. In intergroup conflicts, two or more groups of animals show aggressive behaviour, mostly when competing for territories or other resources. The conflicting groups typically do not provide benefits for each other; therefore, only trying to avoid getting injured limits how serious these fights become. Thus, in many species, including wolves, fights between groups can be lethal (Mech & Boitani, 2003; Wilson & Wrangham, 2003; Mitani et al., 2010). Although within-group aggression also arises due to competition over limited resources, like food, mating partners, and resting places, it involves members of the same group (Brown, 1964; Archer, 1988; Saito et al., 1998; Feddersen-Petersen, 2004). As such, it is usually assumed that benefits of group living (e.g., cooperative hunting or breeding, increased defence against predators) outweigh intra-group competition (Schaik van, 1989). Consequently, within-group aggression is usually strongly inhibited, and animals solve conflicts without killing or physically harming each other or harming the relationship the aggressor has with its opponent (Hinde, 1970; Feddersen-Petersen, 2004). As such, within-group aggression is often highly ritualized, and injuries are avoided. This behaviour has been described in wolves, for instance, where aggression can range from staring intently at another animal through barking or growling to chasing, pushing away, and finally snapping, and only ultimately to real fighting with actual physical contact that may in extreme cases cause an injury or death (Mech, 1970; Mech & Boitani, 2003). Displays of dominance and submission as well as aggressive threats and fleeing are important devices for conflict management to avoid incurring high costs to either competitor (Preuschoft & van Schaik, 2000).

### 2.4.2.3 Comparing Social Interactions in Dog and Wolf Packs

Luckily, in this respect, we can rely on actual observations of dogs and wolves that have compared their within-group aggression. Previous, often forgotten, dog–wolf comparisons introduced at the beginning of this chapter provide the most information.

When raising their malamute and wolf pup groups in identical conditions, Frank and Frank (1982) noted that the "malamute pups exhibited unrestrained fighting from about 2 weeks of age, and we suspect that injury was avoided only because they lacked the powerful jaw musculature that was already evident in the wolf pups and because, unlike the wolf pups, the malamutes were born with a heavy winter coat and subcutaneous fat layer" (p. 513). Based on observations that the malamutes showed earlier and more intense aggression and were able to engage in agonistic play at a later age than the wolves did, they suggested that selection against aggression had relaxed in dogs due to the buffering effect of food provisioning and health care by humans (Frank & Frank, 1982).

Feddersen-Petersen (1991), by analysing dyadic interactions of wolves and dogs living either in separate (16 wolves, 20 poodles) or in mixed groups (6 wolves and 7 poodles) of 5 to 13 individuals, confirmed these differences also in older animals. She found that agonistic interactions appeared in poodle groups earlier and more often than among wolves that showed increased aggression only during breeding seasons (Figure 2-4). Additionally, the ritualized agonistic behaviours described earlier were typical only for the wolves but not for

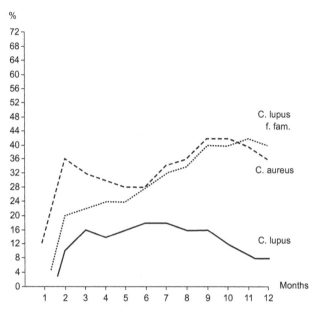

**FIGURE 2-4**   Frequency of agonistic interactions in 16 wolves (*Canis lupus*), 20 poodles (*C. lupus f. fam.*), and 12 golden jackals (*C. aureus*) kept in captive packs during their first year of life, as a percentage of all dyadic interactions observed (from Feddersen-Petersen, 1991). Note that from their sixth month on, dogs appeared to be similarly aggressive to the jackals adapted to a more solitary life. (*Reprinted from Feddersen-Petersen, 1991.*)

the dogs. For instance, attacks launched by a dominant poodle male escalated in 70% of the observed cases into grabbing and bite shaking, regardless of the opponent's reaction. In dogs, she often saw group aggression, when all group members joined a collective attack on one animal. In general, the dominance hierarchy was a lot steeper in dogs than in wolves, resulting in a large social distance between the high-ranking animal(s) and the rest of the group. Interestingly, in the mixed group of young (< 1 year) poodles and wolves, the frequency of agonistic behaviours exhibited by the poodles further increased, probably because most playing attempts of the more playful wolves were answered in this way. In this group, by their third or fourth months of age, the male wolves were dominated by the male poodles (Figure 2-1), the latter obtaining priority of access to food or favoured places.

Similar observations on many more dogs that all lived in packs confirmed that not only poodles but several other breeds, such as German shepherds, Alaskan malamutes, Fila Brasileiros, bull terriers, and Labrador retrievers display aggression towards group members earlier than wolves do, and they maintain a high level of aggression longer than wolves in their first year, albeit there is a great variability in this respect across dog breeds (Feddersen-Petersen, 2004) (Figure 2-5). Although German shepherds belong to those breeds that can cope relatively well with group living, even their threat displays, having the same

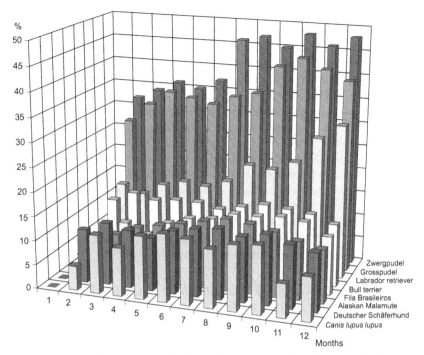

**FIGURE 2-5**   Development of aggressive interactions (dyadic, ritualised, and non-ritualised agonistic interactions in proportion of all social interactions) during the first year of European wolves and different dog breeds, all living in groups (Feddersen-Petersen, 2004). See color plate section. *(Reprinted from Feddersen-Petersen, 2004.)*

length before and after escalated conflicts, differ from those of wolves (Feddersen-Petersen, 2007).

Based on recording the social behaviour of approximately 200 canines, including ritualized and non-ritualized forms of agonistic behaviours, play, and other affiliative behaviours, Feddersen-Petersen (2007) suggested that, in contrast to wolves, dogs have difficulties in cooperating with each other even in a very basic manner of simply doing things together. According to her, in dogs, agonistic interactions often reach high levels of aggression because dogs lack most strategies that wolves commonly use to solve conflicts, such as appeasing, animating, and inhibiting their opponents. She attributed this incompetence of dogs to two factors: partly to their impaired visual communication due to their reduced visual (facial as well as bodily) expression caused by their altered morphology (fur colouring and length, head shape, hanging ears, lack of tail, etc.) (see also Goodwin et al., 1997), and partly to their being adapted to living with humans. This latter factor likely includes motivational changes (e.g., she found that toy poodles interrupted any interaction with conspecifics whenever humans appeared) as well as the adaptation of dogs' communicative skills to interact with humans. It seems that dogs are better adapted to interpret and respond to human signals in agonistic interactions. For instance, when encountering humans in socially ambiguous situations such as being approached by a stranger or playing an object-guarding game, dogs readily adjust to the behaviour of an unfamiliar person even if she frequently switches between acting in a threatening or friendly/playful manner (Gácsi et al., 2013b).

It is important to realise that these observations are in sharp contrast with a widespread view that considers dogs as more docile and affectionate, juvenile versions of wolves (Frank & Frank, 1982; Lindsay, 2008). Many have argued for paedomorphic changes in dogs, and dogs are believed to retain several characteristics of young wolves into their adulthood and to behave in many respects as 4- to 6-month-old wolves (Zimen, reported by Fox, 1971; Frank & Frank, 1982). The second version of the emotional reactivity hypothesis, suggesting reduced aggressiveness and higher tolerance in dogs compared to wolves, is in line with such a view of dogs, but both are in contradiction with the actual data currently available about aggression in dog and wolf packs.

## 2.5 EARLIER ORIGINS OF DOG–HUMAN COOPERATION: CANINE COOPERATION HYPOTHESIS (RANGE ET AL., 2012; RANGE & VIRÁNYI, 2013, 2014)

We propose that, in contrast with the second version of the emotional reactivity hypothesis, wolves do not lag behind dogs in their species-typical tolerance and cooperativeness. As reviewed in the previous section, Feddersen-Petersen's work has already demonstrated the fine-tuned communication and high social competence of wolves that enable their pack life, even if these packs have been constructed artificially and are to some extent forced to live together in captivity. Free-living wolf packs seem to be characterized by even less aggression and more affiliative interactions than captive wolf packs, and it has been proposed that wolf packs should rather be described as a family than a group of animals structured by their dominance hierarchy (Mech, 1999; Packard, 2003; Miklósi, 2008). Wolf packs are dependent on the close action coordination of the pack members in regard to defending territory, raising the offspring, and hunting large game (Mech & Boitani, 2003; Mech, 1970). Consequently, wolves can be expected to pay attention to details of the others' behaviour and also to be tolerant and cooperative, to some extent similarly to humans. Our first results at the Wolf Science Centre confirm these expectations. Our wolves that were hand-raised in peer groups and live in packs in large enclosures at the Gamepark Ernstbrunn follow the gaze of their conspecifics (Range & Viranyi, 2011; Werhahn et al., submitted) and imitate the action of a conspecific demonstrator in a manipulative, two-action task (Range & Virányi, 2014). Moreover, because these animals were similarly well socialized with humans and with conspecifics, they also follow human gaze into distant space (Range & Viranyi, 2011), and profit from a human demonstration in a local enhancement task (Range & Virányi, 2013). Accordingly, we propose that wolves possess most of the skills that have been suggested to be preconditions of successful cooperation, such as high social attentiveness and high tolerance. If so, dog–human cooperation could have evolved on the foundation of wolf–wolf cooperation, potentially by dogs losing their fear of humans and thus being able to extend their relevant social skills to interactions with them (as suggested by the first version of the emotional reactivity hypothesis). At the same time, however, as described in the previous section, domestication has apparently also affected the intraspecific interactions of dogs in a way that makes them less tolerant and less cooperative compared to wolves (in contrast with the implications of the first version of the emotional reactivity hypothesis). As such, the canine cooperation hypothesis presents a complementary or alternative way to reason about the evolutionary origins of dog–human cooperation in addition to domestication, and calls for more careful analyses regarding in what way dog and wolf cognition and social behaviour in intra- and interspecific contexts really differ and, consequently, to what extent dog–human cooperation and communication builds on wolf–wolf cooperation and in what way it has been shaped by domestication.

## 2.6 HUMAN-LIKE BEHAVIOUR IN DOGS BUT NOT IN WOLVES: PART 2

The canine cooperation hypothesis is in contradiction with the second version of the emotional reactivity hypothesis but nicely fits together with its first version. Finding similarities between wolf and human cooperation seems to support the idea that dog–human cooperation occurred when dogs lost their fear of humans and, thus, became able to use their species-typical, tolerant, and cooperative skills with them. Some findings, however, this evolutionary scenario cannot explain either.

### 2.6.1 Do Dogs Interact with Humans as with Conspecifics?

According to the first version of the emotional reactivity hypothesis, by losing their fear of humans, dogs (and also foxes selected for tameness) have become able to interact with humans as with their conspecifics (Hare et al., 2005). This question has been explicitly addressed in the foxes selected either for tameness or for increased aggression, after the researchers recognized that certain vocalisations towards humans clearly differentiated the two lines: it was observed that in the presence of humans, tame foxes produced cackles and pants but never coughed or snorted, whilst aggressive foxes produced coughs and snorts but never cackled or panted (Gogoleva et al., 2010). To investigate whether the foxes interact with humans and conspecifics in the same way, Gogoleva and colleagues (2010) brought together animals from both lines with conspecifics and recorded their vocalisations. They found that in this situation all four kinds of vocalisations occurred in both lines, indicating that the foxes do not consider humans as their conspecifics.

Also dogs respond differently to humans and other dogs in various situations. Moreover, some of the behaviours dogs show specifically towards humans, such as attachment (Topál et al., 2005) or asking for help (Miklósi et al., 2003), do not seem to occur in human-raised wolves. These dog–wolf differences, in contrast to the ones found in the pointing studies, have been addressed neither by extensive data collection nor by evolutionary hypotheses yet. This will be necessary, however, in order to come up with a comprehensive theory of dog domestication.

### 2.6.1.1 Attachment to a Human Caretaker

In stressful situations, human infants use their mother as a safe haven where they can calm down, and as a secure base from where they can start to explore their environment and where they can return to for protection and support (Ainsworth & Wittig, 1969). Adult pet dogs have been shown to benefit from their owners' presence in a similar way (Palmer & Custance, 2008; Gácsi et al., 2013a; but see Prato-Previde et al., 2003, and Chapter 6 in this volume). In the so-called Strange Situation Test, dogs—similarly to human infants—show a preference for their caretaker over an unfamiliar person: at an unknown place, they seek for more contact with the caretaker and show more playful behaviours in her presence than with a stranger (Topál et al., 1998). At the age of 4 months, dogs have already built up this relationship with their owners, whereas hand-raised wolf pups of the same age do not show such a preference for their raiser (Topál et al., 2005). This difference is paralleled by increased help seeking or referencing behaviour of the same dogs compared to the wolves when facing an unsolvable problem (Miklósi et al., 2003), suggesting a higher (and probably rather individualized) dependency on humans in dogs than in wolves. One has to be cautious, however, because at the age of 4 to 5 weeks, in a preference test, the same wolf and dog pups showed a similar (relatively weak) preference for their hand-raiser when they had the choice to spend time with her or with an unfamiliar experimenter after 5 minutes of isolation (Gácsi et al., 2005). Thus, further research is needed to investigate in what way the relationship dogs and wolves develop with their human caretakers differs and changes across the development of the animals.

Nevertheless, based on the currently available data, it has been proposed that the attachment young and adult dogs display towards their owner is a newly derived capacity that evolved during the course of domestication and that, in turn, it serves as an organising background for all further interactions and relationships individual dogs develop with humans (Topál et al., 2009b; Miklósi & Topál, 2013). An important part of arguing for a role of domestication is that apparently conspecifics—even the mother—cannot alleviate a dog's stress response as well as a human being can, and because puppies show no preference for their mother over an unfamiliar bitch (Pettijohn et al., 1977;

Tuber et al., 1996; but see also Chapter 6 in this volume). Accordingly, it seems that dogs can build up a special relationship with humans but not with conspecifics. But again, we need to investigate more carefully whether a comparable relationship can be found in a wolf family.

### 2.6.1.2 Learning from Humans

Similarly, there seems to be a difference in the reaction of dogs to conspecifics and humans when they have the chance to learn from them. Pongrácz and colleagues (2004) found that dogs readily learn from observing another dog that is detouring a V-shaped fence separating the dog from some reward. They do not learn from a human, however, if she/he demonstrates the detour without calling the subjects' attention and having eye contact with them, albeit this form of demonstration is effective in case of a conspecific model. To learn from a human, the model needs to use attention-calling, communicative cues during the demonstration (Pongrácz et al., 2004).

Importantly, also in an A-not-B error task, where the subjects have to search for a hidden object, dogs and human infants are sensitive to the communicative cues given by the experimenter, whereas wolves are not (Topál et al., 2009a). For infants, these communicative cues seem to signal when information conveyed by a person can be generalized and, thus, indicate an opportunity to acquire culturally shared knowledge (Csibra & Gergely, 2009). In dogs, their effects do not generalise across different people but can still help to detect when the behaviour of a person is relevant for the dogs and, thus, facilitate efficient learning and information transfer from human to dog (Topál et al., 2009a; Chapter 11 in this volume). The lack of a similar influence in wolves has been interpreted as evidence for convergent evolution of human and dog cognition (see also Kaminski & Nitzschner, 2013). This question needs further confirmation, however, because the adult wolves tested in this study had considerably less experience with humans than the pet dogs.

In sum, while it is clear that due to their domestication, dogs are predisposed to lose their fear of humans even if minimally socialized, most likely other adaptations have also taken place that prepared dogs to interact with humans, to pay attention to their behaviour and adjust to it, or even to develop social skills enabling them to better adapt to the human environment. However, attributing any of these skills of dogs to domestication has to be preceded and justified by careful wolf–dog comparisons, which to date has rarely been the case. We suggest that this shortcoming has probably led to a false image of dogs and wolves reflected, for instance, by the second version of the emotional reactivity hypothesis: the dog is seen as a tame, tolerant, and cooperative species in contrast to its aggressive ancestor, the wolf. In contrast with this view, the canine cooperation hypothesis proposes that wolves are highly tolerant and cooperative with their pack mates and are characterised by emotional and cognitive skills that might have been a good basis for their interactions with humans when their domestication started. Further research needs to examine, however, to what extent dog–human and wolf–wolf cooperation relies on the same mechanisms. Based on our current knowledge, we suggest that dog–human cooperation likely relies on the leading role of humans enhanced by the dependency of dogs on humans and by a steeper dominance hierarchy that characterises dogs in comparison to wolves.

## 2.7  PRACTICAL RELEVANCE

Wolf–dog comparisons and the resulting domestication hypotheses have important practical implications as well. The dog–wolf differences discussed in the previous section show that dogs have most probably evolved special adaptations to interact with humans. This means that, despite their close relatedness to wolves, modeling wolf packs is not a sufficient approach to understanding the relationship, communication, and co-working of dog–owner pairs or bigger groups of dogs and humans living together. The limitations of this 'lupomorph' approach (Miklósi, 2008) are further emphasized by the fact that, as already indicated by Feddersen-Petersen's work described previously, domestication apparently also changed the within-species social life of dogs. Despite the fact that free-ranging dogs live in pack-like social groups and display differentiated social relationships with each other (Bonanni et al., 2010; Cafazzo et al., 2010; Chapter 3 in this volume), in contrast with wolves, female dogs usually raise their pups alone (Daniels & Bekoff, 1989; Boitani & Ciucci, 1995) or may receive limited support by the fathers in some populations (Pal, 2005). Hunting large prey in groups also seems to be less effective (Boitani & Ciucci, 1995), or is simply unnecessary because feral dogs have a lot of other and less exhaustive possibilities to find food (Berman & Dunbar, 1983). Overall, it seems that most dog packs might not be such close-knit units as the usual wolf pack, which makes keeping dogs in packs more challenging than wolves, at least in captive settings.

**Future Directions**

- Test the two versions of the emotional reactivity hypothesis: The first version of the emotional reactivity hypothesis predicts that (1) the social behaviour of dogs and wolves is similar in intraspecific contexts, but (2) dogs accept humans as social partners easier than wolves do, and (3) dogs interact similarly with humans and conspecifics (if identical socialisation with both is given). In contrast with this, the second version predicts that dogs are less aggressive and more cooperative with humans as well as with conspecifics than wolves are.
- Investigate the plasticity of aggression and cooperation with humans and conspecifics in dogs and wolves using animals that have been raised in different ways.

- Elucidate the differences and similarities in the human–animal bond compared to the animal–animal relationship.
- Test for special adaptations of dogs such as their sensitivity towards attention-calling, communicative cues in social interactions in samples of wolves and dogs that were raised and kept in the same way.
- Track the development of social skills important for cooperation to better understand the influence of experience and individual learning both in regard to differently socialized dogs and in regard to wolf–dog comparisons.

# REFERENCES

Agnetta, B., Hare, B., Tomasello, M., 2000. Cues to food location that domestic dogs (*Canis familiaris*) of different ages do and do not use. Aggress. Behav. 3 (2), 107.

Ainsworth, M.D.S., Wittig, B.A., 1969. Attachment and exploratory behavior of one-year-olds in a strange situation. In: Foos, B.M. (Ed.), Determinants of infant behavior. Methuen, London, pp. 111–136.

Albert, F., Shchepina, O., Winter, C., Rompler, H., Teupser, D., Palme, R., et al., 2008. Phenotypic differences in rats selected for tameness and aggression are associated with differences in stress response. Horm. Behav. 53, 413–421.

Archer, J., 1988. The behavioural biology of aggression, vol. 1. Cambridge University Press, Cambridge.

Belyaev, D.K., 1969. Domestication of animals. Sci. Justice. 5 (1), 47–52.

Bensky, M.K., Gosling, S.D., Sinn, D.L., 2013. The world from a dog's point of view: a review and synthesis of dog cognition research. Adv. Study Behav. 45, 209–406.

Berman, M., Dunbar, I., 1983. The social behaviour of free-ranging suburban dogs. Appl. Anim. Ethol. 10 (1), 5–17.

Bilkó, Á., Altbäcker, V., 2000. Regular handling early in the nursing period eliminates fear responses toward human beings in wild and domestic rabbits. Dev. Psychobiol. 36 (1), 78–87.

Boitani, L., Ciucci, P., 1995. Comparative social ecology of feral dogs and wolves. Ethol. Ecol. Evol. 49–72.

Boitani, L., Francisci, F., Ciucci, P., 1995. Population biology and ecology of feral dogs in central Italy. In: Serpell, J. (Ed.), The domestic dog: its evolution, behaviour and interactions with people. Cambridge University Press, Cambridge, pp. 218–245.

Bonanni, R., Cafazzo, S., Valsecchi, P., Natoli, E., 2010. Effect of group size, dominance rank and social bonding on leadership behaviour in free-ranging dogs. Anim. Behav. 79, 981–991.

Brown, J.L., 1964. The evolution of diversity in avian territorial systems. Wilson Bull. 76, 160–169.

Cafazzo, S., Valsecchi, P., Bonanni, R., Natoli, E., 2010. Dominance in relation to age, sex and competitive contexts in a group of free-ranging domestic dogs. Behav. Ecol. 21 (3), 443–455.

Clarke, A.S., Mason, W.A., Moberg, G.P., 1988. Differential behavioral and adrenocortical responses to stress among three macaque species. Am. J. Primatol. 14 (1), 37–52.

Clutton-Brock, J., 1992. The process of domestication. Mammal Rev. 22 (2), 79–85.

Coppinger, R., Coppinger, L., 2001. Dogs: a startling new understanding of canine origin, behavior & evolution. Scribner, New York.

Csibra, G., Gergely, G., 2009. Natural pedagogy. Trends Cogn. Sci. 13 (4), 148–153.

Daniels, T.J., Bekoff, M., 1989. Population and social biology of free-ranging dogs, *Canis familiaris*. J. Mammal. 70, 754–762.

Duffy, D.L., Hsu, Y., Serpell, J.A., 2008. Breed differences in canine aggression. Appl. Anim. Behav. Sci. 114 (3), 441–460.

Feddersen-Petersen, D., 1991. The ontogeny of social play and agonistic behaviour in selected canid species. Bonner Zoologische Beiträge 42 (2), 97–114.

Feddersen-Petersen, D., 2000. Vocalization of European wolves (*Canis lupus lupus* L.) and various dog breeds (*Canis lupus* f. fam.). Archiv für Tierzucht 43 (4), 387–397.

Feddersen-Petersen, D., 2004. Hundepsychologie: Sozialverhalten und Wesen, Emotionen und Individualität. Kosmos Verlag, Stuttgart.

Feddersen-Petersen, D., 2007. Social behaviour of dogs and related canids. In: Jensen, P. (Ed.), The behavioural biology of dogs. Cromwell Press, Trowbridge, UK, pp. 105–119.

Feddersen-Petersen, D., 1994. Social behavior of wolves and dogs. Vet. Q. 16 (suppl), 51–52.

Fitch, W.T., Huber, L., Bugnyar, T., 2010. Social cognition and the evolution of language: constructing cognitive phylogenies. Neuron 65, 795–814.

Fox, M.W., 1971. The behaviour of wolves, dogs and related canids. Jonathan Cape, London.

Frank, H., 2011. Wolves, dogs, rearing and reinforcement: complex interactions underlying species differences in training and problem-solving performance. Behav. Genet. 41 (6), 830–839.

Frank, H., Frank, M., 1987. The University of Michigan canine information-processing project (1979–1981). In: Frank, H. (Ed.), Man and wolf. Advances, issues, and problems in captive wolf research. Dr. W. Junk Publishers, Dordrecht, pp. 143–167.

Frank, H., Frank, M.G., 1982. On the effects of domestication on canine social-development and behavior. Appl. Anim. Ethol. 8 (6), 507–525.

Frank, H., Frank, M.G., Hasselbach, L.M., Littleton, D.M., 1989. Motivation and insight in wolf (*canis-lupus*) and Alaskan malamute (*canis-familiaris*)—visual-discrimination learning. Bull. Psychonomic Soc. 27 (5), 455–458.

Frank, M.G., Frank, H., 1988. Food reinforcement versus social reinforcement in timber wolf pups. Bull. Psychonomic Soc. 26 (5), 467–468.

Gácsi, M., Győri, B., Miklósi, Á., Virányi, Z., Kubinyi, E., Topál, J., et al., 2005. Species-specific differences and similarities in the behavior of hand raised dog and wolf puppies in social situations with humans. Dev. Psychobiol. 47, 111–122.

Gácsi, M., Győri, B., Virányi, Z., Kubinyi, E., Range, F., Belényi, B., et al., 2009a. Explaining dog wolf differences in utilizing human pointing gestures: selection for synergistic shifts in the development of some social skills. PLoS One 4 (8), e6584.

Gácsi, M., Kara, E., Belényi, B., Topál, J., Miklósi, Á., 2009b. The effect of development and individual differences in pointing comprehension of dogs. Anim. Cogn. 12 (3), 471–479.

Gácsi, M., Maros, K., Sernkvist, S., Faragó, T., Miklósi, Á., 2013a. Human analogue safe haven effect of the owner: behavioural and heart rate response to stressful social stimuli in dogs. PLoS One 8 (3), e58475.

Gácsi, M., Vas, J., Topál, J., Miklósi, Á., 2013b. Wolves do not join the dance: sophisticated aggression control by adjusting to human social signals in dogs. Appl. Anim. Behav. Sci. 145 (3), 109–122.

Galibert, F., Quignon, P., Hitte, C., André, C., 2011. Toward understanding dog evolutionary and domestication history. C. R. Biol. 334 (3), 190–196.

Gogoleva, S.S., Volodin, I.A., Volodina, E.V., Kharlamova, A.V., Trut, L.N., 2010. Vocalization toward conspecifics in silver foxes (*Vulpes vulpes*) selected for tame or aggressive behavior toward humans. Behav. Processes 84 (2), 547–554.

Goodwin, D., Bradshaw, J.W.S., Wickens, S.M., 1997. Paedomorphosis affects agonistic visual signals of domestic dogs. Anim. Behav. 53, 297–304.

Gulevich, R., Oskina, I., Shikhevich, S., Fedorova, E., Trut, L., 2004. Effect of selection for behavior on pituitary–adrenal axis and proopiomelanocortin gene expression in silver foxes (*Vulpes vulpes*). Physiol. Behav. 82 (2), 513–518.

Hare, B., Brown, M., Williamson, C., Tomasello, M., 2002. The domestication of social cognition in dogs. Science 298 (5598), 1634–1636.

Hare, B., Plyusnina, I., Ignacio, N., Schepina, O., Stepika, A., Wrangham, R., et al., 2005. Social cognitive evolution in captive foxes is a correlated by-product of experimental domestication. Curr. Biol. 15 (3), 226–230.

Hare, B., Rosati, A., Kaminski, J., Brauer, J., Call, J., Tomasello, M., 2010. The domestication hypothesis for dogs' skills with human communication: a response to Udell et al. (2008) and Wynne, et al. (2008). Anim. Behav. 79 (2), E1–E6.

Hare, B., Tomasello, M., 2005. Human-like social skills in dogs? Trends Cogn. Sci. 9 (9), 439–444.

Hare, B., Wobber, V., Wrangham, R., 2012. The self-domestication hypothesis: evolution of bonobo psychology is due to selection against aggression. Anim. Behav. 83 (3), 573–585.

Hemmer, H., 1990. Domestication. The decline of environmental appreciation. Cambridge University Press, Cambridge.

Hinde, R., 1970. Aggression in animals. Proc. R. Soc. Med. 63 (2), 162.

Hsu, Y., Sun, L., 2010. Factors associated with aggressive responses in pet dogs. Appl. Anim. Behav. Sci. 123 (3), 108–123.

Kaminski, J., Nitzschner, M., 2013. Do dogs get the point? A review of dog–human communication ability. Learn. Motiv. 44 (4), 294–302.

Klinghammer, E., Goodmann, P.A., 1987. Socialization and management of wolves in captivity. In: Frank, H. (Ed.), Man and wolf: advances, issues and problems in captive wolf research. Dr. W. Junk Publishers, Dordrecht, pp. 31–61.

Kotrschal, K., Hemetsberger, J., Dittami, J., 1993. Food exploitation by a winter flock of greylag geese: behavioral dynamics, competition and social status. Behav. Ecol. Sociobiol. 33 (5), 289–295.

Kubinyi, E., Viranyi, Z., Miklósi, Á., 2007. Comparative social cognition: from wolf and dog to humans. Comp. Cogn. Behav. Rev. 2, 26–46.

Künzl, C., Sachser, N., 1999. The behavioral endocrinology of domestication: a comparison between the domestic guinea pig (Cavia aperea f.porcellus) and its wild ancestor, the cavy (Cavia aperea). Horm. Behav. 35 (1), 28–37.

Lakatos, G., Soproni, K., Doka, A., Miklósi, Á., 2009. A comparative approach to dogs' (Canis familiaris) and human infants' comprehension of various forms of pointing gestures. Anim. Cogn. 12 (4), 621–631.

Liinamo, A.-E., van den Berg, L., Leegwater, P.A., Schilder, M.B., van Arendonk, J.A., van Oost, B.A., 2007. Genetic variation in aggression-related traits in Golden Retriever dogs. Appl. Anim. Behav. Sci. 104 (1), 95–106.

Lindsay, S.R., 2005. Handbook of applied dog behavior and training, procedures and protocols. Iowa State University Press, Ames.

Macdonald, D., Carr, G., 1995. Variation in dog society: between resource dispersion and social flux. In: Serpell, J. (Ed.), The domestic dog: its evolution, behaviour, and interactions with people. Cambridge University Press, Cambridge, pp. 199–216.

Mech, D., 1970. The wolf: the ecology and behaviour of an endangered species. Natural History Press, Garden City, NY.

Mech, L.D., 1999. Alpha status, dominance, and division of labor in wolf packs. Can. J. Zool.- Revue Canadienne De Zoologie 77 (8), 1196–1203.

Mech, L.D., Boitani, L., 2003. Wolf social ecology. In: Mech, L.D., Boitani, L. (Eds.), Wolves: behavior, ecology, and conservation. The University of Chcago Press, Chicago, London, pp. 1–35.

Miklósi, Á., 2008. Dog behaviour, evolution, and cognition. Oxford University Press, Oxford.

Miklósi, Á., Kubinyi, E., Topál, J., Gácsi, M., Viranyi, Z., Csanyi, V., 2003. A simple reason for a big difference: Wolves do not look back at humans, but dogs do. Curr. Biol. 13 (9), 763–766.

Miklósi, Á., Soproni, K., 2006. A comparative analysis of animals' understanding of the human pointing gesture. Anim. Cogn. 9, 81–93.

Miklósi, Á., Topál, J., 2011. On the hunt for the gene of perspective taking: pitfalls in methodology. Learning Behav. 39 (4), 310–313.

Miklósi, Á., Topál, J., 2013. What does it take to become 'best friends'? Evolutionary changes in canine social competence. Trends Cogn. Sci. 17 (6), 287–294.

Miklósi, Á., Topál, J., Csanyi, V., 2004. Comparative social cognition: what can dogs teach us? Anim. Behav. 67, 995–1004.

Mitani, J.C., Watts, D.P., Amsler, S.J., 2010. Lethal intergroup aggression leads to territorial expansion in wild chimpanzees. Curr. Biol. 20 (12), R507–R508.

Moyer, K.E., 1968. Kinds of aggression and their physiological basis. Commun. Behav. Biol. 2 (2), 65–87.

Naumenko, E., Popova, N., Nikulina, E., Dygalo, N., Shishkina, G., Borodin, P., et al., 1989. Behavior, adrenocortical activity, and brain monoamines in Norway rats selected for reduced aggressiveness towards man. Pharmacol. Biochem. Behav. 33 (1), 85–91.

Nelson, R.J., Trainor, B.C., 2007. Neural mechanisms of aggression. Nat. Rev. Neurosci. 8 (7), 536–546.

Nikulina, E.M., Popova, N.K., 1986. Serotonin's influence on predatory behavior of highly aggressive CBA and weakly aggressive DD strains of mice. Aggress. Behav. 12 (4), 277–283.

Oskina, I., 1996. Analysis of the functional state of the pituitary-adrenal axis during postnatal development of domesticated silver foxes (*Vulpes vulpes*). Scientifur 20, 159–167.

Packard, J.M., 2003. Wolf behavior: reproductive, social and intelligent. In: Mech, L.D., Boitani, L. (Eds.), Wolves: behavior, ecology and conservation. University of Chicago Press, Chicago, pp. 35–65.

Pal, S., Ghosh, B., Roy, S., 1999. Inter- and intra-sexual behaviour of free-ranging dogs (*Canis familiaris*). Appl. Anim. Behav. Sci. 62 (2), 267–278.

Pal, S.K., 2005. Parental care in free-ranging dogs, Canis familiaris. Appl. Anim. Behav. Sci. 90 (1), 31–47.

Palmer, R., Custance, D., 2008. A counterbalanced version of Ainsworth's Strange Situation Procedure reveals secure-base effects in dog–human relationships. Appl. Anim. Behav. Sci. 109 (2), 306–319.

Pettijohn, T.F., Wong, T., Ebert, P., Scott, J., 1977. Alleviation of separation distress in 3 breeds of young dogs. Dev. Psychobiol. 10 (4), 373–381.

Plyusnina, I.Z., Solov'eva, M.Y., Oskina, I.N., 2011. Effect of domestication on aggression in gray Norway rats. Behav. Genet. 41 (4), 583–592.

Podberscek, A., 2006. Positive and negative aspects of our relationship with companion animals. Vet. Res. Commun. 30, 21–27.

Pongrácz, P., Gácsi, M., Hegedüs, D., Péter, A., Miklósi, Á., 2013. Test sensitivity is important for detecting variability in pointing comprehension in canines. Anim. Cogn. 16 (5), 721–735.

Pongrácz, P., Miklósi, Á., Timar-Geng, K., Csanyi, V., 2004. Verbal attention getting as a key factor in social learning between dog (*Canis familiaris*) and human. J. Comp. Psychol. 118 (4), 375–383.

Popova, N., 2004. The role of brain serotonin in the expression of genetically determined defensive behavior. Russ. J. Genet. 40 (6), 624–630.

Popova, N., Voitenko, N., Kulikov, A., Avgustinovich, D., 1991. Evidence for the involvement of central serotonin in mechanism of domestication of silver foxes. Pharmacol. Biochem. Behav. 40 (4), 751–756.

Popova, N.K., 2006. From genes to aggressive behavior: the role of serotonergic system. Bioessays 28 (5), 495–503.

Popova, N.K., Nikulina, E.M., Kulikov, A.V., 1993. Genetic analysis of different kinds of aggressive behavior. Behav. Genet. 23 (5), 491–497.

Prato-Previde, E., Custance, D., Spiezio, C., Sabatini, F., 2003. Is the dog–human relationship an attachment bond? An observational study using Ainsworth's Strange Situation. Behaviour 140, 225–254.

Preuschoft, S., van Schaik, C.P., 2000. Dominance and communication. Conflict management in various social settings. In: Aureli, F., Waal, F.D. (Eds.), Natural conflict resolution. University of California Press, Berkeley, pp. 77–105.

Price, E.O., 1998. Behavioral genetics and the process of animal domestication. In: Grandin, T. (Ed.), Genetics and the behavior of domestic animals. Academic Press, New York, pp. 31–66.

Price, E.O., 1999. Behavioral development in animals undergoing domestication. Appl. Anim. Behav. Sci. 65 (3), 245–271.

Range, F., Leitner, K., Virányi, Z., 2012. The influence of the relationship and motivation on inequity aversion in dogs. Soc. Justice. Res. 25 (2), 170–194.

Range, F., Viranyi, Z., 2011. Development of gaze following abilities in wolves (Canis Lupus). PLoS One 6 (2), e16888.

Range, F., Virányi, Z., 2013. Social learning from humans or conspecifics: differences and similarities between wolves and dogs. Frontiers Psychol. 4, 868.

Range, F., Virányi, Z., 2014. Wolves are better imitators of conspecifics than dogs. PLoS One 9 (1), e86559.

Reid, P.J., 2009. Adapting to the human world: dogs' responsiveness to our social cues. Behav. Processes 80 (3), 325–333.

Riedel, J., Schumann, K., Kaminski, J., Call, J., Tomasello, M., 2008. The early ontogeny of human–dog communication. Anim. Behav. 75 (3), 1003–1014.

Saetre, P., Lindberg, J., Leonard, J.A., Olsson, K., Pettersson, U., Ellegren, H., et al., 2004. From wild wolf to domestic dog: gene expression changes in the brain. Mol. Brain Res. 126 (2), 198–206.

Saito, C., Sato, S., Suzuki, S., Sugiura, H., Agetsuma, N., Takahata, Y., et al., 1998. Aggressive intergroup encounters in two populations of Japanese macaques (Macaca fuscata). Primates 39 (3), 303–312.

Schaik van, C.P., 1989. The ecology of social relationships amongst female primates. In: Standen, V., Foley, R.A. (Eds.), Comparative socioecology. The behavioural ecology of humans and other mammals. Blackwell Scientific Publications, Oxford, pp. 195–218.

Scott, J.P., Fuller, J.L., 1965. Genetics and the social behavior of the dog. University of Chicago Press, Chicago.

Thierry, B., 2000. Covariation of conflict management patterns across macaque species. In: Aureli, F., de Waal, F.B.M (Eds.), Natural conflict resolution. University of California Press, Berkeley, pp. 106–128.

Toates, F.M., 1995. Stress: conceptual and biological aspects. Wiley, New York.

Topál, J., Gácsi, M., Miklósi, Á., Viranyi, Z., Kubinyi, E., Csanyi, V., 2005. Attachment to humans: a comparative study on hand-reared wolves and differently socialized dog puppies. Anim. Behav. 70, 1367–1375.

Topál, J., Gergely, G., Erdohegyi, Á., Csibra, G., Miklósi, Á., 2009a. Differential sensitivity to human communication in dogs, wolves, and human infants. Science 325 (5945), 1269–1272.

Topál, J., Miklósi, Á., Csanyi, V., Doka, A., 1998. Attachment behavior in dogs (*Canis familiaris*): a new application of Ainsworth's (1969) Strange Situation Test. J. Comp. Psychol. 112 (3), 219–229.

Topál, J., Miklósi, Á., Gácsi, M., Doka, A., Pongrácz, P., Kubinyi, E., et al., 2009b. The dog as a model for understanding human social behavior. Adv. Study Behav. 39, 71–116.

Trut, L., 1980. The role of behavior in domestication-associated changes in animals as revealed with the example of Silver Fox, Doctoral (Biol.) Dissertation. Inst. Cytol. Genet, Novosibirsk.

Trut, L., 1999. Early canid domestication: the farm-fox experiment. Foxes bred for tamability in a 40-year experiment exhibit remarkable transformations that suggest an interplay between behavioral genetics and development. Am. Sci. 87 (2), 160–169.

Trut, L., Oskina, I., Kharlamova, A., 2009. Animal evolution during domestication: the domesticated fox as a model. Bioessays 31 (3), 349–360.

Trut, L., Plyusnina, I., Oskina, I., 2004. An experiment on fox domestication and debatable issues of evolution of the dog. Russ. J. Genet. 40 (6), 644–655.

Trut, L., Plyusnina, I., Oskina, I., Prasolova, L., 2000. Phenotypic diversity of domestic animals and temporal developmental parameters. Biodiversity Dynamics Ecosystems N. Eurasia 1, 119–123.

Tsigos, C., Chrousos, G.P., 2002. Hypothalamic–pituitary–adrenal axis, neuroendocrine factors and stress. J. Psychosom. Res. 53 (4), 865–871.

Tuber, D.S., Hennessy, M.B., Sanders, S., Miller, J.A., 1996. Behavioral and glucocorticoid responses of adult domestic dogs (*Canis familiaris*) to companionship and social separation. J. Comp. Psychol. 110 (1), 103.

Udell, M.A.R., Dorey, N.R., Wynne, C.D.L., 2008. Wolves outperform dogs in following human social cues. Anim. Behav. 76 (6), 1767–1773.

Udell, M.A.R., Dorey, N.R., Wynne, C.D.L., 2010. What did domestication do to dogs? A new account of dogs' sensitivity to human actions. Biol. Rev. 85 (2), 327–345.

Udell, M.R., Dorey, N., Wynne, C.L., 2011. Can your dog read your mind? Understanding the causes of canine perspective taking. Learning Behav. 39 (4), 289–302.

Virányi, Z., Gácsi, M., Kubinyi, E., Topál, J., Belényi, B., Ujfalussy, D., et al., 2008. Comprehension of human pointing gestures in young human-reared wolves and dogs. Anim. Cogn. 11, 373–387.

Werhahn, G., Barrera, G., Virányi, Z., & Range, F. (submitted). Wolves and dogs follow their packmates' gaze into distant space.

Wilson, M.L., Wrangham, R.W., 2003. Intergroup relations in chimpanzees. Annu. Rev. Anthropol. 363–392.

Woolpy, J.H., Ginsburg, B.E., 1967. Wolf socialization: a study of temperament in a wild social species. Am. Zool. 7 (2), 357–363.

Zimen, E., 1987. Ontogeny of approach and flight behavior towards humans in wolves, poodles and wolf-poodle hybrids. In: Frank, H. (Ed.), Man and wolf. Advances, issues, and problems in captive wolf research. Dr. W. Junk Publishers, Dordrecht, pp. 275–292.

# Social Behaviour

# Chapter 3

# The Social Organisation of a Population of Free-Ranging Dogs in a Suburban Area of Rome: A Reassessment of the Effects of Domestication on Dogs' Behaviour

Roberto Bonanni[1] and Simona Cafazzo[2]

[1]*Department of Neuroscience, University of Parma, Italy,* [2]*Wolf Science Center, Ernstbrunn, Austria*

## 3.1 INTRODUCTION

Analysis of several genetic markers, together with assessment of archaeological remains, has shown that dogs were domesticated from Eurasian wolves (*Canis lupus*) some 15,000–35,000 years BP, although there is still considerable debate about the exact timing and location of the domestication event, how many wolf populations were involved, and the selective mechanism by which wolves were turned into dogs (Vilà et al., 1997; Bokyo et al., 2009; Pang et al., 2009; vonHoldt et al., 2010; Ding et al., 2012; Larson et al., 2012; Wang et al., 2013; Druzhkova et al., 2013). Traditionally, it was thought that dogs had been domesticated only through artificial selection, i.e., by capturing wild wolves and by selectively breeding those bearing desirable traits such as tameness (reviewed in Price, 1984; Clutton-Brock, 1995; Coppinger & Schneider, 1995). Partial support for this view is provided by artificial selection experiments on captive foxes (*Vulpes vulpes*). In this species, selection for a single behavioural trait, i.e., 'tameness towards humans', led to the appearance, through pleiotropic effects, of several morphological and physiological characters that are typically observed in dogs, such as piebald coat, floppy ears, earlier sexual maturation, and dioestrus breeding cycle (Trut, 1999). So, it seems conceivable that similar processes also might have operated during the evolution of dogs. However, more recently, researchers have suggested that natural selective forces also probably

*The Social Dog.* http://dx.doi.org/10.1016/B978-0-12-407818-5.00003-6

contributed to the evolution of wolves into domestic dogs (e.g., Morey, 1994; Clutton-Brock, 1995; Coppinger & Coppinger, 2001; Zeder, 2012; Wang et al., 2013). Actually, one problem with the hypotheses of dog domestication based entirely on artificial selection is that, although genetic studies indicate that several hundred wolves were probably involved in the domestication process (Savolainen, 2007; Vilà & Leonard, 2007; Pang et al., 2009; Ding et al., 2012), there is currently no archaeological evidence that Mesolithic humans artificially bred such a large number of wolves (Coppinger & Coppinger, 2001). For these and other reasons, Coppinger and Coppinger (2001) hypothesised that domestication of dogs was initiated through natural selection when some wolves started scavenging on food leftovers around the first permanent human settlements. According to this hypothesis, wolves that had a shorter flight distance from humans (or higher tameness) had a natural selective advantage over shyer individuals because they were more efficient in exploiting the new available food source provided by humans. At a later stage, once wolves had already evolved into tamer primitive dogs, humans began selecting dog breeding types to meet their requirements and to make dogs suitable for performing specific tasks. The hypothesis of dog self-domestication by natural selection (Coppinger & Coppinger, 2001) seems to be partially supported by two recent genetic studies: one showing that domestication of dogs was accompanied by a positive selection on genes involved in starch metabolism (Axelsson et al., 2013), which suggests an adaptation of dogs to a different ecological niche, and a parallel switch from a strictly carnivorous diet to a more generalised, omnivorous diet; the other one suggesting that during domestication dogs underwent a much milder genetic bottleneck if compared to other domesticated animals (Wang et al., 2013), indicating that domestication was most probably a continuous and dynamic process.

Whichever the mechanisms that led to dog domestication, it is clear that dogs have evolved in association with humans for a very long time (Axelsson et al., 2013; Wang et al., 2013), and that in this new 'domestic environment' they were subjected to selective pressures (both natural and artificial) that were quite different from those experienced by wolves in their original environment of adaptation. Consequently, it is expected that important behavioural differences between dogs and wolves also evolved in the meantime, some of these probably reflecting adaptations to different ecological niches, and others resulting from artificial selection on dogs. Moreover, during their long association with humans, dogs have very often formed heterospecific social groups with them, and the complexity of the human social system may have provided further selective pressures leading to the evolution of new social skills in these animals (Miklósi et al., 2004; Miklósi & Topál, 2013). Most studies aimed at investigating the effect of domestication on the behaviour of dogs have focussed on the dog–human relationship, and results suggest that dogs may have evolved a higher ability to understand human communicative gestures in comparison to wolves, possibly due to their reduced emotional reactivity and higher capacity for attention towards human beings (Miklósi et al., 2003; Hare & Tomasello,

2005; Gacsi et al., 2009; Topál et al., 2009; Hare et al., 2010). However, other researchers have stressed that at least some of the wolf–dog differences with respect to interspecific communication can be greatly diminished when wolves are provided with intense socialisation with humans (Udell & Wynne, 2010; Udell et al., 2010a; Udell et al., 2012). But what about the effects of domestication on dogs' intraspecific social relationships? Although there are far fewer studies on this topic, it is usually believed that artificial selection for retention of juvenile traits (morphological and behavioural) into adulthood caused a reduction in the capacity of dogs to communicate visually with conspecifics and to perform ritualised agonistic behaviour, at least in some breeds (Frank & Frank, 1982; Bradshaw & Nott, 1995; Goodwin et al., 1997; Feddersen-Petersen, 2007). Moreover, there are frequent statements that intraspecific relationships would be less relevant to dogs than relationships with humans, given that many dogs actually spend more of their lifetime in association with humans than with other dogs (e.g., Frank & Frank, 1982; Tuber et al., 1996; Gacsi et al., 2005; Miklósi, 2007a,b; Nitzschner et al., 2012; Brauer et al., 2013). However, the fact that dogs evolved in a 'domestic environment' does not necessarily imply that they can live *only* integrated into a human family. Dogs vary greatly in their degree of association/dependence on human beings (Udell et al., 2010b; Miklósi & Topál, 2013): at one extreme, some dogs develop life-long attachments to their owners (Topál et al., 1998, 2005), whereas, at the opposite extreme, other dogs can even attack and kill humans as prey (Borchelt et al., 1983; Avis, 1999). Moreover, many dogs live in a free-ranging state (Coppinger & Coppinger, 2001); i.e., they do not have any constraints placed by humans on their activities, either because they are unowned or because they are unsupervised by their owners. Free-ranging dogs typically live around human dwellings, scavenging on human refuse (Coppinger & Coppinger, 2001), and have higher opportunities for social interactions with conspecifics than restrained dogs. In spite of the common belief that intraspecific relationships are not so relevant for dogs (see previous references), current estimates suggest that free-ranging dogs represent about 76%–83% of the global dog population (Hughes & Macdonald, 2013; Lord et al., 2013). Furthermore, even reproductive events involving pet dogs are often not planned by human beings (New et al., 2004; Lord et al., 2013), indicating that activities of these animals are probably influenced by conspecifics to a higher degree than expected. Consequently, in order to obtain a comprehensive picture of dogs' evolution and biology, we need to increase our knowledge of dogs' intraspecific relationships and of the social behaviour of free-ranging dogs. This leads directly to the question of whether free-ranging dogs are really social animals and, if so, to what extent they retain features of the wolves' social organisation after their prolonged evolution in a domestic environment. Several studies on free-ranging dogs concluded that those living in urban areas are often solitary or form only small temporary associations with conspecifics (usually 2–3 individuals) indeed (Beck, 1973, 1975; Rubin & Beck, 1982; Berman & Dunbar, 1983; Daniels, 1983a; Daniels &

Bekoff, 1989a; Ortolani et al., 2009). According to Coppinger & Coppinger (2001), since dogs have evolved to scavenge on human refuse, they do not need to form large packs to cooperate in hunting and in raising pups as wolves do, and so they are semi-solitary carnivores. However, note that this view assumes that pack living in wolves has evolved to allow for cooperative hunting and breeding, which is difficult to demonstrate and may not necessarily be true (Harrington et al., 1983; Packard, 2003; MacNulty et al., 2013). Moreover, cooperation in food and territorial defence is another important functional consequence of pack living in wolves (Harrington & Mech, 1979; Mech & Boitani, 2003; Stahler et al., 2013), and it may be potentially functional also for animals scavenging at dumps, at least under certain ecological conditions. In fact, many other studies on free-ranging dogs, carried out in various environments, found that these animals do form stable social groups (2–12 individuals), and some of these also exhibit cooperation in territorial defence (Scott & Causey, 1973; Nesbitt, 1975; Fox et al., 1975; Causey & Cude, 1980; Borchelt et al., 1983; Gipson, 1983; Font, 1987; Daniels & Bekoff, 1989a,b; Macdonald & Carr, 1995; Boitani et al., 1995; Pal et al., 1998). So, it appears that free-ranging dogs exhibit considerable variation with respect to their social organisation, possibly reflecting differences in ecological conditions, in the degree of human influence on their activities, and also in the methodologies applied by different researchers. Indeed, it must be stressed that free-ranging dogs are a very heterogeneous category of animals. For example, Boitani et al. (2007) made a distinction between 'village dogs', i.e., those free-ranging dogs living in proximity to human dwellings, and 'feral dogs', i.e., those that actively avoid human proximity by living in more natural environments. However, note that both kinds of dogs actually subsist mainly on food provided by humans and, according to this definition, they can comprise animals that differ with respect to the degree of human influence on their behaviour, i.e., both dogs that are socialised to humans and animals that are not (Boitani et al., 2007). So, for the purposes of this chapter, we prefer to classify free-ranging dogs based on their degree of dependency/socialisation to humans in the following categories: (1) free-ranging pets, i.e., dogs that are owned, although they are allowed to roam free for varied amounts of time; (2) abandoned/escaped dogs, i.e., dogs that are socialised to humans although, having lost their owner, they are completely free to roam; and (3) non-socialised free-ranging dogs, i.e., dogs that were born free and have formed social bonds only with conspecifics during the earlier stages of their development. The last two categories can be collectively referred to as 'unowned free-ranging dogs' when it is not possible to discriminate between abandoned and non-socialised animals. As we discuss later, these categories are likely to differ with respect to their intraspecific social skills (Daniels, 1983a; Daniels & Bekoff, 1989a,b).

Since there is evidence that at least some free-ranging dogs form stable social groups, we should ask how domestication has changed their organisation in comparison to wolf packs. Some authors (Boitani & Ciucci, 1995;

Boitani et al., 2007) suggest that groups of feral dogs seem to lack both the hierarchical social structure and the strong social bonds that are typical of wolf packs, and that this lack of a clear organisation places an upper limit to the number of individuals that can effectively cooperate as a social unit. Moreover, recently, there have been many claims that dogs' social relationships cannot be described in terms of a dominance–subordination paradigm (Coppinger & Coppinger, 2001; Semyonova, 2003; van Kerkhove, 2004; Bradshaw et al., 2009; Eaton, 2011; McGreevy et al., 2012). However, we stress that most previous studies on free-ranging dogs focussed mainly on ecological aspects (e.g., population censuses, demography, spacing pattern, diet), and only very few of the previously cited papers actually published quantitative data useful to describe the pattern of social interactions among group members, and how these social interactions may potentially affect cooperation. Consequently, we believe that the refusal of a hierarchical model to describe the social relationships of dogs is not fully justified on the available data, and that much more information is needed to assess how domestication has changed dogs' intraspecific social behaviour.

In this chapter, we summarise the main results of our 6-year research on the social organisation of a large population of free-ranging dogs living in a suburban area of Rome (Italy). We focus particularly on quantitative analyses of the social interactions among group companions and between dogs of different groups, given that these aspects have received little attention in the past. Moreover, we set our results in the general framework of the existing literature on dogs and compare the social organisation of dogs with that of wolves. Finally, in the last section of the chapter, we discuss the factors that are likely to explain variation in the social organisation of free-ranging dogs.

## 3.2 DOG POPULATION

Our study area was located at the southwestern periphery of Rome (a district traditionally called 'Muratella') and covered a surface of about 300 ha. It was delimited to the north, west, and south sides by roads with intense traffic and to the east side by cultivated areas. The area was crossed by a central road that split it into two clearly distinct sectors: the southwest sector was urbanised, although not densely populated, whereas the northeast sector was occupied by a nature reserve called 'Tenuta dei Massimi'. The habitat in the reserve consisted mainly of open grasslands with interspersed wooded areas (for a more detailed description of the environment, see Bonanni et al., 2010b; Cafazzo et al., 2010; Bonanni et al., 2011). Free-ranging dogs were free to move across the entire area, although they usually had their resting sites and dens within the reserve. However, in the early morning, they frequently approached the central road crossing the study area to feed on the food brought there by voluntary dog caretakers. These people drove along the central road every morning and placed food (mainly meat from a slaughterhouse) and water at specific feeding sites located alongside the road itself.

We regularly monitored the dogs of this population from April 2005 to October 2011. Detailed population censuses were carried out periodically in 2005–2006, 2007–2008, and again in 2010–2011, by enumerating all individually recognised dogs that approached the road to feed. Moreover, we conducted an intensive behavioural study of a 27-member pack from April 2005 to May 2006, and another similar study focussed on three packs (ranging in size from 3 to 15 individuals) from May 2007 to September 2008 (for a detailed description of packs, see Bonanni et al., 2010b; Cafazzo et al., 2010). The studied packs were selected because they inhabited sectors of the study area characterised by good visibility. These behavioural studies were based mainly on direct observation and recording (*ad libitum* and *focal sampling*; Altmann, 1974) of affiliative, agonistic, and sexual interactions among dogs belonging to the same or to different packs. Additionally, we performed direct observation and recording of the location of animals (on a 1:1,250 scaled map of the area, to the nearest 20–30 m), of scent marking events (e.g., raised leg urinations; Bekoff, 1979), and of intergroup conflicts. Demographic data (number of individuals born, dead, dispersing, immigrating, etc.) were also collected, although those concerning infant survival were meaningful only with respect to the period 2005–2006. The reason is that in 2006 a management programme was started by the Rome Municipality, consisting in periodically trapping, neutering, and releasing back in the area adult dogs and in removing most puppies. All the behavioural results presented in the next sections of the chapter refer to dogs that were intact or sterilised at least 6 months before data collection, unless specified otherwise.

Population size was relatively stable across years. Censuses revealed that about 90–100 adult dogs were living in the area, leading to an estimate of density of about 30 animals/km². This density is much lower than that recorded for dog populations living in urban areas, although still much higher than that recorded in natural environments for both free-ranging dogs and wolves (see Boitani et al., 1995, for review). Sex ratio in adult dogs was male biased, similar to other populations (reviewed in Ortolani et al., 2009), although by a small margin (about 1.3–1.4:1).

Almost all dogs of the population lived in packs. We defined a pack as a distinct and stable unit of individuals that travelled, fed, and defended resources together. Packs were identified by observing animals continuously and for several weeks. Mean pack size ± SD was 12.6 ± 7.4 in 2006 (range: 6–27, N = 7), 9.0 ± 5.2 in 2008 (range: 2–16, N = 10), and 7.6 ± 6.6 in 2011 (range: 3–24, N = 12). Although the size of these groups is quite unusual if compared to that reported for other populations (see previous references), it should be noted that groups of 11–25 free-ranging dogs have already been observed both in North America and in Italy (Beck, 1973; Gipson, 1983; Borchelt et al., 1983; Boitani, 1983). The percentage of solitary dogs in the overall population was in the range 0.011–0.043 across years.

We stress that the composition of our studied packs showed remarkable stability over time and thus was very different from the temporary associations described by several authors (see previous references). Although undoubtedly several individuals could join or leave a given group on a daily or monthly basis, each pack usually contained a building block of animals that stayed together for several years. For example, in 2008, the 'Corridoio pack' comprised 9 dogs that had been together for at least 3 years, plus 2 members that had been born in this pack at the beginning of 2006. Although most members of this group eventually died in 2010, due to an epidemic, the 3 remaining dogs were still together in 2011, which means that they had shared pack membership for at least 6 years. Another example was the 'Fused pack': it comprised 13 individuals at the end of 2007, and 9 out of these were still group companions during the census we carried out 4 years later.

Several of the packs we closely monitored were undoubtedly composed of relatives at least to some extent, given that newborn puppies were usually recruited into their natal group. Survival of puppies was recorded for the 'Corridoio pack' in 2005–2006, and it was about 50% to 1 year of age. Notably, this figure is much higher than the 0%–5% reported by Beck (1975) and by Boitani et al. (1995), respectively, although comparable to the 15%–40% survival reported by many other authors (Scott & Causey, 1973; Nesbitt, 1975; Oppenheimer & Oppenheimer, 1975; Gipson, 1983; Daniels & Bekoff, 1989a,b; Macdonald & Carr, 1995; Butler & Bingham, 2000; Pal, 2001).

All free-ranging dogs of the studied population were medium- or large-sized mongrels (height at the withers 55–80 cm, weight 20–50 kg), and it was not possible to identify any predominant breeding type. Most animals had floppy ears and lacked a pronounced 'stop' between the muzzle and the forehead, whereas their coat colour was highly variable, ranging from uniform tawny/grey to black and tan, with some white/black piebald.

We are reasonably confident that almost all dogs of the population were 'non-socialised' to humans, although we could directly ascertain this only for those animals that were born during our study (e.g., 20 out of 39 dogs for which we have behavioural data were born free during the study). However, even dogs that were born before the initiation of our research never showed any sign of affiliation towards humans, and instead they usually displayed fear/aggressive responses to people. At the beginning of the study, they were very elusive, avoiding human presence at about 50 m of distance. Therefore, to be able to observe and follow them in different parts of the studied area, we spent 2–3 months to get them used to us by following them every day and by decreasing little by little the distance of observation. After this period, all dogs accepted our presence within about 20–50 m of distance, losing interest in us.

The few dogs that showed obvious signs of socialisation to humans, during our behavioural studies, were two 'free-ranging pets' that spent virtually all their time integrated into two different packs. That they had not been abandoned, but rather had spontaneously opted for a 'free-ranging life', was clearly indicated by the fact that, after losing their status as pack members, they went back to their respective human families and they were well received by them

(R. Bonanni, personal observation). Moreover, after the end of our behavioural studies, few other dogs socialised to humans appeared in the area, including some abandoned dogs and a pack affiliated to a gipsy family.

## 3.3 WITHIN GROUP RELATIONSHIPS

### 3.3.1 The 'Dominance Debate'

Before we discuss the applicability of the concept of dominance to domestic dogs, it is useful, first of all, to revise briefly its ethological meaning, as well as the methodological issues concerning its assessment, and the applicability of the concept itself to the dogs' wild ancestors, since this has also been questioned in the past (e.g., Lockwood, 1979).

According to Drews (1993), dominance is an attribute of a social relationship between two individuals in which one of two (the dominant one) emerges as the consistent winner of repeated agonistic interactions, whereas the other (the subordinate) usually defers without escalation. From a functional perspective, social dominance can be interpreted as a 'convention' allowing animals to resolve social conflicts in a relatively peaceful manner; i.e., when two animals with a dominant/subordinate relationship compete for a given resource, usually the dominant one gets it and no physical conflict takes place (Hand, 1986). More in general, 'dominance' refers to the asymmetric distribution of a given behaviour between two individuals (e.g., submissive gestures directed from subordinate to dominant animals and aggressive/dominance gestures displayed by dominant animals towards subordinates; see van Hoof & Wensing, 1987). When dominance relationships characterise all or most dyads in a social group of animals, then it may be possible to describe the social structure of that group as a 'linear dominance hierarchy'. However, in a stringent sense, a linear hierarchy model can be applied only when dominance relationships within the group are 'transitive' (Appleby, 1983). This means that if individual A is dominant over B and B is dominant over C, then A has to be dominant over C as well. Conversely, relationships in which C dominates A are called 'circular triads', and their presence may indicate that the organisation of the group under consideration does not fit a 'linear hierarchy model' (Appleby, 1983). Quantitatively, a dominance hierarchy can be described using indexes of linearity (e.g., Landau's $h$, 1951), whose values usually range from 0, indicating complete absence of transitivity (or no hierarchy), to 1, indicating a perfect linear hierarchy. However, in natural animal societies, perfect linear hierarchies are rare, and lower scores for the linearity indexes will be obtained, the higher the proportion of circular triads and of unknown and tied relationships (i.e., dyads that never interact or that exchange equal rates of the behaviour used to assess dominance, respectively) within the group (Appleby, 1983). However, it is possible to obtain an improved estimate of the Landau's linearity index (h') that takes into account both unknown and tied relationships (de Vries, 1995). So, in summary, the best way to demonstrate that the social structure of a group fits a 'linear hierarchy model' consists in applying appropriate statistical methods (de Vries, 1995) to test whether the number of circular triads is lower than that expected by chance. However, statistical significance of linearity can be demonstrated only with hierarchies comprising a minimum of six individuals (Appleby, 1983).

Sometimes, even when dominance is transitive, the structure of a group may be better described as a 'pyramidal hierarchy', rather than as a linear one (van Hoof & Wensing, 1987). This happens, for example, when dominance relationships are clearly manifested between top-ranking animals and subordinates, but not among animals ranking below the top.

Although definitions like that of Drews (1993) undoubtedly place emphasis on the role of dominance in competitive interactions, it must be stressed that dominance relationships can also be expressed in the context of affiliative bonding (de Waal, 1986). For example, some highly social primates have evolved 'formal submissive gestures', i.e., signals that unambiguously communicate the acceptance of a subordinate status, and that allow the subsequent development of affiliative relationships between dominant and subordinate animals, thus resulting in enhanced group cohesion (de Waal, 1986). It is conceivable that such a formalisation of dominance relationships also exists in canids, if we think that Schenkel (1967) defined submission in the wolf and dog as 'the effort of the inferior to attain friendly or harmonic social integration'. Actually, although the first studies on the social behaviour of captive wolves conveyed the impression that the life of these animals involved frequent fighting to achieve the top-rank (or alpha) position in a dominance hierarchy (e.g., Schenkel, 1947; Zimen, 1975, 1976), more recent works in the wild have actually emphasised the cohesive nature of wolf packs (Mech, 1999). In particular, it has been stressed that whereas many captive wolf groups consisted of assemblages of unrelated individuals, or of orphaned siblings, in the wild most wolf packs are families comprising a breeding pair and their offspring of various ages (Mech, 1999, 2000; Packard, 2003). Within these family groups, there is a natural age-based dominance order in which offspring submit to parents, and puppies submit to both parents and older siblings (Mech, 1999). Around the age at which they attain sexual maturity (2 years), offspring disperse from their natal pack to start a new family elsewhere, and so they usually do not need to fight in order to achieve a dominant-breeder status (Mech, 1999; Mech & Boitani, 2003). However, under some ecological conditions, like those found in Yellowstone National Park, individuals may delay dispersal, and packs will contain several sexually mature individuals (MacNulty et al., 2009, 2012; Stahler et al., 2013). In these cases, competition for dominance and reproduction may be stronger (Mech, 1999), and usually only the top-ranking male and female breed, although in a few cases subordinate females related to the highest-ranking female also succeed in breeding (Peterson et al., 2002; vonHoldt et al., 2008). In any case, using submissive/dominance postures, it is mostly possible to arrange the members of a wolf pack in a dominance hierarchy, whose function in family packs is probably that of regulating access to food resources (Mech, 1999; Packard, 2003). Sometimes this hierarchy will be pyramidal, as when most submissions are directed by offspring to parents, and there are few submission/dominance displays between offspring of 1–2 years of age (e.g., Table 2 in Mech, 1999; Peterson et al., 2002; Packard, 2003). However, at other times the hierarchy will be linear because submissive/dominance signals are exchanged by most members of the group (e.g., van Hoof & Wensing, 1987; Tables 3 through 5 in Mech, 1999; Packard, 2003; Sands & Creel, 2004). Note that the difference between a pyramidal and a linear hierarchy could also be explained in terms of reduced coverage in the former (higher percentage of dyads in which submissive/dominance signals are not observed), which in turn may be due to insufficient sample size concerning social interactions. This point may be especially relevant with respect to wild packs for which prolonged observations of social interactions, and even individual recognition of group members, can be often difficult to achieve (Sands & Creel, 2004). Another point is that not every wolf behavioural pattern seems appropriate to rank group members. For example, van Hoof and Wensing (1987) found an almost perfect linear dominance hierarchy ($h = 0.99$) in a captive family pack based on the direction of the behaviour 'low posture', whereas aggressive behaviour was basically intransitive. On the other hand, Lockwood (1979) found quite low $h$ when using either submissive or aggressive patterns in several small captive packs, although it is unclear if this was due to the presence of circular triads or rather to reduced coverage.

In our studies on free-ranging dogs, we assessed hierarchies using basically the same behavioural patterns described for wolves, i.e., submissive gestures, dominance displays, and aggressive behaviour (see Cafazzo et al., 2010, for a detailed description). Submissive behaviour was recorded both during greeting ceremonies (affiliative submissions) and in response to aggressive and dominance displays (agonistic submissions). Based on the directionality of these behavioural patterns within dyads, it was possible to demonstrate the presence of a statistically significant linear hierarchy in a pack of 27 dogs ('Corridoio pack') studied during a 1-year period (Cafazzo et al., 2010). Data on social interactions collected for this pack during 2005–2006 were sufficiently numerous to show that the dominance rank order was consistent across different contexts (i.e., competition for food; competition for mates; absence of any sources of competition). Although the values obtained for $h'$ were not high (range 0.3–0.65, depending on the behaviour used to assess the hierarchy and on competitive context), this was clearly due to the relatively low coverage and not to the presence of circular triads (Table 3-1). As in wolves, submissive gestures proved to be more useful to assess the hierarchy when compared to dominance and aggressive gestures because they had higher directional consistency (i.e., higher frequency with which the behaviour occurred in its more frequent direction within dyads relative to the total number of times the behaviour occurred; see van Hoof & Wensing, 1987) and higher coverage (Cafazzo et al., 2010; Table 1). Moreover, affiliative submissions showed complete unidirectionality

**TABLE 3-1** Submissive Behavioural Acts Recorded in the 'Corridoio Pack' in 2005–2006, Both in Affiliative and Agonistic Contexts*

Receiver

| Signaler | Mer | Gas | Pip | Leo | Gol | Lan | May | Nan | Iso | Dia | Sim | Pon | Sem | Kim | Mor | Ste | Han | Cuc | Mam | Dot | Gon | Gre | Bro | Eol | Emy | Mag | Pis |
|---|---|---|---|---|---|---|---|---|---|---|---|---|---|---|---|---|---|---|---|---|---|---|---|---|---|---|---|
| Mer | - | | | | | | | | | | | | | | | | | | | | | | | | | | |
| Gas | 95 | - | | | | | | | | | | | | | | | | | | | | | | | | | |
| Pip | 37 | 49 | - | | | | | | | | | | | | | | | | | | | | | | | | |
| Leo | 13 | 3 | | - | | | | | | | | | | | | | | | | | | | | | | | |
| Gol | 24 | 13 | 10 | 6 | - | | 1 | | 2 | 3 | | | 1 | | | | | | | | | | | | | | |
| Lan | 7 | 10 | 14 | 5 | | - | | | | | | | | | | | | | | | | | | | | | |
| May | 5 | 2 | 5 | 1 | 1 | 3 | - | | | | | | 1 | | | | | | | | | | | | | | |
| Nan | 8 | 1 | 1 | | | 3 | | - | | | | | | | | | | | | | | | | | | | |
| Iso | 2 | 2 | 1 | 3 | 5 | 1 | 2 | | - | 1 | | | | | | | | | | | | | | | | | |
| Dia | 1 | | | | | | | | | - | | 2 | 1 | | | | | | | | | | | | | | |
| Sim | 59 | 34 | 25 | 16 | 19 | 18 | 8 | 6 | 1 | 3 | - | 12 | | | | 1 | | | | | | | | | | | |
| Pon | 14 | 30 | 26 | 4 | 13 | 9 | 7 | 1 | 3 | 1 | 13 | - | 1 | 3 | 2 | | 1 | | | | | | | | | | |
| Sem | 20 | 23 | 34 | 26 | 18 | 10 | 3 | 23 | 98 | 8 | | | - | 5 | | | | | | 3 | | | | | | | |
| Kim | 1 | 5 | 10 | 3 | 3 | 10 | 16 | 3 | 2 | 3 | 4 | 5 | | - | 1 | | | | | | | | | | | | |
| Mor | 15 | 5 | 7 | 1 | 7 | 7 | 3 | 7 | 1 | 1 | 3 | 1 | | | - | | | | | | | | | | | | |

Continued

**TABLE 3-1** Submissive Behavioural Acts Recorded in the 'Corridoio Pack' in 2005–2006, Both in Affiliative and Agonistic Contexts*—cont'd

| Signaler | Mer | Gas | Pip | Leo | Gol | Lan | May | Nan | Iso | Dia | Sim | Pon | Sem | Kim | Mor | Ste | Han | Cuc | Mam | Dot | Gon | Gre | Bro | Eol | Emy | Mag | Pis |
|---|---|---|---|---|---|---|---|---|---|---|---|---|---|---|---|---|---|---|---|---|---|---|---|---|---|---|---|
| Ste | 11 | 2 | | | 2 | 3 | 10 | 8 | | | | 1 | | 3 | | – | | | | 1 | 3 | | | | | | 1 |
| Han | 13 | 7 | 10 | 11 | 4 | 5 | 7 | 7 | 1 | 9 | 6 | | | 3 | 3 | 2 | – | 2 | | 1 | 3 | | | 1 | | | |
| Cuc | 5 | 4 | 3 | 3 | 5 | 8 | 16 | 9 | | 5 | | | 15 | 9 | 11 | | | – | | | | | | | | | |
| Mam | 4 | 5 | 3 | | 6 | 9 | 11 | 8 | | 1 | | 2 | 11 | 3 | 6 | 1 | 4 | 1 | – | | 2 | | | 2 | | | |
| Dot | 12 | 9 | 2 | | 17 | 31 | 10 | 12 | | 4 | 4 | 1 | 22 | 2 | 4 | 1 | | 7 | 9 | – | | | | | | 4 | |
| Gon | 12 | 1 | 8 | 1 | 5 | 27 | 10 | 7 | 3 | 13 | 13 | 4 | 13 | 6 | 1 | | | 1 | | 4 | – | | | | | 1 | 1 |
| Gre | 11 | | | | 2 | 4 | 4 | 12 | | 2 | | 5 | | 1 | | | | | 1 | 1 | | – | | | | | |
| Bro | 12 | 4 | 4 | | 9 | 4 | 9 | 8 | | 4 | 4 | 1 | 7 | 1 | 7 | 1 | 4 | 3 | 1 | 7 | 9 | – | – | 5 | 2 | 2 | |
| Eol | 16 | 5 | 6 | 3 | 12 | 19 | 11 | 19 | | 4 | 4 | 1 | 8 | 4 | 6 | 3 | 4 | 4 | 1 | 5 | 5 | 5 | 5 | – | 7 | 7 | 5 |
| Emy | 2 | | 1 | | 11 | 10 | 5 | 12 | | 4 | 4 | | 10 | | 4 | 2 | 1 | 1 | 1 | 1 | 2 | | 2 | 7 | – | 11 | |
| Mag | 3 | 3 | 1 | | 3 | 11 | 8 | 15 | | 3 | 3 | 1 | 33 | 2 | 3 | 1 | 2 | 2 | | 4 | 1 | 1 | 2 | 7 | 11 | – | |
| Pis | 6 | 2 | 3 | | 6 | 5 | 11 | 8 | | 2 | 2 | 2 | 8 | 7 | 7 | 1 | 2 | 4 | 4 | 1 | 1 | 1 | 2 | 5 | 5 | | – |

Receiver

*Performers on the vertical axis and receivers on the horizontal one. Improved linearity test (de Vries, 1995): h' = 0.65, P = 0.00001. Directional consistency index (DCI) = 0.96. The dominance order was built following de Vries (1998). (This table has been modified from Cafazzo et al., 2010.)

(i.e., the behaviour was expressed by only one individual in all dyads), and thus they could indicate formal acceptance of subordination (for a similar argument in pet dogs, see Bauer & Smuts, 2007; Trisko, 2011).

We studied the 'Corridoio pack' again in 2007–2008, when its size had shrunk to 11 individuals, and found that it was still possible to detect a significant linear hierarchy in it, based on the direction of submissive behaviour (Bonanni, 2008). On the other hand, for another group of 9 dogs studied in the same period, the 'Curva pack', the distribution of submissive signals resulted in a hierarchy whose linearity was just marginally significant (Bonanni, 2008). However, this lack of significant linearity was undoubtedly due to low coverage. In fact, this pack was studied in an undisturbed manner for just 5 months before its composition changed substantially (it was joined by another group of 5 dogs to form the 'Fused pack'), and several members were sterilised by the Rome Municipality. Nonetheless, by summing all submissive interactions recorded before and after sterilisation, we achieved higher coverage and found a significant linear hierarchy in this pack as well (Bonanni et al., unpublished data). Finally, the distribution of submissive behaviours recorded in a group of 3 dogs, the 'Piazza pack', allowed to arrange its members in a perfect linear hierarchy ($h = 1$; Bonanni, 2008) although, due to the small size of this pack, it was not possible to demonstrate that linearity could not have arisen by chance (see Appleby, 1983).

Overall, in our packs the dominance rank order (built following de Vries, 1998) was positively correlated to age, with older dogs being dominant over younger ones (Bonanni et al., 2010a; Cafazzo et al., 2010). Moreover, there was a tendency for males being dominant over females of similar age, although females were usually dominant over younger males (Bonanni et al., 2010a; Cafazzo et al., 2010). Both results suggest that the structure of our dog packs was relatively similar to that of wolf family packs, in which the dominance order is also based on age and males tend to dominate females within a given age class (van Hoof & Wensing, 1987; Mech, 1999; Packard, 2003).

We stress that dominance hierarchies in dogs are not restricted to our study area. Work carried out by our collaborators, using our same methods, has found two significant linear hierarchies in a pack of six free-ranging dogs living in a different district of Rome (Abis, 2004) and in a group of eight dogs housed in a shelter in Southern Italy (Barillari, 2004). Furthermore, a significant linear hierarchy has recently been found in a large group of pet dogs housed at a daycare facility in North America (Trisko, 2011).

In an influential review, Bradshaw et al. (2009) re-examined a data set previously published by Pal et al. (1998) concerning the pattern of social interactions in two packs of Indian free-ranging dogs, and presented new data on dominance interactions in neutered male dogs. They applied David's score (Gammel et al., 2003) in an attempt to arrange the dogs of all these groups in a rank order and, after finding some apparent inconsistencies, they rejected the usefulness of dominance as a construct to explain dog social behaviour. First of all, it should be noted that, although David's score is a valuable tool in order to rank the members of a social group, it does not represent a test of the linearity of a hierarchy at all. Consequently, it cannot be used to reject a linear hierarchy model as applied to groups of domestic dogs. Second, we re-analysed the data published by Pal et al. (1998) using an improved linearity test developed by de Vries (1995), and found the following results: in the pack of eight dogs, outcomes of submissive interactions (see Pal et al., 1998, p. 337, Table 3, LIG group) could be arranged in a hierarchy whose linearity was highly significant ($h' = 0.90$, $P = 0.0009$), thus contradicting the findings by Bradshaw et al. (2009); in the pack of five dogs, submissive interactions (see Pal et al., 1998, p. 337, Table 3, HIG group) again resulted in a hierarchy whose degree of linearity was even higher ($h' = 0.95$) although, because this group comprised less than six individuals, there was no statistical power to demonstrate that this linearity could not have arisen by chance (see Appleby, 1983).

In summary, the finding of seven statistically significant linear hierarchies in large packs of domestic dogs, and of two hierarchies with very high indexes of linearity in small packs for which statistical testing of linearity was not applicable, suggests that hierarchies in this species are common, although it cannot probably be expected that every dog group will fit a hierarchy model. Hierarchies can be detected only by applying appropriate linearity tests, and mostly only after prolonged recording of social interactions necessary to achieve sufficient coverage. Our results strongly contradict previous claims about the presumed lack of a clear hierarchical structure in dog groups (e.g., Boitani & Ciucci, 1995; Boitani et al., 2007; Bradshaw et al., 2009; McGreevy et al., 2012) and suggest instead that evolution in a domestic environment has not substantially altered the ability of dogs to form structured packs with conspecifics.

### 3.3.2 Leadership

Social animals have a strong tendency to do what their group companions are doing (Epple & Alveario, 1985; Glickman et al., 1997; Ferrari et al., 2005, 2009), and this holds for domestic dogs as well (Ross & Ross, 1949; Scott & Marston, 1950; Range et al., 2011). This coordination of group activities is fundamental in order to maintain group cohesion and eventually allow animals to get the benefits of social living (Conradt & Roper, 2005). However, group members are likely to vary in the degree to which they affect the behaviour of companions. For example, some individuals (usually referred to as 'leaders') can be more likely to make decisions about which activity the group will perform at a given time and companions will simply accept their decisions (Conradt & Roper, 2005).

In family groups of wolves, collective activities are usually led by the dominant breeders given that the offspring tend naturally to follow the initiatives of their parents (Mech, 1999, 2000). Parents usually make decisions about awakening at the den, initiation of foraging, and travelling direction, and also take initiatives in hunting, territorial defense, etc. (Mech, 2000). In small family packs, leadership is shared by both the breeding male and the breeding female (Mech, 2000). However, in large family packs, containing multiple sexually mature individuals, dominant breeders usually lead movements during 60%–90% of travel time (Peterson et al., 2002), meaning that in a non-negligible minority of cases, subordinate offspring can also provide leadership. From a functional perspective, it is usually believed that young subordinates can benefit from accepting their parents' decisions because the latter are more experienced and thus have higher knowledge about the location of resources such as prey and refuges within the territory (Mech, 1999; Peterson et al., 2002; Packard, 2003). On the other hand, sharing leadership in large packs may allow dominant animals to take advantage of pooled knowledge of a territory, and also to reduce both energy expenditure and the risks associated with travelling at the front of the pack (Peterson et al., 2002).

Although leader–follower relationships have been observed and described also for free-ranging dogs (e.g., Beck, 1973, 1975; Nesbitt, 1975; Fox et al., 1975; Boitani et al., 1995), until recently quantitative data on social interactions in these animals were too scarce to allow any detailed comparison with the pattern found in their wild ancestors. In our study, we often observed free-ranging dogs of all packs engaging in coordinated collective movements (Figure 3-1) that usually occurred in correspondence with group activity shifts (e.g., from resting to travelling, from resting to feeding/drinking, etc.; see Bonanni et al., 2010a). This provided us with the opportunity to investigate leadership as a mechanism for promoting pack coordination, and also to assess how the individual tendency

**FIGURE 3-1**    The 'Corridoio pack' on the move. Dogs were often observed moving in a single file. See color plate section. *(Photo by Simona Cafazzo.)*

to lead the pack was influenced by social relationships. In particular, we focussed on leadership at group departure for two reasons: (1) it was not always possible to follow the group to its travel destination, and (2) studies on other species (Byrne, 2000) suggest that animals walking at the front of the group are *not necessarily* those who decide the direction of travel. Operationally, we defined a 'leader' as the first dog that, after leaving all other group members behind, moved along a direction followed by a minimum of two companions ('followers') within 10 minutes (see Bonanni et al., 2010a, for more details). Note that wolf researchers in the past have measured leadership both at group departure and as the frequency/duration of staying at the front of the pack during travelling, and that these two different measures have provided very similar results (Mech, 2000; Peterson et al., 2002). So, for this reason and others (see below), we are quite confident that our definition of leadership provides a useful measure to allow a comparison with studies on wolves.

We found that in our studied packs, leadership was shared among group members, although not equally (Bonanni et al., 2010a). In other words, although every dog of at least 1 year of age could behave as a leader sometimes, i.e., it could successfully initiate a collective movement involving a minimum of three animals, each pack contained a limited number of individuals that could be classified as 'habitual leaders' (i.e., individuals that behaved more frequently as leaders than they behaved as followers). For example, in the largest pack studied ('Corridoio pack', 2005–2006), only 6 out of 27 dogs could be classified as 'habitual leaders', whereas there was just 1 habitual leader in the 'Corridoio pack', 2007–2008, and in the 'Curva pack', that comprised 11 and 9 dogs, respectively (Bonanni et al., 2010a). Habitual leaders could be both male and female, and overall, gender had no significant effect on the frequency of leading. Instead, the individual frequency of leading was strongly influenced by a linear combination of age and measures of dominance (e.g., proportion of submissive group companions; Bonanni et al., 2010a). Specifically, dogs that were old and high-ranking were much more likely to behave as leaders than young, low-ranking individuals (Figure 3-2). However, a multivariate statistical analysis revealed that receiving affiliative submissions, or formal recognition of dominance (Bauer & Smuts, 2007; Cafazzo et al., 2010; Trisko, 2011), was overall a better predictor of leadership than receiving agonistic submissions. Specifically, dogs that more frequently behaved as leaders usually received both affiliative and agonistic submissions by many companions, whereas dogs that received submissions by many companions *only* in agonistic contexts rarely behaved as leaders (Bonanni et al., 2010a). These results suggest that the development of an affiliative relationship between dominant and subordinate dogs could play a fundamental role in eliciting follower behaviour. In any case, the two kinds of submissions were highly and positively correlated during the period 2005–2006, but not in 2007–2008. This difference between the two phases of the research was partly due to the fact that the dog scoring highest for 'received agonistic submissions' in 2007–2008 actually scored 'zero' for

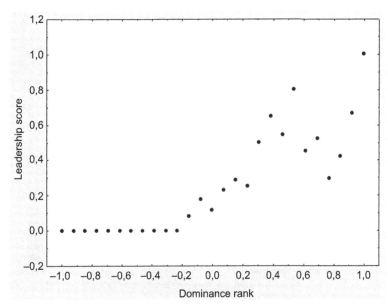

**FIGURE 3-2**   The relation between 'leadership score', as defined in Bonanni et al. (2010a), and 'dominance rank' for the members of the Corridoio pack during 2005–2006. Dominance rank is based on the directionality of both affiliative and agonistic submissions. Highest rank scored as +1 and lowest rank scored as −1. Spearman correlation between the two variables: $rs = 0.92$, $P < 0.00001$.

'received affiliative submissions' (Bonanni et al., 2010a). This animal was a young adult male that was attaining higher positions in the dominance hierarchy of the 'Corridoio pack', although he rarely behaved as a leader (Figure 3-3). This might have been simply due to insufficient time passing from the achievement of a high-rankings status and hence a lack of recognition of his leadership role by the other members of the pack. Alternatively, this animal may not have possessed the social skills necessary to develop affiliative relationships with his potential followers (Bonanni et al., 2010a). The latter seems to us the most likely explanation based on the observations we carried out on this pack in 2009–2010. Moreover, the affiliative nature of leader–follower relationships is also emphasised by our finding that, during resting times, followers spent more time in close proximity with the 'habitual leaders' of their pack than with other followers (Bonanni et al., 2010a). This suggests that leaders were usually regarded by companions as more 'attractive' social partners than non-leaders.

From a proximate perspective, collective movements in free-ranging dogs may arise from the effort of subordinates to maintain proximity with dominant individuals. A further interpretation is that followers tend to do what the leaders do (e.g., go to a given place when the leader goes to that place) because they pay higher attention to the leaders' actions than to the actions of other companions, having a stronger affiliative relationship with the former. Consistent

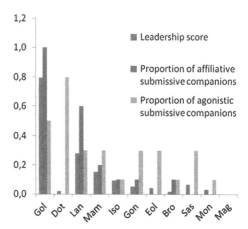

**FIGURE 3-3** Distribution of 'leadership score', 'proportion of affiliative submissive companions', and 'proportion of agonistic submissive companions', respectively, for the members of the Corridoio pack during 2007–2008. Spearman's correlation between 'leadership score' and 'affiliative/agonistic submissive companions': $rs = 0.75$, $P < 0.008$; $rs = 0.47$, $P = 0.14$, respectively. The latter failed to be statistically significant due to the behaviour of Dot. See color plate section. *(Modified from Bonanni et al., 2010a.)*

with this hypothesis, recent studies have shown that dogs pay more attention to the actions of their owners than to the actions of other familiar humans (Horn et al., 2013), indicating that in these animals attention is enhanced by affiliation. Moreover, Lockwood (1979) also proposed that the social organisation of his captive wolf packs could be described in terms of the differential attention that dominant animals received by companions.

So, based on our results, it can be argued that leadership in the free-ranging dogs of our population is mainly provided by old and dominant individuals (note that age and dominance are positively correlated), a pattern that seems to be relatively similar to that found in wolf family packs, in which parents lead activities and they are older than offspring and usually dominant over them (Mech, 2000; Peterson et al., 2002). However, from a functional perspective, it is unclear why dogs have retained the wolf trait of following the movements of experienced individuals because, unlike wolves, they subsist on food resources that are quite predictable in time and space. However, it remains possible that leaders may provide effective protection against common enemies (King et al., 2008) or know the better strategies and refuges to avoid them.

In wolves, the expression of both dominance and leadership closely corresponds to a 'breeding status', given that parents are in most cases the only breeding pack members, and they tend to suppress reproductive activity in their sexually mature offspring (Packard et al., 1985; Derix et al., 1993; von-Holdt et al., 2008). Conversely, packs of free-ranging dogs are known for containing often multiple breeding individuals of both sexes (Daniels & Bekoff,

1989a,b; Boitani et al., 1995; Macdonald & Carr, 1995; Pal et al., 1999; but see Gipson, 1983), and it has been argued that they lack those social mechanisms that allow dominant wolves to control the reproductive activity of subordinates (Boitani & Ciucci, 1995; Bradshaw et al., 2009). However, the latter is clearly a very questionable point because few data on dominance in free-ranging dogs have been published until the recent past and, consequently, the effect of social relationships on reproductive activities of group members has rarely, if ever, been tested. Our dog packs were similar to many others in that they contained multiple breeding individuals of both sexes. However, even if all sexually mature females mated and eventually gave birth, and several males of each group mated with oestrus females, in the pack for which analysis was possible (Corridoio pack, 2005–2006), both female reproductive success and male copulation rate were strongly influenced by leadership and dominance status (Cafazzo et al., submitted). In particular, high-ranking females that led the pack more frequently had a higher number of puppies surviving to sexual maturity, and high-ranking males that led the pack more frequently enjoyed a much higher copulation rate (Cafazzo et al., submitted). So, it seems that, at least in the Corridoio pack, leadership and dominance had a strong relationship with reproductive activity. Whereas in wolf packs there are usually just two 'breeding' individuals that consistently lead group activities, in our Corridoio pack multiple breeders could be ranked along a continuum from those that led frequently and had the highest reproductive success to those that rarely/never led and had the lowest reproductive success. Is the difference in the number of breeding individuals between wolf and dog packs an 'adaptive' consequence of the evolution of dogs in a domestic environment? We believe that this may be the case, and there are at least two possible reasons that packs of free-ranging dogs often contain multiple breeding individuals. First, as already pointed out by Lord et al. (2013), availability of food resources provided directly or indirectly by humans throughout the year allows dogs to reproduce in their first year of life. Conversely, wolves are much more food-limited than dogs and usually cannot breed before their second year of life (Lord et al., 2013). This means that young dogs can reproduce in their natal pack, whereas young wolves usually reproduce after dispersal. Second, abundant food resources in the domestic environment would presumably decrease feeding competition and nutritional stress in free-ranging dogs, and this may have favored the evolution of a social system in which dominant individuals would put less pressure over subordinates and would allow them to breed. This is also supported by the observation that wolf packs with multiple breeders are found where food resources are unusually abundant (Mech & Boitani, 2003). So, although much more data on the social regulation of reproduction in dogs are clearly needed, the interdependence of variables such as leadership, dominance, age, and reproductive activity, found both in studies of wolves and in our study on free-ranging dogs, suggests that some common organising mechanism may shape the social organisation of both species.

### 3.3.3 Cooperation

Cooperation can be defined as a joint action by two or more individuals to achieve a common goal (Boesch & Boesch, 1989). Moreover, in a broader sense, cooperation can also refer to prosocial behaviour by which an animal acts to benefit another one (e.g., de Waal et al., 2008). In wolves, social bonds among pack members have several functional consequences that include cooperation in hunting large prey (Mech, 1975; MacNulty et al., 2009, 2012), cooperation in territorial defence (Harrington & Mech, 1979; Mech, 1993; Mech & Boitani, 2003), and cooperation in breeding (Packard et al., 1992; Mech, 1999). Notably, wolf family packs are characterised by monogamous pair bonding, and by a division of *labour* system in which the breeding male leads hunts at places located several kilometers away from the den where the female remains to attend puppies (Mech, 1999, 2000). Subsequently, the breeding male and the older offspring come back to the den and provision the female and the puppies with food, by carrying pieces of prey and/or by regurgitating partially digested food (Mech, 1999, 2000). Conversely, many authors have stressed that domestication seems to have caused a reduction in cooperative tendencies in dogs, particularly because cooperative breeding and hunting seem to be greatly reduced in dogs relative to wolves (Boitani & Ciucci, 1995; Coppinger & Coppinger, 2001; Kubinyi et al., 2007; Miklósi, 2007a; Range et al., 2009; Brauer et al., 2013). For example, unlike wolves, free-ranging dogs have a primarily promiscuous mating system (Daniels, 1983b; Ghosh et al., 1984; Boitani et al., 1995; Pal et al., 1999; Pal, 2003), they rarely form monogamous pairs (for exceptions, see Gipson, 1983; Pal, 2005), and puppies are rarely fed by group members other than their mother (Macdonald & Carr, 1995; Boitani et al., 1995; Lord et al., 2013; but see Pal, 2005). According to Boitani et al. (1995), the lack of paternal care is one of the main reasons that feral dog populations suffer from very high infant mortality and are probably not self-sustaining. However, it should be stressed that many other authors have reported infant mortality rates for populations of free-ranging dogs, living in various environments, that fell within the range reported for wild populations of wolves (Scott & Causey, 1973; Nesbitt, 1975; Oppenheimer & Oppenheimer, 1975; Gipson, 1983; Daniels & Bekoff, 1989a,b; Butler & Bingham, 2000; Pal, 2001). This suggests that the apparent reduction in cooperative breeding in dogs is not necessarily a maladaptive trait. For example, Lord et al. (2013) have recently emphasised that the availability of food at dumps allows dog puppies to start foraging independently of parents at a much earlier age than wolf puppies (that have to learn to hunt). This might have favored the evolution in dogs of a decreased parental investment and of a consequent higher fertility (Lord et al., 2013). In populations that are close to the habitat carrying capacity, high fertility would lead to high infant mortality (e.g., Beck, 1973; Daniels, 1983a) but, on the other hand, it would allow fast recovery after increases in adult mortality (Lord et al., 2013). We add to these arguments the following considerations. First of all, human dumps are likely to be much more predictable in time and space than mobile prey on which wolves subsist. This usually allows lactating female dogs to place their den in the vicinity of human food sources (e.g., Gipson, 1983; Boitani et al., 1995; Macdonald & Carr, 1995), and to feed without the need of waiting the return of any hunting companions while, at the same time, minimising the time in which puppies are left unattended. Consequently, in our opinion, allofeeding of mothers and puppies in free-ranging dogs does not happen frequently because it is not required by mothers, and it is probably not necessary in order to increase their reproductive success. Second, free-ranging dog populations are often characterised by much higher density than any wild canid species and by the lack of a clear reproductive season (e.g., Beck, 1973; Daniels, 1983a,b; Boitani et al., 1995; Butler & Bingham, 2000). These conditions undoubtedly provide male dogs with the opportunity to court multiple oestrus females throughout the year while, at the same time, distracting them from any eventual caring for puppies. Consequently, these factors may have favoured the evolution from a monogamous mating system with prominent paternal care (like that of wolves) to a promiscuous mating system with reduced paternal care (like that of dogs). This seems to be supported by the fact that some wild canid species can switch from a monogamous to a polygynous/promiscuous mating system when population density increases (reviewed in Hennessy et al., 2012).

In our study population, we had limited possibilities to assess the level of cooperative breeding because dens with puppies were often well concealed in the dense vegetation and not easy to locate. On one occasion we did observe a neutered male dog regurgitating food for puppies that were past their weaning age. Moreover, on a number of occasions, we observed several members of a given pack gathering around a den location, or a male associating with a lactating female (Cafazzo, 2007); hence, we cannot rule out the possibility that certain adult dogs, other than the mother, regurgitated food to puppies. However, lactating females were frequently observed coming to the feeding stations to feed together with their group companions and then going back to the nature reserve to feed their pups without being followed by the rest of the pack. Another point is that we also observed some pregnant females leaving the Corridoio pack before giving birth and never coming back (Cafazzo, 2007). Notably these females, which presumably raised their puppies alone, had a reproductive success that was lower than that of the top-ranking female of the Corridoio pack, although still higher than that of non-dispersing subordinate females (Cafazzo et al., submitted). So, although these data are very limited, we hypothesise that (1) either any potential cooperative breeding effort concentrated mainly on the top-ranking female, or (2) it had no major impact on female reproductive success.

However, the fact that male dogs do not frequently regurgitate food to their companions does not, in our opinion, imply that dogs are not cooperative carnivores or that they are unable to display prosocial behaviour towards conspecifics. For example, free-ranging dogs, including males, can provide parental care by guarding puppies against enemies (Nesbitt, 1975; Pal, 2003, 2005; Cafazzo, 2007; Guenther Bloch, personal communication), or can defend collectively an injured companion against approaching humans (Macdonald & Carr, 1995). Moreover, large dogs have been sometimes observed knocking over garbage cans to make food available for smaller dogs (Beck, 1973). Furthermore, in our opinion, cooperation should not be seen as a unitary trait. To give an example, cooperative breeding in wolves seems to be promoted by a spring peak in the secretion of prolactin (Kreeger et al., 1991; Asa, 1997; Asa & Valdespino, 1998). However, wolves also cooperate in hunting and in territorial defence at times other than spring (Mech & Boitani, 2003; MacNulty et al., 2012), indicating that different proximate mechanisms may underlie different kinds of cooperation. So, although group-living free-ranging dogs seem to show an overall reduction in cooperative breeding and hunting relative to wolves, they frequently engage in cooperative defence of territory and food resources against rival dogs (Fox et al., 1975; Font, 1987; Daniels & Bekoff, 1989b; Macdonald & Carr, 1995; Boitani et al., 1995; Pal et al., 1998). This is usually accomplished by several pack members joining forces to threaten or chase dogs belonging to stranger packs and often do not involve physical contact because the weaker opponents retreat without a fight. In our population, intergroup competition for food and space was very frequent. We were interested in assessing which variables influenced cooperation among group companions and individuals' active participation in conflicts against stranger packs. Active participation was defined as approaching opponents aggressively (e.g., barking, snarling, staring at the opponents with a tense body posture while keeping the tail high) by moving forward a minimum distance of 10 m, although we also recorded which dogs came closest to the opponents during an intergroup conflict or eventually in aggressive physical contact (e.g., biting, scratching, jumping upon) with them (see Bonanni et al., 2010b, for more details). Our results suggest that affiliative relationships between pack members promoted cooperation. Specifically, individuals that were more likely to actively participate in intergroup contests were those with the higher proportion of adult companions with whom they exchanged affiliative gestures (Figure 3-4; Bonanni et al., 2010b). From a functional perspective, cooperation with affiliative partners may ensure long-term reciprocation of social support and limit exploitation by cheaters (de Waal, 2008; Schino & Aureli, 2009; Berghanel et al., 2011). Surprisingly, we found that high-ranking dogs tended to participate more only in contests against groups that were larger than their own, although they did not stay at the front of the pack more often than subordinates (Bonanni et al., 2010b). Importantly, the coordinated aggressive behaviour exhibited by group companions appeared to be functional with respect to the competitive outcome because larger packs (those containing the higher number of participating individuals) were more likely to elicit a retreat response in the opposing groups (Bonanni et al., 2011).

From a socio-cognitive perspective, it would be interesting to assess whether these coordinated threatening displays observed in free-ranging dogs actually

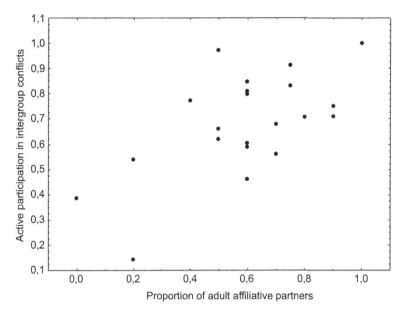

**FIGURE 3-4**   The relation between 'relative frequency of active participation in intergroup conflicts' and 'proportion of adult affiliative companions' for the members of three packs studied in 2007–2008. Pearson's correlation between the two variables: $r = 0.63$, $P = 0.002$. *(Modified from Bonanni et al., 2010b.)*

represent cooperation in a strict sense (see Boesch & Boesch, 1989). In other words, do dogs adjust their behaviour to that of companions so as to make coordination possible, or are they simply reacting to strangers simultaneously and independently of each other? For example, recently Brauer et al. (2013) found that even if pet dogs successfully coordinated their actions in a problem-solving task, there was no obvious indication that they attended to each other during the experiment, although the authors admitted that monitoring the partner may not have been necessary to solve the task. Although a compelling demonstration that dogs take their partner's role into account should be based on controlled studies, our results provide some indications that free-ranging dogs were *not* acting independently of each other when engaging in coordinated aggressive displays. First of all, in our study, dogs were more likely to behave aggressively when their own pack outnumbered the opposing group (Bonanni et al., 2011), which suggests that dogs took their companions' presence into account as if they were expecting to receive support from them during conflicts. Second, the previously mentioned correlation between individual participation in conflicts and measures of affiliative integration is, in our opinion, unlikely to emerge unless dogs are also attending to which companions participated in conflicts, and not just reacting to the opponents. For example, the first dogs that detect the presence of strangers will usually start challenging them (conditionally to an assessment of relative group size), and this may facilitate the activation of similar motor patterns and similar emotional states in their companions (sensu Preston & de Waal, 2002); the latter, in turn, may happen with a higher probability, the closer the social relationship between the animals involved. Affiliative partners either may be more motivated to act together or simply pay higher attention to the actions of companions.

Moreover, during intergroup conflicts, we observed dogs displaying behavioural patterns indicating that they could take into account the behaviour of partners. For instance, we frequently observed dogs glancing at companions as if they were checking their position, although systematic recording of this behavioural pattern was prohibitively difficult because typically many dogs moved fast all together. However, on some occasions when dogs approached opponents slowly, we could observe some individuals monitoring the position of companions (e.g., by turning their heads backwards) and then stopping the approach until other dogs also started moving forward. So, we had the feeling that looking at companions could be used to check whether they would join the collective action and/or as a way to recruit them (see Fox et al., 1975, for a similar description). Finally, when groups of dogs chasing a lone opponent succeeded in reaching the opponent, they frequently encircled it, a behaviour indicating that dogs were adjusting their movements to those of companions.

In contrast, we almost never observed our studied dogs engaging in cooperative hunting, aside from an unsuccessful attempt at chasing a red fox (*Vulpes vulpes*), although they were suspected to have killed a crested porcupine (*Hystrix cristata*), some sheep (*Ovis aries*), and a donkey (*Equus africanus asinus*). It is probable that dogs had low opportunities to find suitable prey in the area, and the possibility of feeding on food provided by humans may have decreased their motivation to hunt and hence to cooperate in hunting. More generally, dogs have smaller teeth and less powerful jaws than wolves (Coppinger & Coppinger, 2001), and thus, they are likely to be less effective than wolves in killing large prey. Even so, in our opinion, evolution in a domestic environment does not necessarily imply that dogs have lost their predatory motivation or the ability to hunt cooperatively. Notably, this environment also contains several other species that were commensal of humans and potential prey for dogs during their evolution, including several ungulates (Zeder, 2012), and this may have contributed to maintain selective pressures for predatory motivation. Studies on free-ranging dogs' diet confirm that most of them subsist primarily on human waste and carrions, but they can also prey on small- and medium-sized animals such as reptiles (e.g., *Amblyrynchus cristatus*), birds (e.g., *Apterix australis*), mice (*Peromyscus* spp.), rabbits (*Sylvilagus* spp.), hares (*Lepus* spp.), white-tail deer (*Odocoileus virginianus*), roe deer (*Capreolus capreolus*), gazelles (e.g., *Procapra gutturosa, Gazella gazella*), impala (*Aepyceros melampus*), goats (*Capra hircus*), and sheep (Scott & Causey, 1973; Nesbitt, 1975; Kreeger, 1977; Lowry & MacArthur, 1978; Causey & Cude, 1980; Gipson, 1983; Boitani, 1983; Butler et al., 2004; Vanak & Gompper, 2009; Young et al., 2011). Although systematic recording of how many dogs were involved in killing these prey was almost never carried out, these studies report that on a number of occasions two to four dogs participated in killing (e.g., Kreeger, 1977; Butler et al., 2004), indicating that free-ranging dogs can hunt cooperatively. Moreover, we were personally shown a series of pictures, taken by a game warden in Italy, showing a group of six free-ranging dogs chasing a herd of wild boars (*Sus scrofa*) and succeeding in isolating one of the juveniles and in killing it. In our view, it is realistic to suggest that most domestic dogs are less specialised than wolves in preying on large herbivores, but they can easily hunt medium-sized animals, including some ungulates, and can do this cooperatively. Although it has been suggested that free-ranging pets kill ungulates more frequently than unowned dogs (e.g., Kreeger, 1977; Lowry & MacArthur, 1978), this may be simply due to the fact that the former were more numerous in the areas where those predation events were recorded. It is plausible that unowned dogs are more likely to engage in cooperative hunting where suitable prey are available and food resources provided by humans are less abundant.

To summarise this part, our belief is that ecological pressures experienced by dogs in the domestic environment have led to the evolution of a promiscuous mating system characterised by a reduction in cooperative breeding relative to wolves, and of a feeding strategy primarily based on scavenging and predation of small- or medium-sized animals, still involving some degree of cooperative hunting. However, defending food resources collectively against conspecifics seems to be still advantageous in dogs, at least when these are abundant enough to allow pack living, and this might have provided selective pressures for the ability to develop a complex social structure, relatively similar to that of wolves.

## 3.4 INTERGROUP RELATIONSHIPS: SPACING PATTERN

A territory can be defined as a defended area from which competitors are actively excluded (reviewed in Maher & Lott, 1995, 2000). Wolves are usually depicted as highly territorial carnivores because packs live in more or less exclusive areas usually showing little overlap with areas occupied by neighbouring packs (Peters & Mech, 1975; Harrington & Mech, 1979; Harrington & Asa, 2003; Mech & Boitani, 2003). Moreover, such areas are defended both directly, i.e., through attacks against trespassing wolves, and indirectly through scent marking and howling (Peters & Mech, 1975; Harrington & Mech, 1979; Harrington & Asa, 2003; Mech & Boitani, 2003). Scent marking is accomplished mainly through raised-leg urinations (in males), flexed-leg urinations (in females), and ground scratching, all of which are usually performed only by the dominant breeders (Peterson et al., 2002; Harrington & Asa, 2003). Scent marks are released at higher rates along the boundaries of territories than in the interior parts, and there is some evidence that they can deter intruders from trespassing, and that trespassing wolves usually suspend marking (Peters & Mech, 1975; Mech & Boitani, 2003). Howling can also be used to advertise a pack's presence in a given area, although it seems to be independent of location (Harrington & Mech, 1983). In any case, both marking and howling appear to be functional in order to space different packs and to reduce the probability of interpack encounters. Nevertheless, territorial competition in wolves can become more intense during the mating season, and interpack conflicts are actually one of the main causes of wolf mortality (Mech & Boitani, 2003).

However, the degree of territoriality exhibited by a given species is known to vary with ecological conditions (Maher & Lott, 2000). For instance, wolf territories may overlap more when food resources become more abundant, population density increases, and the habitat becomes saturated (Peterson, 1979; Mech & Boitani, 2003). More in general, when population density increases above a given threshold, intruder pressure may become so intense that territories would no longer be economically defendable, leading animals to switch to alternative competitive strategies (Brown, 1964; Maher & Lott, 2000). So, since free-ranging dogs often subsist on abundant food resources and, as a consequence, live at much higher densities than wolves, it may be expected that overall they will be less territorial than their wild ancestors. For example, in some high-density urban populations, free-ranging pets seem to defend their household as a territory but, when outside, their small home ranges greatly overlap with those of other dogs, indicating that they are not defending exclusive areas (e.g., Berman & Dunbar, 1983; Daniels, 1983a; Daniels & Bekoff, 1989a). Conversely, packs of unowned free-ranging dogs living in low-density rural areas have often been described as territorial (Nesbitt, 1975; Gipson, 1983; Daniels & Bekoff, 1989a,b; Macdonald & Carr, 1995; Boitani et al., 1995). On the other hand, there are also reports of territorial pack-living dogs in urban areas (Fox et al., 1975; Font, 1987; Pal et al., 1998) and of non-territorial packs in low-density rural areas (e.g., Scott & Causey, 1973). So, it appears that between-population variation in territoriality is affected by multiple variables in free-ranging dogs, which could possibly include density, abundance, distribution, and predictability of food resources as well as the degree of relatedness to certain breeds selected for higher/lower territoriality, etc. (see next section).

Moreover, free-ranging dogs' territories are usually much smaller than those of wolves, the former ranging from 0.1 ha in urban areas (Daniels, 1983a) to 70 km² in Alaska (Gipson, 1983), and the latter ranging from 30 to more than 6,000 km² (Mech & Boitani, 2003). Clearly, the size of wolf territories will be inversely related to prey biomass (Mech & Boitani, 2003), whereas dogs' territories presumably just need to embrace human food sources and are unlikely to be strongly influenced by highly mobile prey. There is no evidence that free-ranging dogs rely on howling to space each other as wolves do, although there is some indication that they can use collective barking to intimidate rival packs even at considerable distance and to keep them away from dumps (Daniels & Bekoff, 1989b; Macdonald & Carr, 1995). Overall, it appears that free-ranging dogs are usually less spaced than wolves (because they live at higher densities, in smaller territories), and that interpack conflicts are much more frequent, although rarely escalating into fatal aggression (Daniels & Bekoff, 1989b; Boitani et al., 1995;

Bonanni et al., 2010b, 2011; for exceptions, see Gipson, 1983; Macdonald & Carr, 1995).

There are somewhat ambiguous indications that dogs in our study population were defending territories. For example, based on the locations of intergroup contests, the Corridoio pack in 2005–2006 appeared to be defending an area of about 22 ha, comprised within a larger home range of about 60 ha (Cafazzo et al., 2012). Members of this pack scent marked (using the same behavioural patterns described previously for wolves) more frequently along the territorial boundary and within the territory than outside (Cafazzo et al., 2012). Moreover, they marked more frequently during interpack than during intrapack agonistic interactions (Cafazzo et al., 2012). Unlike wolves, in dogs marking was not restricted to the highest-ranking male and female, although there was a very high positive correlation between marking rate and dominance rank both in males and females (Figure 3-5; Cafazzo et al., 2012; Bonanni et al., unpublished data). Nevertheless, in 2007–2008 we found that the areas marked with raised-leg urinations by the members of three competing packs overlapped to a considerable extent (45%–75% of the total area), indicating that marking was not really effective in deterring strangers from entering a given area (Bonanni et al., 2011). However, the larger packs were usually able to keep the smaller ones away from a given feeding site at least temporarily (Bonanni et al., 2011), and the smaller packs often could feed at a given site only when the larger groups were elsewhere (Bonanni et al., unpublished data). It is likely that the spacing pattern observed in our population was a reflection of the way in which people distributed the food, that led several packs to concentrate their activities around the central road. However, the situation might have been different in 2005–2006, when the Corridoio pack was much bigger than all the other groups in the area, and might have been also more effective in keeping strangers away from its territory.

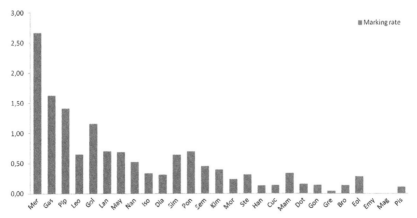

**FIGURE 3-5**   The relation between 'marking rate' and 'dominance rank' for the members of the Corridoio pack during 2005–2006. Spearman rank correlation between the two variables: $r_s = 0.92$, $P = 0.0001$. Individuals are ordered from the highest (on the left) to the lowest dominance rank.

Whereas in wolves marking and territorial defence seem to increase during mating periods (Mech & Boitani, 2003), the opposite seems to hold for free-ranging dogs. During the oestrus period, male dogs can frequently trespass into other packs' territories and succeed in mating with females (Boitani et al., 1995; Pal et al., 1998, 1999; Cafazzo et al., submitted). Consistent with these observations, in our population, males of the Corridoio pack marked less frequently in the presence of oestrus females than at other times (Cafazzo et al., 2012). Furthermore, the oestrus period in dogs can also provide opportunities to recruit strangers into resident packs (Boitani et al., 1995; Macdonald & Carr, 1995; Cafazzo, 2007). It has been suggested (Macdonald & Carr, 1995) that recruitment of immigrant dogs would be more probable during periods in which the structure of packs is highly perturbed by human-caused mortality (e.g., car accidents, capturing, killing). Conversely, when stable affiliative and kinship ties develop among the members of a pack, the chances of unrelated immigrants being recruited will probably diminish (Macdonald & Carr, 1995). More generally, we believe that recruitment of immigrant dogs would be functional for relatively small packs that need to increase in size to face intergroup competition for resources.

Overall, it seems that the pattern of abundance and distribution of food resources in the domestic environment has led to an increase in the frequency of interpack encounters and to a higher degree of interpack tolerance in dogs relative to wolves.

## 3.5 VARIATION IN DOG SOCIETY

Our studies have shown that a population of free-ranging dogs living in a suburban area of Rome exhibited a complex degree of social organisation. Nevertheless, the social organisation of our dogs apparently was very different from that reported by several other authors who claimed that free-ranging dogs either rarely form stable social groups (Beck, 1973, 1975; Rubin & Beck, 1982; Berman & Dunbar, 1983; Daniels, 1983a; Coppinger & Coppinger, 2001; Butler et al., 2004; Ortolani et al., 2009) or, if they do, they lack a clear hierarchical structure (Boitani & Ciucci, 1995; van Kerkhove, 2004; Boitani et al., 2007; Bradshaw et al., 2009). Consequently, it appears of paramount importance to unravel which variables are likely to explain the observed variation in the social organisation of dogs, and to assess to what extent the findings from our population can be extended to other dog populations.

First of all, differences in the results of different studies on free-ranging dogs can partly arise from differences in the methodologies employed. For example, in our opinion, previous claims about the presumed lack of social structure in dog packs were not justified due to very limited quantitative data on social interactions (see the previous discussion about dominance). Moreover, methodology may also affect whether or not dogs are depicted as solitary *versus* social animals. In our study, we concluded that dogs were living in stable packs after observing the animals continuously for several weeks. For example, in 2005 we took about 6 weeks to recognise individually all members of the 'Corridoio' pack and to conclude that they belonged to the same group, although the time necessary to assess the composition of a given pack will inevitably vary depending on group size, population density, and visibility. On the other hand, in most studies claiming that urban free-ranging dogs are semi-solitary carnivores (Beck, 1973; Berman & Dunbar, 1983; Daniels, 1983a; Daniels & Bekoff, 1989a; Ortolani et al., 2009), sociality was assessed by recording the frequencies of group sizes during population surveys, and then by comparing them with the expected frequencies generated by a zero-truncated Poisson distribution, under the hypothesis of random grouping. This method always led to the finding that solitary dogs comprised 50%–80% of the total population and that they were more frequent than expected. However, as pointed out by Font (1987), this method does not involve prolonged and continuous observations of animals, and it can lead to an overestimation of the proportion of solitary individuals in cases in which members of social groups are solitary foragers. Actually, prolonged focal observations of free-ranging dogs living in urban areas and in villages have revealed that they can form stable social groups that defend territories collectively, although they spend considerable time foraging alone (Font, 1987; Macdonald & Carr, 1995). Consequently, the applicability of the previously described method to investigate dog sociality may be questionable.

Whether free-ranging dogs forage alone or in groups or, more in general, whether they live solitary or in packs is likely to be affected by local ecological conditions, particularly by the abundance and distribution of food resources. Theories concerning the ecology of carnivore sociality predict that the size of social groups will be positively correlated to food biomass during periods of minimum availability (Macdonald, 1983). For example, it has been argued that where food resources are dispersed in small packages (e.g., handouts, scraps from restaurants, dustbins), as in many urban areas and villages, that are usually insufficient to satiate more than one individual at a time, dogs will often forage alone (Daniels & Bekoff, 1989a; Macdonald & Carr, 1995; Coppinger & Coppinger, 2001). Conversely, when food resources provided by humans are clumped in large packages (e.g., garbage dumps), then group foraging becomes possible, and living in large packs probably allows dogs to increase access to food by outcompeting single dogs and smaller groups (Daniels & Bekoff, 1989a,b; Macdonald & Carr, 1995). In our study site, food dispensed by humans at each feeding site was probably abundant enough to support packs of an unusual size (we always saw group members feeding together). However, accurate measurements of food abundance in all these studies were not carried out, and so this hypothesis still awaits testing.

Free-ranging dogs that live in urban areas and in villages have been classified as 'village dogs' by several authors (Macdonald & Carr, 1995; Coppinger & Coppinger, 2001; Boitani et al., 2007; Ortolani et al., 2009). However, this category is very heterogeneous because it comprises both owned and unowned dogs and, in our opinion, it is very important to distinguish dogs on the basis of both their ownership status and their degree of socialisation to people. For example, Daniels (1983a) found that free-ranging pets spent less time in association with other dogs, and had smaller home ranges, than unowned free-ranging dogs. Moreover, it should be noted that in virtually all the studies concluding that free-ranging dogs are semi-solitary animals, conducted both in urban and in rural environments, the population comprised a considerable proportion of free-ranging pets (Beck, 1973, 1975; Rubin & Beck, 1982; Berman & Durban, 1983; Daniels, 1983a; Daniels & Bekoff, 1989a; Butler et al., 2004; Ortolani et al., 2009). Conversely, all studies in which free-ranging dogs formed stable groups in urban (Fox et al., 1975; Font, 1987; Pal et al., 1998), suburban (Bonanni et al., 2010a,b; Cafazzo et al., 2010, 2012), and rural/wild environments (Scott & Causey, 1973; Nesbitt, 1975; Causey & Cude, 1980; Gipson, 1983; Daniels & Bekoff, 1989b; Boitani et al., 1995; Macdonald & Carr, 1995) concerned unowned animals that were probably to a large extent non-socialised to humans (based on their avoidance/aggressive response to people) or with a minority of abandoned/escaped dogs recruited into packs. So, why do free-ranging pets seem to be less likely to live in packs than unowned free-ranging dogs? The question seems particularly intriguing because many pet dogs undoubtedly possess the potential to engage in complex social interactions with conspecifics (Bauer & Smuts, 2007; Cools et al., 2008; Ward et al., 2008, 2009; Trisko, 2011). We propose the following potential explanations: one possibility is that some free-ranging pets may actually form groups, but these may go undetected because these dogs can forage alone for considerable amounts of time (see previous discussion); a second possibility is that, depending on the degree of restraint placed by their owners, free-ranging pets may have limited chances of coordinating their actions with those of potential companions, hence reducing the likelihood of group formation; third, depending on the timing and amount of socialisation to humans relative to timing and amount of socialisation to conspecifics, some free-ranging pets may be less motivated to develop social relationships with other dogs (see Scott & Fuller, 1965), considering humans their primary relationship partners; finally, free-ranging pets that do not share their home site with other dogs may have limited experience/social skill in interactions with conspecifics (see Daniels & Bekoff, 1989a). Note that, if one of the last two hypotheses is correct, one should expect that even 'abandoned dogs', which are socialised to humans, will be on average less skilled at interacting with conspecifics than 'non-socialised dogs'. In partial support for this, Daniels and Bekoff (1989b) found that abandoned dogs had fewer social contacts with conspecifics, and much smaller home ranges, than non-socialised dogs, although some of them were also recruited into packs of non-socialised animals.

Some authors have stressed that the behaviour of free-ranging dogs is likely to be influenced also by the effects of artificial selection (e.g., Boitani & Ciucci, 1995; Macdonald & Carr, 1995). Although 'pure breed' dogs are rarely found in free-ranging groups (Boitani & Ciucci, 1995), it is believed that differential genetic relatedness with specific breeds may contribute to explain the observed variation in free-ranging dog behaviour. For example, Boitani and Ciucci (1995) suggested that higher territoriality in a given group/population may depend on relatedness of its members with breeds artificially selected to guard livestock and other human property. Moreover, differential relatedness with specific breeds can also affect body size and physiology which, in turn, will affect the energetic and ecological requirements of a given population of free-ranging dogs and eventually their fitness in a given environment. For example, a pack of free-ranging dogs studied in Alaska showed excellent survival at extreme low temperatures, and most of its members apparently were related to breeds specifically selected in cold latitudes (Gipson, 1983). On the one hand, we agree that the effects of artificial selection will complicate the adaptive interpretations of dog behaviour and will probably increase its variability relative to that of their wild ancestor. On the other hand, we believe that predicting the behaviour of free-ranging dogs based on their degree of relatedness to specific breeds is a very difficult task. First of all, hybridisation studies suggest that the behaviour of crossbred dogs usually differs from that of both parental breeds (Scott & Fuller, 1965). Moreover, free-ranging dogs could differ from genetically related breeds, not just with respect to their genetic sequences, but also with respect to patterns of gene expression mediated by different environmental stimuli (sensu Saetre et al., 2004; Crews, 2011). Aside from all these considerations, we also caution against arguing that the complex sociality observed in some free-ranging dogs is a mere consequence of artificial selection. We stress, instead, that some of the salient features of dog sociality that emerged in our and in other studies on free-ranging dogs (e.g., dominance, cooperation) appear to be shared by multiple breeds and also by the primitive Indian dogs, suggesting a possible derivation by a common ancestor. For example, a recent study found a linear dominance hierarchy in a large group of pet dogs comprising 18 different breeds (Trisko, 2011). Moreover, cooperative behaviours such as coalitions to intimidate/attack common targets, observed by us and other researchers in free-ranging dogs, have been described also in several breeds of both large and small size (Scott & Fuller, 1965; Borchelt et al., 1983). Most important, Indian free-ranging dogs, which are presumably genetically closer to the most primitive dogs than Western free-ranging dogs (Clutton-Brock, 1995; Gonzalez, 2012; Crapon de Caprona & Savolainen, 2013), display a social organisation that shares several features with the dogs of our population, i.e., stable social groups with hierarchical structure, cooperation in intergroup conflicts, and a promiscuous mating system (Pal et al., 1998, 1999; see also the previous section on dominance).

In summary, we suggest that all the previously cited factors could contribute to explain variation in the social organisation of free-ranging dogs. However, we predict that stable packs are more likely to develop when unowned free-ranging dogs, and especially those that are not socialised to humans, can subsist on abundant and non-dispersed food resources.

## 3.6 CONCLUSIONS

Our results suggest that domestic dogs possess the potential to develop long-term, complex social bonds with their conspecifics and not just with humans. Under the appropriate ecological conditions, free-ranging dogs can form stable packs whose organisation resembles that of wolf packs with respect to several aspects, including age-based dominance hierarchies, leader–follower relationships, social regulation of reproductive activities, and cooperation in resource defence. Differences between their social organisation and that of wolves can be reasonably explained as an adaptation to the domestic environment. Finally, the fact that some dogs do not form stable groups should not be interpreted as an indication that domestication has drastically reduced their intraspecific sociality, but rather as a flexible response to specific ecological conditions and higher dependency on humans.

### Future Directions

- Our finding that affiliative relationships seem to promote coordination and coalition formation in free-ranging dogs is based on correlational evidence only. However, the existence of a causal link between affiliative and cooperative interactions could be tested by administering substances that in other species are related to both behaviours (e.g., 'oxytocin'; De Vries et al., 2003; Madden & Clutton-Brock, 2011) to domestic dogs.
- What is the effect of kinship on social relationships in dogs? We currently do not know whether the social organisation of packs composed primarily by relatives differs from that of packs made up of unrelated individuals.
- Why do male dogs rarely regurgitate food for puppies? In wolves, cooperative breeding seems to be promoted by a seasonal prolactin peak, coinciding with the birth of puppies (Kreeger et al., 1991; Asa, 1997; Asa & Valdespino, 1998). However, little is known about variations of prolactin and parental behaviour in dogs, especially in males.
- Is pack living in dogs a trait maintained by natural selection? Although it has been shown that larger packs of free-ranging dogs outcompete smaller packs and lone individuals over access to resources (Daniels & Bekoff, 1989b; Macdonald & Carr, 1995; Bonanni et al., 2011), there are currently no data about the effect of pack size on reproductive success of free-ranging animals.
- It is fundamental to replicate studies on the social organisation of free-ranging dogs using methods based on continuous observations of animals (e.g., *focal animal sampling*; Altmann, 1974), which would be also useful in order to assess the degree of human influence on their behaviour (e.g., ownership, socialisation to people).
- It is useful to conduct studies on the intraspecific social behaviour of pure breeds. In the absence of more information about this topic, any attempt at interpreting the behaviour of free-ranging dogs on the basis of their genetic relatedness to specific breeds would be highly speculative.

## ACKNOWLEDGEMENTS

We wish to thank Sarah Marshall-Pescini and Juliane Kaminski for inviting us to write this chapter. Thanks also to the people who provided relevant scientific papers: Kerstin Wilhelm, Sarah Marshall-Pescini, Kathryn Lord, Monique Udell, Cheryl Asa, Joelene Hughes, Carles Vilà, Huw Nolan, Marc Bekoff, Rolf Peterson, and Jane Packard. Special thanks to Claudio Carere for useful suggestions about the text and the figures, and to Gabriele Schino for suggestions about the mechanisms of animal reciprocity. Finally, we are grateful to a number of colleagues, friends, and dog lovers who provided useful discussions about dog behaviour by sharing their experience with us: Eugenia Natoli, Paola Valsecchi, Alberto Fanfani, Lucia Paolini, Michele Minunno, Guenther Bloch, Claudio Marcolli, Carles Vilà, Maurizio Guiducci, Manuela Rossetti, and Claudia Strinati.

## REFERENCES

Abis, A., 2004. Analisi della struttura gerarchica di un gruppo di cani vaganti (*Canis familiaris*) in una zona periferica di Roma in presenza e in assenza di fonti di competizione. M.S. Thesis, University of Rome La Sapienza, Rome, Italy.

Altmann, J., 1974. Observational study of behavior: sampling methods. Behaviour 48, 227–265.

Appleby, M.C., 1983. The probability of linearity in hierarchies. Anim. Behav. 31, 600–608.

Asa, C.S., 1997. Hormonal and experiential factors in the expression of social and parental behavior in canids. In: Solomon, N.G., French, J.A. (Eds.), Cooperative breeding in mammals. Cambridge University Press, Cambridge, pp. 129–149.

Asa, C.S., Valdespino, C., 1998. Canid reproductive biology: an integration of proximate mechanisms and ultimate causes. Am. Zool. 38, 251–259.

Avis, S.P., 1999. Dog pack attack: hunting humans. Am. J. Foren. Med. Path. 20, 243–246.

Axelsson, E., Ratnakumar, A., Arendt, M.L., Maqbool, K., Webster, M.T., Perloski, M., Liberg, O., Arnemo, J.M., Hedhammar, A., Lindblad-Toh, K., 2013. The genomic signature of dog domestication reveals adaptation to a starch-rich diet. Nature 495, 360–364.

Barillari, E., 2004. Le relazioni sociali tra i membri di un gruppo stabile di cani domestici, ospitati presso un'oasi canina, con particolare riferimento alla gerarchia di dominanza. M.S. Thesis, University of Rome La Sapienza, Rome, Italy.

Bauer, E.B., Smuts, B.B., 2007. Cooperation and competition during dyadic play in domestic dogs. Anim. Behav. 73, 489–499.

Beck, A.M., 1973. The ecology of stray dogs: a study of free-ranging urban animals. York Press, Baltimore, MD.

Beck, A.M., 1975. The ecology of "feral" and free-roving dogs in Baltimore. In: Fox, M.W. (Ed.), The wild canids. Van Nostrand Reinhold, New York, pp. 380–390.

Bekoff, M., 1979. Scent-marking by free-ranging domestic dogs. Biol. Behav. 4, 123–139.

Berghänel, A., Ostner, J., Schröder, U., Schülke, O., 2011. Social bonds predict future cooperation in male Barbary macaques, *Macaca sylvanus*. Anim. Behav. 81, 1109–1116.

Berman, M., Dunbar, I., 1983. The social behaviour of free-ranging suburban dogs. Appl. Anim. Ethol. 10, 5–17.

Boesch, C., Boesch, H., 1989. Hunting behavior of wild chimpanzee in the Tai National Park. Am. J. Phys. Anthropol. 78, 547–573.

Boitani, L., Francisci, F., Ciucci, P., Andreoli, G., 1995. Population biology and ecology of feral dogs in central Italy. In: Serpell, J. (Ed.), The domestic dog: its evolution, behaviour and interactions with people. Cambridge University Press, Cambridge, pp. 217–244.

Boitani, L., 1983. Wolf and dog competition in Italy. Acta Zoologica Fennica 174, 259–264.

Boitani, L., Ciucci, P., 1995. Comparative social ecology of feral dogs and wolves. Ethol. Ecol. Evol. 7, 49–72.

Boitani, L., Ciucci, P., Ortolani, A., 2007. Behaviour and social ecology of free-ranging dogs. In: Jensen, P. (Ed.), The behavioural biology of dogs. CAB International, Wallingford, UK, pp. 147–165.

Bonanni, R., 2008. Cooperation, leadership and numerical assessment of opponents in conflicts between groups of feral dogs. Ph.D. thesis, University of Parma, Italy.

Bonanni, R., Cafazzo, S., Valsecchi, P., Natoli, E., 2010a. Effect of affiliative and agonistic relationships on leadership behaviour in free-ranging dogs. Anim. Behav. 79, 981–991.

Bonanni, R., Natoli, E., Cafazzo, S., Valsecchi, P., 2011. Free-ranging dogs assess the quantity of opponents in intergroup conflicts. Anim. Cogn. 14, 103–115.

Bonanni, R., Valsecchi, P., Natoli, E., 2010b. Pattern of individual participation and cheating in conflicts between groups of free-ranging dogs. Anim. Behav. 79, 957–968.

Borchelt, P.L., 1983. Attacks by packs of dogs involving predation on human beings. Appl. Public Health Rep. 98, 57–68.

Boyko, A.R., Boyko, R.H., Boyko, C.M., Parker, H.G., Castelhano, M., Corey, L., Degenhardt, J.D., Auton, A., Hedimbi, M., Kityo, R., Ostrander, E.A., Schoenebeck, J., Todhunter, R.J., Jones, P., Bustamante, C.D., 2009. Complex population structure in African village dogs and its implications for inferring dog domestication history. Proc. Natl. Acad. Sci. U.S.A., Early Ed. 106, 13903–13908.

Bradshaw, J.W., Blackwell, E.J., Casey, R.A., 2009. Dominance in domestic dogs—useful construct or bad habit? J. Vet. Behav. Clin. Applic. Res. 4, 135–144.

Bradshaw, J.W.S., Nott, H.M.R., 1995. Social and communication behaviour of companion dogs. In: Serpell, J. (Ed.), The domestic dog: its evolution, behaviour and interactions with people. Cambridge University Press, Cambridge, pp. 115–130.

Bräuer, J., Bös, M., Call, J., Tomasello, M., 2013. Domestic dogs (Canis familiaris) coordinate their actions in a problem-solving task. Anim. Cogn. 16, 273–285.

Brown, J.L., 1964. The evolution of diversity in avian territorial systems. Wilson Bull. 76, 160–169.

Butler, J.R.A., Bingham, J., 2000. Demography and dog–human relationships of the dog population in Zimbabwean communal lands. Vet. Rec. 147, 442–446.

Butler, J.R.A., du Toit, J.T., Bingham, J., 2004. Free-ranging domestic dogs (Canis familiaris) as predators and prey in rural Zimbabwe: threats of competition and disease to large wild carnivores. Biol. Conserv. 115, 369–378.

Byrne, R.W., 2000. How monkeys find their way: leadership, coordination, and cognitive maps of African baboons. In: Boinski, S., Garber, P.A. (Eds.), On the move: how animals travel in groups. University of Chicago Press, Chicago, IL, pp. 491–518.

Cafazzo, S., 2007. Dinamiche sociali in un gruppo di cani domestici (Canis lupus familiaris) liberi in ambiente suburbano. Ph.D. Thesis, University of Parma, Italy.

Cafazzo, S., Bonanni, R., Valsecchi, P., Natoli, E. Social variables affecting mate preferences, copulation and reproductive outcome in a pack of free-ranging dogs. Submitted.

Cafazzo, S., Natoli, E., Valsecchi, P., 2012. Scent-marking behaviour in a pack of free-ranging domestic dogs. Ethology 118, 1–12.

Cafazzo, S., Valsecchi, P., Bonanni, R., Natoli, E., 2010. Dominance in relation to age, sex and competitive contexts in a group of free-ranging domestic dogs. Behav. Ecol. 21, 443–455.

Causey, M.K., Cude, C.A., 1980. Feral dog and white-tailed deer interaction in Alabama. J. Wildl. Manage. 44, 481–484.

Clutton-Brock, J., 1995. Origins of the dogs: domestication and early history. In: Serpell, J. (Ed.), The domestic dog: its evolution, behaviour and interactions with people. Cambridge University Press, Cambridge, pp. 199–216.

Conradt, L., Roper, T.J., 2005. Consensus decision making in animals. Trends Ecol. Evol. 20, 449–456.

Cools, A.K.A., Van Hout, A.J.M., Nelissen, M.H.J., 2008. Canine reconciliation and third-party initiated post-conflict affiliation: do peacemaking social mechanisms in dogs rival those of higher primates? Ethology 114, 53–63.

Coppinger, R., Coppinger, L., 2001. Dogs: a new understanding of canine origin, behavior and evolution. Chicago University Press, Chicago, IL.

Coppinger, R., Schneider, R., 1995. Evolution of working dogs. In: Serpell, J. (Ed.), The domestic dog: its evolution, behaviour and interactions with people. Cambridge University Press, Cambridge, pp. 21–47.

Crapon de Caprona, M.D., Savolainen, P., 2013. Extensive phenotypic diversity among South Chinese dogs. ISRN Evol. Biol. 2013, Article ID 621836, 8 pages.

Crews, D., 2011. Epigenetic modifications of brain and behavior: theory and practice. Horm. Behav. 59, 393–398.

Daniels, T.J., 1983a. The social organization of free-ranging urban dogs I. Non-estrous social behavior. Appl. Anim. Ethol. 10, 341–363.

Daniels, T.J., 1983b. The social organization of free-ranging urban dogs II. Estrous groups and the mating system. Appl. Anim. Ethol. 10, 365–373.

Daniels, T.J., Bekoff, M., 1989a. Population and social biology of free-ranging dogs, *Canis familiaris*. J. Mammal. 70, 754–762.

Daniels, T.J., Bekoff, M., 1989b. Spatial and temporal resource use by feral and abandoned dogs. Ethology 81, 300–312.

de Vries, H., 1995. An improved test of linearity in dominance hierarchies containing unknown or tied relationships. Anim. Behav. 50, 1375–1389.

de Vries, A.C., Glasper, E.R., Detillion, C.E., 2003. Social modulation of stress responses. Physiol. Behav. 79, 399–407.

de Vries, H., 1998. Finding a dominance order most consistent with a linear hierarchy: a new procedure and review. Anim. Behav. 55, 827–843.

de Waal, F.B.M., 1986. The integration of dominance and social bonding in primates. Q. Rev. Biol. 61, 459–479.

de Waal, F.B.M., 2008. Putting the altruism back into altruism: the evolution of empathy. Annu. Rev. Psychol. 59, 279–300.

de Waal, F.B.M., Leimgruber, K., Greenberg, A.R., 2008. Giving is self rewarding for monkeys. Proc. Natl. Acad. Sci. U.S.A. 105, 13685–13689.

Derix, R., Van Hooff, J., de Vries, H., Wensing, J., 1993. Male and female mating competition in wolves: female suppression vs. male intervention. Behaviour 127, 141–174.

Ding, Z.L., Oskarsson, M., Ardalan, A., Angleby, H., Dahlgren, L.G., Tepeli, C., Kirkness, E., Savolainen, P., Zhang, Y.P., 2012. Origins of domestic dog in Southern East Asia is supported by analysis of Y-chromosome DNA. Heredity 108, 507–514.

Drews, C., 1993. The concept and definition of dominance in animal behaviour. Behaviour 125, 283–313.

Druzhkova, A.S., Thalmann, O., Trifonov, V.A., Leonard, J.A., Vorobieva, N.V., Ovodov, N.D., Graphodatsky, A.S., Wayne, R.K., 2013. Ancient DNA analysis affirms the canid from Altai as a primitive dog. PLoS One 8, 57754.

Eaton, B., 2011. Dominance in dogs: fact or fiction? Dogwise Publishing, Wenatchee, WA.

Epple, G., Alveario, M.C., 1985. Social facilitation of agonistic responses to strangers in pairs of saddle back tamarins (*Saguinus fuscicollis*). Am. J. Primatol. 9, 207–218.

Feddersen-Petersen, D.U., 2007. Social behaviour of dogs and related canids. In: Jensen, P. (Ed.), The behavioural biology of dogs. CAB International, Wallingford, UK, pp. 105–119.

Ferrari, P.F., Bonini, L., Fogassi, L., 2009. From monkey mirror neurons to primate behaviours: possible "direct" and "indirect" pathways. Phil. Trans. R. Soc. B. 364, 2311–2323.

Ferrari, P.F., Maiolini, C., Addessi, E., Fogassi, L., Visalberghi, E., 2005. The observation and hearing of eating actions activates motor programs related to eating in macaque monkeys. Behav. Brain Res. 161, 95–101.

Font, E., 1987. Spacing and social organization: urban stray dogs revisited. Appl. Anim. Behav. Sci. 17, 319–328.

Fox, M.W., Beck, A.M., Blackman, E., 1975. Behaviour and ecology of a small group of urban dogs. Appl. Anim. Ethol. 1, 119–137.

Frank, H., Frank, M.G., 1982. On the effects of domestication on canine social development and behavior. Appl. Anim. Ethol. 8, 507–525.

Gácsi, M., Gyoöri, B., Miklósi, Á., Virányi, Z., Kubinyi, E., Topál, J., Csányi, V., 2005. Species-specific differences and similarities in the behavior of hand-raised dog and wolf pups in social situations with humans. Dev. Psychobiol. 47, 111–122.

Gácsi, M., Gyoöri, B., Virányi, Z., Kubinyi, E., Range, F., Belényi, B., Miklósi, Á., 2009. Explaining dog–wolf differences in utilizing human pointing gestures: selection for synergistic shifts in the development of some social skills. PLoS One 4, e6584.

Gammell, M.P., de Vries, H., Jennings, D.J., Carlin, C.M., Hayden, T.J., 2003. David's score: a more appropriate dominance ranking method than Clutton-Brock et al.'s index. Anim. Behav. 66, 601–605.

Ghosh, B., Choudhuri, D.K., Pal, B., 1984. Some aspects of sexual behaviour of stray dogs, *Canis familiaris*. Appl. Anim. Behav. Sci. 13, 113–127.

Gipson, P.S., 1983. Evaluations of behavior of feral dogs in interior Alaska, with control implications. Vertebrate Pest Control Manag. Mater. 4th Symp. Am. Soc. Testing Mater. 4, 285–294.

Glickman, S.E., Zabel, C.J., Yoerg, S.I., Weldele, M.L., Drea, C.M., Frank, L.G., 1997. Social facilitation, affiliation, and dominance in the social life of spotted hyenas. Ann. N. Y. Acad. Sci. 807, 175–184.

Gonzalez, T., 2012. The pariah case: some comments on the origin and evolution of primitive dogs and on the taxonomy of related species. Ph.D. Thesis, Australian National University, Canberra, Australia.

Goodwin, D., Bradshaw, J.W.S., Wickens, S.M., 1997. Paedomorphosis affects agonistic visual signals of domestic dogs. Anim. Behav. 53, 297–304.

Hand, J.L., 1986. Resolution of social conflicts: dominance, egalitarianism, spheres of dominance, and game theory. Q. Rev. Biol. 61, 201–219.

Hare, B., Rosati, A., Kaminski, J., Bräuer, J., Call, J., Tomasello, M., 2010. The domestication hypothesis for dogs' skills with human communication: a response to Udell et al. (2008) and Wynne, et al. (2008). Anim. Behav. 79, 1–6.

Hare, B., Tomasello, M., 2005. Human-like social skills in dogs? Trends Cogn. Sci. 9, 439–444.

Harrington, F.H., Asa, C.S., 2003. Wolf communication. In: Mech, L.D., Boitani, L. (Eds.), Wolves: behavior, ecology, and conservation. University of Chicago Press, Chicago, IL, and London, pp. 66–103.

Harrington, F.H., Mech, L.D., 1979. Wolf howling and its role in territory maintenance. Behaviour 68, 207–249.

Harrington, F.H., Mech, L.D., 1983. Wolf pack spacing: howling as a territory-independent spacing mechanism in a territorial population. Behav. Ecol. Sociobiol. 12, 161.

Harrington, F.H., Mech, L.D., Fritts, S.H., 1983. Pack size and wolf pup survival: their relationship under varying ecological conditions. Behav. Ecol. Sociobiol. 13, 19–26.

Hennessy, C., Dubach, J., Gehrt, D., 2012. Long-term pair bonding and genetic evidence for monogamy among urban coyotes (*Canis latrans*). J. Mammal. 93, 732–742.

Horn, L., Range, F., Huber, L., 2013. Dogs' attention towards humans depends on their relationship, not only on social familiarity. Anim. Cogn. 16, 435–443.

Hughes, J., Macdonald, D.W., 2013. A review of the interactions between free-roaming domestic dogs and wildlife. Biol. Conserv. 157, 341–351.

King, A.J., Douglas, C.M.S., Huchard, E., Isaac, N.J.B., Cowlishaw, G., 2008. Dominance and affiliation mediate despotism in a social primate. Curr. Biol. 18, 1833–1838.

Kreeger, T.J., 1977. Impact of dog predation on Minnesota whitetail deer. MN Acad. Sci. 43, 8–13.

Kreeger, T.J., Seal, U.S., Cohen, Y., Plotka, E.D., Asa, C.S., 1991. Characterization of prolactin secretion in gray wolves (*Canis lupus*). Can. J. Zool. 69, 1366–1374.

Kubinyi, E., Virányi, Z., Miklósi, Á., 2007. Comparative social cognition: from wolf and dog to humans. Comp. Cogn. Behav. Rev. 2, 26–46.

Landau, H.G., 1951. On dominance relations and the structure of animal societies: I. Effect of inherent characteristics. Bull. Math. Biophys. 13, 1–19.

Larson, G., Karlsson, E.K., Perri, A., Webster, M.T., Ho, S.Y.W., Peters, J., Stahl, P.W., Piper, P.J., Lingaas, F., Fredholm, M., Comstock, K.E., Modiano, J.F., Schelling, C., Agoulnik, A.I., Leegwater, P.A., Dobney, K., Vigne, J.D., Vilà, C., Andersson, L., Lindblad-Toh, K., 2012. Rethinking dog domestication by integrating genetics, archeology, and biogeography. Proc. Natl. Acad. Sci. U.S.A., Early Ed. 109, 8878–8883.

Lockwood, R., 1979. Dominance in wolves: useful construct or bad habit? In: Klinghammer, E. (Ed.), The behavior and ecology of wolves. Garland STPM Press, New York, pp. 225–244.

Lord, K., Feinstein, M., Smith, B., Coppinger, R., 2013. Variation in reproductive traits of members of the genus *Canis* with special attention to the domestic dog (*Canis familiaris*). Behav. Processes 92, 131–142.

Lowry, D.A., McArthur, K.L., 1978. Domestic dogs as predators on deer. Wildl. Soc. Bull. 6, 38–39.

Macdonald, D.W., 1983. The ecology of carnivore social behavior. Nature 301, 379–384.

Macdonald, D.W., Carr, G.M., 1995. Variation in dog society: between resource dispersion and social flux. In: Serpell, J. (Ed.), The domestic dog: its evolution, behaviour and interactions with people. Cambridge University Press, Cambridge, pp. 199–216.

MacNulty, D.R., Smith, D.W., Mech, L.D., Vucetich, J.A., Packer, C., 2012. Nonlinear effects of group size on the success of wolves hunting elk. Behav. Ecol. 23, 75–82.

MacNulty, D.R., Smith, D.W., Vucetich, J.A., Mech, L.D., Stahler, D.R., Packer, C., 2009. Predatory senescence in ageing wolves. Ecol. Lett. 12, 1347–1356.

Madden, J.R., Clutton-Brock, T., 2011. Experimental peripheral administration of oxytocin elevates a suite of cooperative behaviours in a wild social mammal. Proc. R. Soc. Lond. B. 278, 1189–1194.

Maher, C.R., Lott, D.F., 2000. A review of ecological determinants of territoriality within vertebrate species. Am. Midland Natural. 143, 1–29.

Maher, C.R., Lott, D.F., 1995. Definitions of territoriality used in the study of variation in vertebrate spacing systems. Anim. Behav. 49, 1581–1597.

McGreevy, P.D., Starling, M., Branson, N.J., Cobb, M.L., Calnon, D., 2012. An overview of the dog–human dyad and ethograms within it. J. Vet. Behav. Clin. Appl. Res. 7, 103–117.

Mech, L.D., 1975. Hunting behavior in two similar species of social canids. In: Fox, M.W. (Ed.), The wild canids. Van Nostrand Reinhold, New York, pp. 363–368.

Mech, L.D., 1993. Details of a confrontation between two wild wolves. Can. J. Zool. 71, 1900–1903.

Mech, L.D., 1999. Alpha status, dominance, and division of labor in wolf packs. Can. J. Zool. 77, 1196–1203.

Mech, L.D., 2000. Leadership in wolf, *Canis lupus*, packs. Can. Field-Natural. 114, 259–263.

Mech, L.D., Boitani, L., 2003. Wolf social ecology. In: Mech, L.D., Boitani, L. (Eds.), Wolves: behavior, ecology, and conservation. University of Chicago Press, Chicago, IL, and London, pp. 1–34.

Miklósi, Á., 2007a. Dog behaviour, evolution and cognition. Oxford University Press, Oxford.

Miklósi, Á., 2007b. Human-animal interactions and social cognition in dogs. In: Jensen, P. (Ed.), The behavioural biology of dogs. CAB International, Wallingford, UK, pp. 207–222.

Miklósi, Á., Kubinyi, E., Topál, J., Gacsi, M., Viranyi, Z., Csanyi, V., 2003. A simple reason for a big difference: wolves do not look back at humans, but dogs do. Curr. Biol. 13, 763–766.

Miklósi, Á., Topál, J., 2013. What does it take to become "best friends"? Evolutionary changes in canine social competence. TICS. 17, 287–294.

Miklósi, Á., Topál, J., Csányi, V., 2004. Comparative social cognition: what can dogs teach us? Anim. Behav. 67, 995–1004.

Morey, D.F., 1994. The early evolution of the domestic dog. Am. Sci. 82, 336–347.

Nesbitt, W.H., 1975. Ecology of a feral dog pack on a wildlife refuge. In: Fox, M.W. (Ed.), The wild canids. Van Nostrand Reinhold, New York, pp. 391–395.

New Jr., J.C., Kelch, W.J., Hutchison, J.M., Salman, M.D., King, M., Scarlett, J.M., Kass, P.H., 2004. Birth and death rate estimates of cats and dogs in U.S. households and related factors. J. Appl. Anim. Welfare Sci. 7, 229–241.

Nitzschner, M., Melis, A.P., Kaminski, J., Tomasello, M., 2012. Dogs (*Canis familiaris*) evaluate humans on the basis of direct experiences only. PLoS One 7, e46880.

Oppenheimer, E., Oppenheimer, R., 1975. Certain behavioral features in the pariah dog (*Canis familiaris*) in west Bengal. Appl. Anim. Ethol. 2, 81–92.

Ortolani, A., Vernooij, H., Coppinger, R., 2009. Ethiopian village dogs: behavioural responses to a stranger's approach. Appl. Anim. Behav. Sci. 119, 210–218.

Packard, J.M., 2003. Wolf behavior: reproductive, social and intelligent. In: Mech, L.D., Boitani, L. (Eds.), Wolves: behavior, ecology, and conservation. University of Chicago Press, Chicago, IL, and London, pp. 35–65.

Packard, J.M., Mech, L.D., Ream, R.R., 1992. Weaning in an arctic wolf pack: behavioral mechanisms. Can. J. Zool. 70, 1269–1275.

Packard, J.M., Seal, U.S., Mech, L.D., Plotka, E.D., 1985. Causes of reproductive failure in two family group of wolves (*Canis lupus*). Z. Tierpsychol. 68, 24–40.

Pal, S.K., 2001. Population ecology of free-ranging urban dogs in West Bengal, India. Acta Theriol. 46, 69–78.

Pal, S.K., 2003. Reproductive behaviour of free-ranging rural dogs in West Bengal, India. Acta Theriol. 48, 271–281.

Pal, S.K., 2005. Parental care in free-ranging dogs, *Canis familiaris*. Appl. Anim. Behav. Sci. 90, 31–47.

Pal, S.K., Ghosh, B., Roy, S., 1998. Agonistic behaviour of free-ranging dogs (*Canis familiaris*) in relation to season, sex and age. Appl. Anim. Behav. Sci. 59, 331–348.

Pal, S.K., Ghosh, B., Roy, S., 1999. Inter- and intra-sexual behaviour of free-ranging dogs (*Canis familiaris*). Appl. Anim. Behav. Sci. 62, 267–278.

Pang, J., Kluetsch, C., Zou, X.J., Zhang, A.B., Luo, L.Y., Angleby, H., Ardalan, A., Ekström, C., Sköllermo, A., Lundeberg, J., Matsumura, S., Leitner, T., Zhang, Y.P., Savolainen, P., 2009. mtDNA data indicate a single origin for dogs south of Yangtze River, less than 16,300 years ago, from numerous wolves. Mol. Biol. Evol. 26, 2849–2864.

Peters, R.P., Mech, L.D., 1975. Scent-marking in wolves. Am. Sci. 63, 628–637.

Peterson, R.O., 1979. Wolves of Isle Royale—new developments. In: Klinghammer, E. (Ed.), The behavior and ecology of wolves. Garland STPM Press, New York, pp. 3–18.

Peterson, R.O., Jacobs, A.K., Drummer, T.D., Mech, L.D., Smith, D.W., 2002. Leadership behavior in relation to dominance and reproductive status in gray wolves, Canis lupus. Can. J. Zool. 80, 1405–1412.

Preston, S.D., de Waal, F.B.M., 2002. Empathy: its ultimate and proximate bases. Behav. Brain Sci. 25, 1–72.

Price, E.O., 1984. Behavioral aspects of animal domestication. Q. Rev. Biol. 59, 1–32.

Range, F., Horn, L., Viranyi, Z., Huber, L., 2009. The absence of reward induces inequity aversion in dogs. PNAS 106, 340–345.

Range, F., Huber, L., Heyes, C., 2011. Automatic imitation in dogs. Proc. R. Soc. B. 278, 211–217.

Ross, S., Ross, G.W., 1949. Social facilitation of feeding behavior in dogs: II. Feeding after satiation. J. Gen. Psychol. 74, 293–304.

Rubin, H.D., Beck, A.M., 1982. Ecological behaviour of free-ranging urban pet dogs. Appl. Anim. Ethol. 8, 161–168.

Saetre, P., Lindberg, J., Leonard, J.A., Olsson, K., Pettersson, U., Ellegren, H., Bergström, T.F., Vilà, C., Jazin, E., 2004. From wild wolf to domestic dog: gene expression changes in the brain. Mol. Brain Res. 126, 198–206.

Sands, J., Creel, S., 2004. Social dominance, aggression and faecal glucocorticoid levels in a wild population of wolves, Canis lupus. Anim. Behav. 67, 387–396.

Savolainen, P., 2007. Domestication of dog. In: Jensen, P. (Ed.), The behavioural biology of dogs. CAB International, Wallingford, UK, pp. 21–37.

Schenkel, R., 1947. Expression studies on wolves—captivity observations. Behavior 1, 81–129.

Schenkel, R., 1967. Submission: its features and function in the wolf and dog. Am. Zool. 7, 319–329.

Schino, G., Aureli, F., 2009. Reciprocal altruism in primates: partner choice, cognition, and emotions. Adv. Study Behav. 39, 45–69.

Scott, J.P., Fuller, J.L., 1965. Genetics and the social behavior of the dog. Chicago University Press, Chicago.

Scott, J.P., Marston, M.V., 1950. Social facilitation and allelomimetic behavior in dogs. Behaviour 2, 135–143.

Scott, M.D., Causey, K., 1973. Ecology of feral dogs in Alabama. J. Wildl. Manage. 37, 253–265.

Semyonova, A., 2003. The social organization of the domestic dog; a longitudinal study of domestic canine behavior and the ontogeny of domestic canine social systems. The Carriage House Foundation, The Hague, Netherland.

Stahler, D.R., MacNulty, D.R., Wayne, R.K., vonHoldt, B., Smith, D.W., 2013. The adaptive value of morphological, behavioural and life-history traits in reproductive female wolves. J. Anim. Ecol. 82, 222–234.

Topál, J., Gácsi, M., Miklósi, Á., Virányi, Z., Kubinyi, E., Csányi, V., 2005. Attachment to humans: a comparative study on hand reared wolves and differently socialized dog puppies. Anim. Behav. 70, 1367–1375.

Topál, J., Miklósi, Á., Csányi, A., Dóka, A., 1998. Attachment behaviour in dogs (Canis familiaris): a new application of Ainsworth's (1969) strange situation test. J. Comp. Psychol. 112, 219–229.

Topál, J., Miklósi, Á., Gácsi, M., Dóka, Pongrácz, P., Kubinyi, E., Virányi, Z., Csányi, V., 2009. The dog as a model for understanding human social behavior. Adv. Study Behav. 39, 71–116.

Trisko, R.K., 2011. Dominance, egalitarianism and friendship at a dog daycare facility. Ph.D. Thesis, University of Michigan, Ann Arbor, Michigan.

Trut, L.N., 1999. Early canid domestication: the farm-fox experiment: foxes bred for tamability in a 40-year experiment exhibit remarkable transformations that suggest an interplay between behavioral genetics and development. Am. Sci. 87, 160–169.

Tuber, D., Hennessy, M.B., Sanders, S., Miller, J.A., 1996. Behavioral and glucocorticoid responses of adult domestic dogs (Canis familiaris) to companionship and social separation. J. Comp. Psychol. 110, 103–108.

Udell, M.A.R., Dorey, N.R., Wynne, C.D.L., 2010a. What did domestication do to dogs? A new account of dogs' sensitivity to human actions. Biol. Rev. 85, 327–345.

Udell, M.A.R., Dorey, N.R., Wynne, C.D.L., 2010b. The performance of stray dogs (Canis familiaris) living in a shelter on human-guided object-choice tasks. Anim. Behav. 79, 717–725.

Udell, M.A.R., Spencer, J.M., Dorey, N.R., Wynne, C.D.L., 2012. Human-socialized wolves follow diverse human gestures... and they may not be alone. Int. J. Comp. Psychol. 25, 97–117.

Udell, M.A.R., Wynne, C.D.L., 2010. Ontogeny and phylogeny: both are essential to human-sensitive behaviour in the genus Canis. Anim. Behav. 79, 9–14.

Vanak, A.T., Gompper, M.E., 2009. Dogs Canis familiaris as carnivores: their role and function in intraguild competition. Mammal Rev. 39, 265–283.

van Hooff, J.A.R.A.M., Wensing, J.A.B., 1987. Dominance and its behavioural measures in a captive wolf pack. In: Frank, H.W. (Ed.), Man and wolf. Junk Publishers, Dordrecht, Olanda (Netherlands), pp. 219–252.

van Kerkhove, W., 2004. A fresh look at the wolf-pack theory of companion-animal dog social behavior. J. Appl. Anim. Welfare Sci. 7, 279–285.

Vilà, C., Leonard, J.A., 2007. Origin of dog breed diversity. In: Jensen, P. (Ed.), The behavioural biology of dogs. CAB International, Wallingford, UK, pp. 38–58.

Vilà, C., Savolainen, P., Maldonado, J.E., Amorin, I.R., Rice, J.E., Honeycutt, R.L., Crendall, K.A., Lundeberg, J., Wayne, R.K., 1997. Multiple and ancient origins of the domestic dog. Science 276, 1687–1689.

vonHoldt, B.M., Pollinger, J.P., Lohmueller, K.E., Han, E., Parker, H.G., Quignon, P., Degenhardt, J.D., Boyko, A.R., Earl, D.A., Auton, A., Reynolds, A., Bryc, K., Brisbin, A., Knowles, J.C., Mosher, D.S., Spady, T.C., Elkahloun, A., Geffen, E., Pilot, M., Jedrzejewski, W., Greco, C., Randi, E., Bannasch, D., Wilton, A., Shearman, J., Musiani, M., Cargill, M., Jones, P.G., Qian, Z., Huang, W., Ding, Z.L., Zhang, Y.P., Bustamante, C.D., Ostrander, E.A., Novembre, J., Wayne, R.K., 2010. Genome-wide SNP and haplotype analyses reveal a rich history underlying dog domestication. Nature 464, 898–902.

vonHoldt, B.M., Stahler, D.R., Smith, D.W., Earl, D.A., Pollinger, J.P., Wayne, R.K., 2008. The genealogy and genetic viability of reintroduced Yellowstone grey wolves. Mol. Ecol. 17, 252–274.

Wang, G., Zhai, W., Yang, H., Fan, R., Cao, X., Zhong, L., Wang, L., Liu, F., Wu, H., Cheng, L., Poyarkov, A.D., Poyarkov JR., N.A., Tang, S., Zhao, W., Gao, Y., Lv, X., Irwin, D.M., Savolainen, P., Wu, C., Zhang, Y., 2013. The genomics of selection in dogs and the parallel evolution between dogs and humans. Nat. Commun. 4, 1860.

Ward, C., Bauer, E.B., Smuts, B.B., 2008. Partner preferences and asymmetries in social play among domestic dog, Canis lupus familiaris, littermates. Anim. Behav. 76, 1187–1199.

Ward, C., Trisko, R., Smuts, B.B., 2009. Third-party interventions in dyadic play between littermates of domestic dogs, Canis lupus familiaris. Anim. Behav. 78, 1153–1160.

Young, J.K., Olson, K.A., Reading, R.P., Amgalanbaatar, S., Berger, J., 2011. Is wildlife going to the dogs? Impacts of feral and free-roaming dogs on wildlife populations. Bioscience 61, 125–132.

Zeder, M.A., 2012. Pathways to animal domestication. In: Gepts, P., Famula, T.R., Bettinger, R.L. (Eds.), Biodiversity in agriculture: domestication, evolution and sustainability. Cambridge University Press, Cambridge, pp. 227–259.

Zimen, E., 1975. Social dynamics of the wolf pack. In: Fox, M.W. (Ed.), The wild canids. Van Nostrand Reinhold, New York, pp. 336–362.

Zimen, E., 1976. On the regulation of pack size in wolves. Zeitschrift für Tierpsychologie 40, 300–341.

# Chapter 4

# Social Behaviour among Companion Dogs with an Emphasis on Play

Barbara Smuts

*Department of Psychology, University of Michigan, Ann Arbor, MI, USA*

## 4.1 INTRODUCTION

'Companion dogs' are dogs who are 'owned' by particular individuals or households and who are fed and otherwise cared for by these humans. In the West, most such dogs are the offspring of purebred dogs or mixed-breed descendants of those dogs. Research into domestic dog behaviour has focused on companion dogs because they live in the same societies the scientists inhabit, they are ubiquitous, and they are usually amenable to handling by non-owners.

Although research on dog–dog social interactions has flourished in recent years, it nevertheless constitutes a minute fraction of the behavioural research on dogs, most of which focuses on dog–human interactions. Among the relatively few studies on dog–dog interactions, many focus on social play, several investigate dominance and aggression, and very few consider affiliative relationships. This chapter reflects those differing emphases. Due to space constraints, I focus on quantitative studies whenever they are available. Many books about the lives of individual dogs, which are too numerous to list, contain fascinating accounts of dog–dog interactions (e.g., Marshall Thomas, 2001; Csyani, 2005; Kerasote, 2007). Two recent books contain detailed photographic documentation of dog–dog communication (Aloff, 2005; Handelman, 2008). All these sources of information represent important complements to the data summarised here.

## 4.2 SOCIAL PLAY

Many companion dogs are eager to play with other dogs. This no doubt reflects the fact that humans meet their survival needs, leaving them free time and excess energy that can be devoted to play. Because social play is so common in companion dogs, it has been studied more than any other type of behaviour. In the following sections, I describe how dogs play, competitive and cooperative

*The Social Dog.* http://dx.doi.org/10.1016/B978-0-12-407818-5.00004-8

aspects of play, how play may differ as a function of sex and age, how dogs communicate during play, and triadic interactions during play.

## 4.2.1 How Dogs Play: Ethological Elements

Domestic dog social play (abbreviated here as 'dog play') includes action patterns also seen during agonistic interactions (e.g., chase), sexual interactions (e.g., mounting), and hunting behaviour (e.g., stalking) (Bekoff, 1974; Aloff, 2005; Bauer & Smuts, 2007; Handelman, 2008). It also includes actions unique to play, such as rearing up and jaw sparring. Most often, dogs play in pairs (Adler et al., 2011), and, unless indicated otherwise, *what follows applies only to pair-wise interactions.*

Some play actions are asymmetric, meaning that the roles of play partners differ. These include (1) various forms of *mock attacks,* such as tackling and jumping on, in which one dog tries to force the other to the ground by pushing, slamming, or forceful neck 'biting' (without inflicting harm; Figure 4-1); (2) *chases,* in which one dog runs after the other, sometimes attempting to playfully bite the hindquarters of the fleeing partner; (3) *stalking;* and (4) *mounting.* Symmetric actions include (1) *jaw sparring* (also called jaw wrestling*),* in which two dogs 'fence' with open mouths; this can be done while lying down, sitting, or standing; (2) *wrestling,* in which both dogs are mostly prone, often lying side by side; wrestling can include jaw sparring, pushing with the paws and feet, and rolling around or over the other dog; and (3) *rearing up,* in which two dogs leap or stand vertically, often placing the forelegs on each other's shoulders; while

**FIGURE 4-1**    Bentley, a 3-year-old golden retriever mix, gently bites Lela, a 5-month-old German shepherd, on her neck. Lela shows a play face. See color plate section. *(Copyright Barbara Smuts.)*

vertical they may swipe with their paws and/or bite at one another (Figure 4-2). Play bows (where the dog lowers the forequarters and raises the hindquarters) and play bounces (bobbing up and down or side to side without going into a full play bow) can be symmetric, with both dogs bowing or bouncing at the same time, or asymmetric, with one dog bowing or bouncing at the other. Bows and bounces are typically described as 'play signals' (Bekoff, 1974, 1995) because they are often used to initiate play and sometimes to maintain a play atmosphere (see following section). Although specific action patterns can be fairly stereo-typed (Feddersen-Petersen, 1991), dogs combine these different behaviours (and others, such as paw lifts, spins, and standing over a prone dog) in a wide variety of ways (Bekoff, 1974) such that play overall does not seem stereotyped.

## 4.2.2  Play Fighting versus Real Fighting

The most obvious distinction between play fighting and real fighting involves noticeable restraint during play that is absent from real fighting. During hundreds of hours watching dogs play, I have not witnessed any injuries beyond minor scratches. The mouth often remains open during play 'bites', and the teeth usually fail to make contact with the other dog's skin. Similarly, often the chaser is clearly not trying to catch up with the other dog, or a dog will allow herself to be tackled when she clearly could have escaped.

Another distinction is that during play, actions have different consequences than they do in real fights. For example, after forcing a partner to the ground, a playing dog may quickly release him rather than maintaining a hold on him, or

**FIGURE 4-2**    Bentley and Lela rear up. See color plate section. *(Copyright Barbara Smuts.)*

mock attacks might be interspersed with playful behaviours like play bows or play faces (Figure 4-3).

Two other important differences are the presence of self-handicapping and role reversals during play (Fagen, 1981; Bekoff & Allen, 1998). Self-handicapping (SH) reduces the actor's advantage over the play partner. The restraint mentioned previously can be considered SH. Easier to measure are overt behaviours that clearly put the actor at a disadvantage, such as when a dog voluntarily rolls onto the ground, exposing her abdomen to the play partner. This behaviour differs from passive submission (Schenkel, 1967), in which a dog lies down and reveals the belly to a dominant dog outside of play. In that context, the submissive dog freezes and holds her tail between her legs. During play, a prone dog often shows a play face, wags her tail, and remains active by, for instance, flailing her legs to 'defend' against her partner's mock bites (Figure 4-4). Since animals presumably never place themselves at a disadvantage during a real fight, self-handicapping clearly identifies play fighting as

**FIGURE 4-3**   During a brief pause in play, Bentley and Lela show play faces. See color plate section. *(Copyright Barbara Smuts.)*

**FIGURE 4-4**   Bentley 'defends' against his partner's mock bites. See color plate section. *(Copyright Barbara Smuts.)*

something distinct (Fagen, 1981). A less striking example of how play differs from real fighting involves the way dogs move their heads during jaw sparring. In contrast to a real bite attempt in which the head is thrust forward towards the other dog, when jaw sparring, dogs tend to move their heads back and forth sideways, preventing the teeth from contacting the other dog's face.

Role reversals occur when play partners take turns adopting asymmetric roles. For example, one dog may jump on another, and a moment later the second dog jumps on the first one. Role reversals are clearly related to SH because they can occur only when playing animals sometimes fail to press an advantage (Fagen, 1981; Pellis & Pellis, 1987), offering the partner a chance to gain the upper paw.

Despite these differences, play fighting can get very rough. In some species, subordinates escalate play fighting to try to unseat dominant animals (male rats: Pellis et al., 2005) or to probe another's physical abilities (captive wolves: Zimen, 1981). It is not known whether dogs escalate play for these reasons. However, when one dog gets too rough with another, the disadvantaged partner usually protests by yelping or snapping. Typically, the other dog then immediately desists. In other cases, neither dog resists nor protests as play intensifies. Paradoxically, such rough play seems to occur most often between dogs who are long-term friends ('friends' are dogs who show consistent preferences for each other as affiliative partners, including play partners, for periods of at least 3 months; Smuts, in prep). Presumably, friends can play with mutual abandon because they have established trust (Bekoff, 2001) over time (see section 4.2.6, 'Negotiating and Reconciling during Play').

## 4.2.3 Competition and Cooperation during Dog Play

Play fighting includes inherently competitive actions, since many of its maneuvers involve attempts to catch, bite, or pin the partner while she/he tries to escape. At the same time, a degree of cooperation must exist, or it would not be play at all (Fagan, 1981). In particular, players must exercise restraint to avoid harming each other.

How do animals balance competitive and cooperative tendencies during play fighting? Some researchers (Altmann, 1962; Aldis, 1975) proposed that animals must follow the 50:50 rule so that players adopt 'winning' roles roughly equally. Individuals in some species tend to conform to this 50:50 play rule (e.g., juvenile rats: Pellis & Pellis, 1987), while in other instances, one player in a dyad adopts the offensive role most of the time (e.g., juvenile male rhesus monkeys: Symons, 1978; young kangaroos playing with older ones: Watson & Croft, 1996). Recently, Bauer and Smuts (2007) tested the 50:50 hypothesis in dogs. Based on a variety of clearly defined behaviours coded from videotapes, they calculated how many times each dog in 55 adult dog dyads showed offensive behaviours (such as chases and tackles) and self-handicapping behaviours (such as voluntarily rolling onto the ground). Contrary to prediction, they found that roughly half of the dyads showed significant asymmetry of roles (based on a binomial test). Perhaps most notable, 22% of the dyads showed complete asymmetry of roles yet continued to play over time.

The older individual within dyads adopted offensive roles significantly more and showed self-handicapping significantly less than the younger did, and smaller dogs self-handicapped more often than larger dogs. Sex of the two players did not predict how often a dog showed either behaviour. For a subset of 19 pairs, Bauer and Smuts (2007) used formal status indicators (see section 4.3.1, 'Defining Dominance') to determine relative ranks. In these dyads, the dominant dog performed nearly 80% of the offensive behaviours, whereas the subordinate dog performed about 95% of all self-handicapping behaviours. Since rank co-varied with both age and size (with older, larger dogs tending to be more dominant), Bauer and Smuts (2007) could not test for dominance effects independent of age/size.

These findings refute Miklósi's claim (2007, p. 191) that role reversals *must* occur for play to persist. Instead, older, larger, and/or more dominant dogs appeared to use their age, size, and/or higher rank to press an advantage (e.g., being on top) over the partner. At the same time, in most dyads, these same dogs sometimes allowed their partners to chase and 'attack' them (Figure 4-5), suggesting that status roles were relaxed to a degree during play, and this may have encouraged disadvantaged partners to play (Bekoff & Allen, 1998; Bekoff, 2001). Indeed, play fighting within five pairs of adult dogs living with the author involved more equitable roles (with one dog in the offensive role no more than 60% of the time) despite the fact that in all five pairs one dog was clearly dominant to the other (Smuts, in prep). It is also important to note that the Bauer and Smuts (2007) study did not include any young puppies or closely related adults. Adult dog/puppy play has not been studied, but casual observations indicate that adults very often self-handicap with puppies, often by lying down.

**FIGURE 4-5**   Role reversal: Lela, the smaller, younger dog, adopts the offensive role by biting Bentley's neck. She plays the offensive role less often than Bentley does (Figure 4-1). See color plate section. *(Copyright Barbara Smuts.)*

Ward et al. (2008) used methods similar to those of Bauer and Smuts (2007) to analyse play within four litters of dog puppies (n = 39 dyads) beginning at 3 weeks of age until 7, 8, 23, or 40 weeks of age (depending on the litter). Within litters, offensive behaviours were much more common than self-handicapping, which accounted for only 6% of behavioural acts recorded (play bows were excluded from this total). They found that littermate play roles did not conform to the 50:50 rule at any age, and asymmetry of roles increased as puppies grew older. These findings, combined with those of Bauer and Smuts on adult dogs, suggest that asymmetry of roles is often—but not always—fundamental to dog play.

Identifying factors (e.g., living together) associated with more balanced play roles remains a challenge for future research. We also need to ask whether so-called self-handicapping behaviours really do place the actor in a more vulnerable or less advantageous position during dyadic play. An alternative hypothesis is that they simply function as play signals, helping to communicate playful intent (see the following section). This hypothesis is consistent with the fact that in both studies described previously (Bauer & Smuts, 2007; Ward et al., 2008), within dyads the dog who self-handicapped more also tended to be the one who play bowed more. Researchers also need to identify additional aspects of canine social play (beyond role asymmetries and role reversals) that vary among individuals or dyads.

## 4.2.4 Sex Differences and Changes in Play Behaviours with Age

Bauer and Smuts (2007) did not find any sex differences in amounts of offensive behaviours or self-handicapping during play, but they did find that females played with a wider number of partners (both female and male) than males did. Males seemed particularly reluctant to play with other males. These sex differences need to be replicated.

Bekoff (1974) found no sex difference in the 'roughness' of beagle puppy play, but other sexual behaviours occurred much more often in male puppies than in females. Ward et al. (2008) found that when male and female littermates played together, males initiated play more often than their female partners did (this difference and all others from the Ward et al. study cited in this paragraph and the next were significant). Lund and Vestergaard (1998) also found that in four litters between 3 and 8 weeks of age, male puppies initiated play with females more often than expected. Ward et al. (2008) reported that at all ages females initiated play more often with other females than with males. Males showed no preference for initiating play with either sex with the exception of the one litter studied between 27 and 40 weeks of age; those males initiated play more often with other males.

In mixed-sex puppy dyads, males displayed offensive behaviours more often than their female partners did, and they also exhibited such behaviours more often when playing with females than when playing with other males.

Sex of partner made no difference to how often female littermates showed offensive behaviours. Both sexes self-handicapped at similar rates whether playing with the same sex or opposite sex partner, but in mixed-sex dyads, males self-handicapped more often than their female partners did. Perhaps this helped to compensate for the fact (above) that they also showed more offensive behaviours. Adler et. al. (2011) reported that among 142 dogs at a boarding kennel, female–male play bouts lasted significantly longer than did same-sex play bouts.

Ward et al. (2008) speculated that male puppies initiated play with female littermates more than the reverse in order to develop inter-sexual social skills useful as adults vying for the attention of estrous females. Female puppies may prefer playing with other females compared to males because males showed more offensive behaviours to females than vice versa. It is also possible that female preferences for same-sex partners reflect the importance of female–female relationships. In wolves, female–female competition is intense, but females also form coalitions with same-sex partners when contesting or defending rank positions (Jenks, 2011). In companion dogs, females are more likely to injure one another in fights than are male dogs (Sherman et al., 1996); no data exist on female–female coalitions in dogs.

Bauer et al. (2009) reported variation in the rates at which different behaviours occurred as a function of the players' ages, examining only dyads in which both partners were of similar age (n = 67 dyads). Forced downs and standing over were common in young puppies but decreased as age increased, and in mature adults (older than 3 years), they were quite rare. Voluntary downs occurred at relatively constant rates from early puppyhood until 3 years of age but declined sharply after that. Mounts and chases occurred most often among sub-adults (between 6 and 12 months of age) and were less common before and after that time period. However, rates of chasing remained above two times per minute at all ages. Muzzle licks (considered a self-handicapping behaviour) were also most common among sub-adults. Bauer et al. (2009) never saw chin overs, mounts, or muzzle licks during play between mature adults, but these behaviours have been seen among other mature adult players (personal observation). Ward et al. (2008, Figure 4-3) reported that biting with rapid side-to-side shaking of the head ('bite shakes') disappeared among littermates after the age of 23 weeks. Why various age differences in play behaviours exist is not known, although it is possible that dogs abandon bite shakes once they develop adult dentition in order to avoid injuring their partners.

## 4.2.5 Communication during Play

Researchers have described more than 20 different behaviour patterns that dogs use to initiate play, including pawing at the face, rolling over belly up, barking, stalking, ambushing, and head shaking (Fox, 1970; Bekoff, 1974; Bauer & Smuts, 2007; Horowitz, 2009). Play bows are the most salient action used to initiate play (Bekoff, 1974; Bauer & Smuts, 2007; Ward et al., 2008); they are discussed at the end of this section.

Dog play includes a wide variety of facial expressions, most notably, the open-mouth play face, which is not seen in other social contexts. When Fox (1970) compared the development of play faces in coyotes, wolves, and three fox species, he found that, when compared to foxes, wolves and coyotes showed more variable facial expressions, including expressions that were graded in intensity (p. 59). The dog 'play face' definitely shows such a gradient. At one extreme, the mouth is relaxed and open just a little, revealing only the upper parts of the most forward teeth of the lower jaw (e.g., Aloff, 2005, pp. 284, 343). At a slightly higher intensity, the mouth opens wider so that most or all of the teeth of the lower jaw can be seen (e.g., Aloff, 2005, p. 302). Although no one has studied the contexts in which different degrees of mouth opening occur, the partly open mouth seems most common during play invitations, brief pauses in play (Figure 4-3), and running/chasing, whereas the full open mouth appears to be associated with bite intentions or attempts and efforts to parry bites (personal observation).

Playing dogs also show threatening facial expressions characterised by different degrees of horizontal and vertical lip retraction; the latter produces a fierce-looking face with a puckered muzzle and exposed front teeth (Figure 4-6) (Aloff, 2005; Handelman, 2008). These faces so closely resemble those observed during real agonism that, when the action is frozen, it may be impossible to tell them apart. During play, however, threatening expressions come and go quickly and often occur right before or after 'goofy' behaviours such as bouncing or spinning, suggesting that such facial expressions may be disassociated from the negative emotions that presumably accompany real

**FIGURE 4-6**    Bentley retracts his lips vertically, puckering his muzzle and revealing his teeth to make a fierce face. Lela's play face shows that she does not think this is a real threat. See color plate section. *(Copyright Barbara Smuts.)*

threat faces. Growls emitted by dogs during play with humans are shorter than agonistic growls and higher pitched (Farago et al., 2010). If the same is true of growls during dog–dog play, dogs should be able to easily distinguish between real growls and play growls.

As mentioned previously, both Bauer and Smuts (2007) and Ward et al. (2008) found that in adult dogs and puppy littermates, respectively, the individual who self-handicapped more also gave more play signals, suggesting that self-handicapping may sometimes function as a play signal (Bekoff, 2001).

Dutton et al. (2011) found that dogs at a dog park were more successful at getting other dogs to play when they showed higher rates of attention-getting behaviours during initiation attempts. Horowitz (2009) defined attention-getting behaviours in terms of their effects on the target's sensory field; for example, a dog might bark or make contact to get another dog's visual attention during play. She found that when a dog lost her play partner's attention, she first tried to reclaim it with an attention-getter. If play did not resume, the dog often tried a different attention-getting behaviour. When the partner interacted with someone else, dogs most often used bumping, biting, or pawing, as if they knew that especially salient attention-grabbers were needed. Dogs directed visual play signals, such as play bows, only towards dogs who could see them (Horowitz, 2009; Byosiere & Smuts, in prep). These findings are important because, outside of primates, researchers have rarely studied how the types of signals used might vary as a function of the recipient's attentional state (Horowitz, 2009).

Play bows have received more attention than any other aspect of communication during play. Bekoff (1995) hypothesised that canines might use bows, which are highly stereotyped, to reassure partners that their intentions remained playful, especially when the bower's actions could most easily be misinterpreted as aggressive. Consistent with this hypothesis, he found that play bows occurred most often in captive wolves, captive coyotes, and domestic dogs just before and just after bite shakes, arguably the most aggressive of the play behaviours seen. Since researchers have rarely tested hypotheses about the functions of play signals, this finding is noteworthy. However, bite shakes almost never occur after the juvenile period (see previous text), which suggests that Bekoff's results apply only to the infants in his sample. Byosiere and Smuts (in prep) found that in adult dog dyads, aggressive actions rarely immediately preceded or followed play bows. Instead, play bows most often occurred right after the play partner was sitting or standing still, suggesting that they might help to keep play going during pauses in the action.

## 4.2.6  Negotiating and Reconciling during Play

Researchers (e.g., Fagen, 1981; Bekoff, 2001, 2004) have argued that ongoing negotiations are required to maintain play and to develop trust over time between play partners. Observations of dogs I have lived with support these claims (Smuts, in prep). For example, when adult male Tex first began playing with the adult female Tuna, who was both larger and younger than he, Tex frequently interspersed real growls and charges amongst his play behaviours. The first few times he did this, Tuna behaved submissively and/or ran away, but within hours, she began to ignore Tex's aggressive behaviours and simply continued playing. For a year, Tuna lived in Tex's household, and they played often. During play, Tex continued to occasionally threaten Tuna, but, because she ignored him, his aggression never interfered with their play. I think that Tex used low-level aggression to test whether Tuna might turn on him. When she did not, he could trust that, despite her size advantage, she was safe to play with.

Tex also lived and played with two female dogs who were the same size as him but a few years older. When Tex got too rough or showed aggression towards either female during play, they directed mild aggression back at him and play would cease. In this context, Tex tended to wait a few seconds and then exhibited friendly behaviour (e.g., touching the female's muzzle with his nose; cf. Fagen, 1981, p. 393). These actions seemed to function as reconciliation (Cools et al., 2008) or apologies (Bekoff, 2004), since play nearly always resumed immediately afterward.

## 4.2.7  Triadic Interactions during Play

When two dogs (A and B) play in a group, another dog (C) may intervene by directing playful, affiliative, or aggressive actions at one of them (B, the 'target'). Trisko (2011) studied 193 such interventions at a dog daycare center and evaluated two hypotheses concerning their proximate social benefits. The first hypothesis proposed that C practiced fighting skills and/or developed competitive advantages when he intervened in pair-wise play. If so, C should target B when B is the lower ranking of the two playing dogs, especially if B ranks lower than C (for how rank was determined, see section 4.3.1, 'Defining Dominance'). Also, C should not intervene against a dog who is a preferred affiliative partner (defined by the frequency of several affiliative behaviours, including play). An alternative hypothesis proposed that dogs intervened to initiate play with preferred partners and/or to interrupt their play with another dog, as a means of advancing relationships with preferred partners. It predicts that the relative ranks of the three dogs should not influence which dog C targets, that C will be likely to target preferred partners (especially when it involves directing a play behaviour at the target; Figure 4-7), and that C is more likely to play with the preferred partner rather than the other dog immediately after the intervention. Results were most consistent with the second hypothesis (Trisko, 2011), suggesting that adult dogs often intervene in dyadic play to claim a preferred play partner for themselves.

Ward et al. (2009) investigated third-party interventions during play among puppy littermates. In contrast to the unrelated adult dogs in Trisko's study, related littermates did not preferentially target (nor did they preferentially intervene against) preferred play partners. Instead, puppies most often targeted the

**FIGURE 4-7**   Tex intervenes in dyadic play between Bentley and Lela by mock biting his favorite play partner. See color plate section. *(Copyright Barbara Smuts.)*

puppy in the losing role at the time of the intervention. In many species, when two individuals are fighting, individuals tend to aid the higher ranking of the two (unless a close relative is being attacked). This so-called winner support is thought to benefit the intervener by (1) enhancing his relationship with a valuable ally and/or (2) providing a temporary advantage over the losing animal (de Waal & Harcourt, 1992). Possibly play interventions help puppies become skilled at this pattern of agonistic intervention (a pattern not so far documented in dogs), with the caveat that during play, the supported dog is 'winning' only in that moment. If so, it appears that triadic interventions function differently in young littermates and unrelated adult dogs.

Smuts (in prep) observed a triad of three adult female dogs who played together for over a year. Based on the submissive behaviours they showed to each other, they ranked A > B > C. A and B lived together, and C visited them nearly every day. A, an older dog, frequently intervened in play between B and C, who were both young adults. During her interventions, A always targeted C, usually by directing play bites at her. All three dogs often played together after an intervention (in contrast to the dogs in Trisko's study). During triadic play, B markedly increased offensive behaviours and markedly decreased self-handicapping behaviours towards C compared to when she played with C alone. This is reminiscent of the winner-support coalitions described previously, except that it is unclear whether B's advantage over C in the play context made any difference to her overall relationship with C.

## 4.2.8  Functions of Play

Scientists do not yet know the functional significance of social play in any species, including dogs (Burghart, 2005). Many possibilities have been suggested, including learning social skills, assessing one's abilities compared to others, developing/maintaining or testing social relationships, and learning how to cope with unexpected situations (Spinka et al., 2001). Some or all of these hypotheses could help explain play in companion dogs, but no data exist to test them directly. Play, even within a species, may have more than one function. For example, play in young puppies may have different functions than play in adult dogs. Another difficulty is that play has been studied most intensively in companion dogs and dogs living in other captive situations (e.g., Bekoff, 1974) whose lives likely differ in important ways from those of their ancestors. With the exception of one study on play in juvenile village dogs (Pal, 2010), play has not been studied in free-ranging dogs, whose reproduction, in contrast to that of many companion dogs, is a product of their interactions with other dogs and their success obtaining resources.

Despite these caveats, it is tempting to speculate that social play contributes to bond formation (Bekoff & Allen, 1998; Bekoff, 2001). In particular, Bekoff (1978; 2001) has speculated that social play helps dogs (and other animals) develop the skills necessary for cooperation, such as reciprocity, development of negotiating skills, and learning which individuals to trust as social partners. Based on available data on dog play, this is a very reasonable hypothesis, but it remains to be rigorously tested.

## 4.3 DOMINANCE

In animal behaviour in general (Drews, 1993) and in discussions of dog behaviour, in particular (e.g., Scott & Fuller, 1965; Goddard & Beilharz, 1985), aggression and ritualised signals indicative of formal dominance rank are often confounded. In wild wolves, higher-ranking animals do not show higher rates of aggression (Sands & Creel, 2004), and lower-ranking wolves sometimes show aggression to more dominant wolves (van Hooff & Wensing, 1987). The same holds for feral dogs (Cafazzo et al., 2010). Since, in wolves and dogs, patterns of aggression and patterns of formal dominance are not the same, one cannot use evidence pertaining to one to draw conclusions about the other. For this reason, dominance and aggression are here considered separately.

### 4.3.1 Defining Dominance

Schenkel's (1947; 1967) early descriptions of ritualised interactions in wolves and dogs captured many essential elements of status communication, and all subsequent researchers have built on his scheme. In 'active submission', the subordinate wolf adopts a crouched posture, tucks the tail, and flattens the ears while reaching up to nuzzle or lick the dominant's face. During 'passive submission', the lower-ranking wolf rolls onto the side or back, exposing the abdomen. Again, the ears are flattened and the tail tucked between the thighs. Schenkel (1967) and Fox (1971) considered active submission to be a ritualised form of the face-licking behaviour that wolf puppies use to stimulate adult food regurgitation. They interpreted passive submission to be a ritualisation of the posture a young puppy adopts when her mother licks her anogenital region.

Schenkel (1967, p. 325) wrote that through ritualised submission, an animal offered unambiguous acknowledgement of inferior status in an attempt to elicit another's tolerance. The two different types of submission (active and passive) 'correspond with the nuances in the attitude of the superior [i.e., dominant]' (p. 324). When the dominant wolf or dog shows tolerance and even friendliness, the subordinate tends to offer active submission. When the dominant adopts a more 'severe' (e.g., growling) and/or 'inquisitive' attitude involving sniffing attempts, the subordinate is more likely to show passive submission, offering the genital region for the dominant's inspection. Schenkel emphasised that in both cases (and with forms intermediate between these two extremes), the subordinate's behaviour was an 'often overwhelming offer of friendly affection' rather than an expression of fear (p. 324).

Since Schenkel's time, canine researchers have typically referred to the postures he described as 'ritualised submission' and have focused on how they instantiate status differences, while often neglecting Schenkel's emphasis on their friendly, integrative functions (e.g., Scott & Fuller, 1965; LeBoeuf, 1967). Meanwhile, developments in primatology helped to vindicate Schenkel's perspective. For decades, primatologists had debated the usefulness of the dominance concept (e.g., Bernstein, 1981). Researchers often used winning fights or gaining access to resources as measures of rank, but outcomes of competitive encounters were often inconsistent even over short periods of time (Drews, 1993). De Waal (1986) argued that such inconsistencies ought to occur, to a degree, because individuals are expected to vary in their motivations to compete for particular resources (e.g., a well-fed animal is less motivated to contest a food item than a hungry one). He also pointed out that the presence of allies could alter competitive outcomes. Instead of focusing on winning fights or gaining access to resources, he identified particular submissive behaviours, such as pant-grunts in chimpanzees and the bared-teeth display in macaques, that showed high directional consistency (within dyads) in many different contexts over time. De Waal argued that such behaviours served as unequivocal acknowledgements of subordinate status, and the resultant asymmetric relationships indicated the existence of *formal dominance*. Like Schenkel, de Waal argued that subordinates showed formal submissive signals in order to gain the dominant's favor.

The concept of formal dominance greatly clarified the study of status relationships in many non-human primate species, and soon formal dominance was documented in captive wolves (van Hooff & Wensing, 1987). Like the primate signals described previously, the 'low postures' that characterise active submission and the complementary 'high postures' shown by dominants exhibited very high directionality within dyads across different contexts as well as consistency over time. Displays of passive submission also met the criteria for formal dominance.

Despite the significance of these findings in wolves, until very recently researchers ignored the possibility of formal dominance in dogs. Bradshaw and Nott (1995, p. 124) asserted that, when studying groups of dogs, it was futile to use 'the criteria used to detect dominance in wolves'. Because they observed few ritualised status indicators, these authors instead used displacements (one dog avoids another when one or both move towards a resource or goal) as potential measures of dominance in small, single-breed groups of female Cavalier King Charles spaniels (n = 6) and French bulldogs (n = 5). In the spaniel group, displacements indicated clear-cut alpha and omega positions, but among the 4 dogs in between, relationships were not so clear. In the bulldog group, only some relationships among females were clear-cut, and the single male consistently displaced all 4 females in some contexts but not in others. Bradshaw (Bradshaw et al., 2009) also studied a group of 19 neutered male dogs at a shelter and found inconsistent rank relationships. They used 'David's scores' to assess dominance relationships, a method that depends on a variety of assumptions unlikely to hold for social groups (de Vries, 1998). In addition, they combined measures of aggression, dominance, avoidance, and submission. Neither Bradshaw and Nott (1995) nor Bradshaw et al. (2009) provided raw data on the frequencies or directions of interactions, which makes it hard to evaluate their claims.

In contrast, Cafazzo et al. (2010) used measures of submission similar to those of van Hooff and Wensing (1987) in their study of status in a group of feral dogs. They concluded that behaviours associated with active and passive submission functioned as a good measure of formal dominance in dogs, and that these dogs had a linear dominance hierarchy. Using similar methods, Trisko (2011) reported similar results for the group of daycare dogs mentioned previously. Both authors found that higher-ranking dogs were older, and males and females showed similar rates of submissive behaviour.

Both Cafazzo et al. (2010) and Trisko (2011) reported lower 'coverage' of dyadic dominance interactions (72% and 29%, respectively) compared to van Hooff and Wensing (1987; 98%). Coverage refers to the percentage of dyads in which researchers observed any formal status indicators. Lower coverage by Cafazzo et al. (2010) likely reflects difficulties observing unhabituated feral dogs, but this cannot explain Trisko's even lower coverage. It appears that some companion dog dyads do not exhibit status differentiation, at least under the circumstances of Trisko's daycare center. Dogs attending the center were neutered or spayed and had no resources to compete over except toys and human and canine social attention. Attendees were screened and fight-prone dogs excluded. Conflicts were reduced through counter-conditioning, and highly aroused dogs received brief 'time-outs' until they calmed down. These practices aimed to create a maximally peaceful social environment, and among these daycare dogs, aggression was infrequent and never inflicted serious injury (Trisko, 2011).

Trisko's findings can be interpreted in two ways. On the one hand, because some dyads manifested formal dominance even under such low-conflict conditions, one could argue that rank differences sometimes really matter to companion dogs. On the other hand, since less than one-third of the dyads showed such asymmetries, dominance may not be very important in most pairs. Adoption of either position seems premature. We clearly need studies of dominance in companion dogs under different conditions (e.g., intact dogs, dogs competing for resources), but these can be difficult to achieve for ethical reasons. Two recent studies in non-daycare settings seem to support the importance of dominance relationships in dogs. Van der Borg et al. (2012) studied 16 intact, group-housed dogs ranging in age from puppies to adults. Based on 4,874 dyadic interactions and using the same analytical framework as Trisko (2011) and Cafazzo et al. (2010), they concluded that components of Schenkel's active and passive submission, especially 'low posture', had very high directional consistency within dyads and resulted in a linear dominance hierarchy. In another recent study at the Wolf Science Center in Austria, a pack of wolves and one of domestic dogs were reared and kept under similar conditions (Ritter et al., 2012). The authors used dominant and submissive behaviours during group feeding to assign ranks to group members. Then they observed the behaviours of wolf pairs and dog pairs who were placed in a pen with a large bone. They found that in dogs, but not wolves, high-ranking animals showed more aggression than lower-ranking ones and monopolised the food. They concluded that wolves were more tolerant than dogs during feeding competition and that dogs developed a 'steeper' hierarchy that inhibited aggression directed towards higher-ranking group members. The two studies just summarised are published as abstracts, so important details are lacking. As more information becomes available, it will be important to know how measures of formal dominance compare between these three packs (two dog packs and one wolf pack) and whether, in the Ritter et al. (2012) study, dogs exhibit a steeper hierarchy than wolves when data from the everyday group contexts are compared.

## 4.3.2 Development of Dominance Relationships

Prior to the current emphasis on formal dominance, most researchers typically studied canine competition and dominance through pair-wise 'bone-in-pen' tests like that used by Ritter et al. (2012). If one dog, A, maintained possession of the bone most of the time and could re-possess the bone when it was given to dog B, A was considered dominant to B (Scott & Fuller, 1965; Beach, 1970; Wright, 1980). Problems with using access to resources as a measure of dominance were mentioned previously, and even in these early studies, researchers seemed to realise this. For example, Fox (1969, p. 245) found that, in pairs of infant coyotes and infant wolves, which animal won fights did not predict priority of access to food, and Scott and Fuller (1965) acknowledged that such tests, by inciting competition, helped to create the dominance asymmetries that they were trying to measure.

For all these reasons, some researchers (Bradshaw & Nott, 1995; Serpell & Jagoe, 1995) are not convinced by claims (Scott & Fuller, 1965; Beach, 1970) that domestic dog littermates develop clear-cut dominance relationships by the age of 12 weeks. It is possible that rank differences can be detected during puppy play (Bradshaw & Nott, 1995; Serpell & Jagoe, 1995). Puppies do show behaviours associated with dominance and submission in adult dogs, such as standing over a prone partner or muzzle licking, respectively. Ward et al. (2008) observed such behaviours only during play, and by the age of 8–10 weeks, no dominance hierarchies were apparent. Their study, however, was based on one litter each of Labrador retrievers, Doberman pinschers, and malamutes, plus one mixed-breed litter. It is possible that in some other breeds (or even in different litters of the studied breeds), puppies exhibit formal dominance asymmetries early in life. The Scott and Fuller (1965) bone-in-pen tests indicated striking breed differences in competitive interactions by 11 weeks (e.g., basenjis competed intensively for the bone, whereas beagles and cocker spaniels were much less aggressive), and breed differences in formal dominance relationships also seem likely. These possibilities need to be tested in studies focusing on how and when formal dominance develops among littermates and among adult dogs meeting for the first time.

## 4.4 DOG–DOG AGGRESSION AND RECONCILIATION

A number of studies describe aspects of dog–dog aggression based on questionnaires and/or veterinary case reports (Borchelt, 1983; Sherman et al., 1996; Goodloe & Borchelt, 1998; Guy et al., 2001; Rugbjerg et al., 2003; Fatjo et al., 2007; Duffy et al., 2008; Haug, 2008; Casey et al., 2013), but only a few observational studies exist on dog–dog aggression. Bradshaw and Lea (1992, p. 252) were struck 'by the rarity of even mildly aggressive behaviours' during 292 pair-wise interactions between off-leash dogs meeting in two open areas. Shyan et al. (2003) used focal sampling to assess aggression at a 2-acre dog park in Indianapolis. In 72 hours of observation, they recorded 28 possible conflicts, a conservative number, since the authors thought that half of them could have been rough play. Based on the conservative number, 7% of the 177 dogs observed showed aggression. None of these episodes produced injuries. Capra et al. (2011) videotaped 127 incidents of dog–dog aggression in a dog-park-like setting during 241 group sessions that included from 2 to 25 dogs. Eighty females and 60 males participated in the study. All aggression involved dogs not very familiar with each other, and none resulted in injuries. Carrier et al. (2013) videotaped focal samples of 55 dogs at a dog park in Newfoundland. On average, only 6 other dogs were around during focal samples, and 57% of the sampled dogs did not know any of those present. Twenty-two percent of the dogs showed at least one agonistic behaviour, and 83% played or displayed play signals. Salivary cortisol levels were unrelated to involvement in agonism but were positively related to the frequency with which a dog showed 'lowered or hunched posture' (p. 103). The authors did not report how often lowered postures occurred in response to the proximity or behaviour of another dog. Petak (2013) studied communication among

12 neutered male dogs, all more than 6 years old, living in a large outdoor enclosure at a shelter. Interactions involving neutral approaches followed by sniffing occurred most often, and aggressive interactions occurred least often. A few dyads accounted for most of the aggression, and dogs who initiated more aggression also received more. Some pairs appeared to have unresolved relationships, even after living together for some time. In the last two studies described here, authors made no mention of injuries inflicted by other dogs, suggesting that none occurred.

Santos et al. (2013) used a carefully designed, multistep protocol to try to pair unfamiliar dogs together at a shelter. Twenty-two pairs were formed without any aggression. Another five pairs showed unidirectional aggression but no biting and were successfully paired. The three remaining pairs showed bidirectional aggression (again, no biting) and could not live together, but those dogs were later successfully grouped with other dogs.

In the six studies summarised in the preceding paragraphs, no dogs were wounded, despite the fact that the dogs often did not know each other. By itself, this suggests that dogs are a remarkably peaceful species. However, the dogs observed in these studies represent a biased sample of companion dogs, since people whose dogs are often aggressive towards other dogs are undoubtedly less likely to bring their dogs to off-leash parks. In addition, these dogs were presumably not competing for food, space, or mates. Casey et al. (2013) analysed questionnaire responses from a 'convenience sample' of 3,874 dogs whose owners were recruited at places dog owners tend to frequent, such as dog shows and veterinary clinics. Twenty-two percent of their sample reported that their dog had shown aggression towards another dog outside the household (OHA) at some point. The authors thought that the real percentage of OHA is probably higher than this, since, as hypothesised previously, aggressive dogs mostly stay home. Without information on how many dogs are observable in public and how many are not and the aggressive tendencies of each, it is not possible to reconcile the very different findings of the observational studies and the Casey et al. (2013) questionnaire study.

Casey et al. (2013) found that only 8% of dogs had ever shown *within* household inter-dog aggression (WHA), indicating that dogs familiar with one another tend to be less aggressive. Only 13% of dogs with OHA showed WHA, but 41% of the dogs with WHA also showed OHA, suggesting that WHA has a higher threshold than OHA. From an evolutionary perspective, this makes sense, since wolves and free-ranging dogs show serious aggression towards strangers much more often than towards fellow pack members (Mech & Boitani, 2003; chapter on feral dogs). None of the factors examined in the Casey et al. (2013) study, such as age of dog, sex, or breed, explained much of the variance in either OHA or WHA.

Cools et al. (2008) studied how domestic dogs raised in pairs at a food research facility coped with conflicts that occurred soon after researchers placed them together in larger groups (for ethical reasons, serious aggression was interrupted by the observers). They found that domestic dog opponents were significantly more likely to interact in a friendly manner soon after a conflict than were the same dogs at a neutral point in time. Such behaviour, termed 'reconciliation' (de Waal & van Roosmalen, 1979), has been documented in numerous primates and a few non-primates, including captive wolves (Cordoni & Palagi, 2008). As in many other species, reconciliations occurred most often between individuals who had important relationships (in the dog study, previous cage mates). Cools et al. (2008) also found that, soon after a conflict, an uninvolved third party approached and directed friendly contact to the victim more often than expected. Palagi and Cordoni (2009) showed that, in a captive wolf pack, similar contacts were offered most often by animals closely affiliated with the victim. In wolves, such contacts reduced redirected aggression by the victim and increased her friendly contact with other group members, but similar data are not available for dogs.

The evidence reviewed in the preceding paragraphs indicates that, at least among companion dogs socialised well enough to appear in public, injurious aggression is rare. When dogs meet for the first time or after a time apart, they usually greet. Dog greetings are familiar to most people because of the associated sniffing of anal and genital regions, but dogs also show a wide range of body language and facial expressions during greetings, some of them quite subtle. Through information conveyed during greetings, dogs likely communicate their standing relative to one another, making it unnecessary to resolve the relationship through aggression. In addition, reconciliation and third-party friendly contacts often occur after even minor conflicts. If these behaviours have the same effects in dogs that they do in wolves, the likelihood of further conflict will decline. Finally, as mentioned previously, dogs meeting in public typically are not competing for resources. All these factors combined make serious aggression rare.

Nevertheless, questionnaire studies indicate that some companion dogs can be very aggressive towards other dogs. In many cases, aggression reflects inadequate socialisation and/or a fear of other dogs, manifested as defensive aggression (Lindsay, 2001). Ethologists make a clear distinction between offensive and defensive aggression in wolves and dogs, but the average dog owner probably does not. If we want a further understanding of aggression in companion dogs, therefore, future questionnaire- or interview-based studies must include details about socialisation, the contexts in which aggression occurs, whether a resource (including a social partner) was at stake, and the precise behaviours shown by the dogs. To gain accurate information on dog behaviours, people with aggressive dogs might first need to take a short course in how to read canine body language. There may well be dog owners willing to participate in such studies, given the serious consequences of dog–dog aggression. Once more detailed and accurate information on dog–dog aggression becomes available, we will be in a much better position to reduce its frequency.

## 4.5 FRIENDLY BEHAVIOUR

Although informal observations indicate that dogs show individual preferences when they have opportunities to choose their friends, few studies of such preferences are available. Trisko (2011) studied affiliative behaviour in 24 dogs at a daycare facility (the same dogs mentioned earlier in studies of triadic play and dominance). She developed a composite index of affiliation based on time spent playing with particular individuals and the frequency with which different dogs showed friendly behaviours, like muzzle licking and nose nudges, to other dogs. The affiliation index controlled for individual variation in dogs' tendencies to show friendly behaviours. To understand how affiliation and dominance might interact, Trisko divided dyads into four possible types of relationships: (1) formal: dyad exhibited affiliation and unidirectional submission (22%); (2) egalitarian: dyad showed affiliation and bidirectional submission or no submission (21%); (3) agonistic: dyad showed unidirectional submission and no affiliation (8%); and (4) non-interactive: dyad showed neither affiliation nor submission (50%). Play and affiliation were significantly more common in female–male dyads than in same-sex dyads, and in female–male dyads, egalitarian relationships were significantly more common than were the other relationship types. The mean affinity indices did not differ between formal versus egalitarian relationships, but in the latter, affiliation within dyads tended to be distributed more equally, whereas in formal relationships, affiliation tended to be shown mostly by one individual—the lower-ranking one. However, dominant animals sometimes reciprocated friendly behaviours. Neither age differences nor weight differences within dyads were associated with the different relationship types.

It is not surprising that egalitarian relationships were most likely to occur in mixed-sex dyads, since in most mammals, including wolves (Derix et al., 1993), competition is typically more intense within each sex than between the sexes; this follows from the fact that same-sex animals tend to compete for the same resources. However, Trisko (2011) also found some male–male and female–female egalitarian relationships. Her research also indicated that a lot of potential dyads never interacted at all beyond an initial greeting, sometimes involving sniffing only. No obvious factors, such as being similar in size, predicted which dogs avoided each other. Perhaps certain dogs just dislike others, and such dislike is frequently expressed through avoidance (personal observation).

Further research is clearly needed to determine if these different types of relationships exist in other groups of dogs, and, if so, what factors (beyond being of opposite sex) are involved. One possibility is that dogs of some breeds (or mixes of certain breeds) are simply less inclined towards competition than others. The difference between Labrador retrievers, most of whom would rather play than posture over dominance, and German shepherds, many of whom seem preoccupied with status, come to mind. However, surprisingly, no published observational studies exist demonstrating that such breed differences occur.

Ward et al. (2008) demonstrated that play partner preferences occurred in four litters of dogs. Over the course of time, these preferences tended to become stronger. However, we do not know whether patterns of puppy play affect dogs' future relationships. For example, in feral dogs, if within-litter play partner preferences exist, do they predict which puppies, if any, are most likely to disperse together?

Wolf and dog researchers alike have tended to focus more on dominance and aggression than on affiliative behaviours, but Trisko's research, as well as dog owners' many anecdotes about dog friends (Smuts & Ward, 2010), indicates that affiliation deserves equal emphasis in future studies.

## 4.6 CONCLUSIONS

In modern societies, familiarity with dogs is based mainly on public encounters with dogs in human custody and/or companion dogs owned by people or their friends. This applies to scientists, too, of course. A natural outgrowth of these contexts is the assumption, made explicit by some researchers, that human society is the natural habitat of the domestic dog (Csanyi, 2005; Miklósi 2007, p. 10). This assumption is correct; dogs evolved from wolves as a result of association with humans (Vila et al., 1997). However, this is only half of the story because it ignores the evidence that historically (and pre-historically), most dogs simultaneously maintained social relationships with other dogs. Of course, companion dogs often interact with other dogs, but only at the owner's discretion. In contrast, in the developing world and in (or near) major cities all over the world, many dogs *choose* to associate with other dogs while maintaining only a loose association with humans (see Chapter 3 on feral dogs, this volume). I refer to these animals as free-ranging dogs (FRDs) to emphasise their ability to choose their associates.

Until recently, it was thought that FRDs formed only ephemeral bonds with other dogs, lived mostly alone or in pairs, and lacked behavioural adaptations of the kind shown by wolves (e.g., Beck, 1973; Boitani & Ciucci, 1995; but see Font, 1987). Recent work has shown, in contrast, that such dogs typically live in packs that share many features of wolf society, including a fission–fusion social organisation, hostility to non-group members, cooperative territorial defense, linear dominance hierarchies, differential reproductive success as a function of rank, and sometimes pair-bonds with paternal investment (Chapter 3 on feral dogs). Because companion dogs have been common only in the last 200 years or so (Sampson, 2006), it seems likely that, throughout their history as domesticates, the vast majority of dogs lived in canine societies like these, while also associating with humans. Indeed, accounts of dogs in the 19th and 20th centuries indicate that even in Western societies, many dogs freely associated with other dogs and reproduced with them before leash laws became common.

Thus, although over the course of domestication, dogs *gained* the ability to relate to humans, they never *lost* the ability to interact skillfully with conspecifics. While papillons and mastiffs may not look much like dingoes or Bangkok street dogs, they are nearly identical genetically (Parker et al., 2010). As we consider studies reviewed in this chapter and think about the behaviour of the dogs we know, it is important to keep this perspective in mind because it may help us to understand much of what dogs do together. That understanding will, I hope, lead to more opportunities for dogs to enjoy the company of their own species.

**Future Directions**

We are left with these questions:

- How do opportunities for companion dogs to have frequent, friendly interactions with other dogs affect their mental and physical health?
- How and when do dominance asymmetries develop in groups of dogs, and how do dogs avoid and resolve conflicts?
- How does dog–dog social behaviour differ between intact dogs and dogs who have been spayed/neutered?
- How does conspecific social behaviour vary among different dog breeds?
- How does dog social behaviour in groups compare with that of wolves?

## ACKNOWLEDGEMENTS

I am grateful to the organisers and participants from the working group on *Play, Evolution & Sociality* (sponsored by NIMBios, The National Institute for Mathematical & Biological Synthesis, with support from the National Science Foundation, USA) for many stimulating conversations about play. I also thank the former graduate students whose research is central to my understanding of dog play, Dr. Erika Bauer, Dr. Rebecca Trisko, and Dr. Camille Ward, as well as the many dog guardians, dogs, and undergraduate research assistants who make this research possible.

## REFERENCES

Adler, C., Mackensen-Friedrichs, I., Franz, C., Crailsheim, K., 2011. Social play behavior of group housed domestic dogs (*Canis familiaris*). J. Vet. Behav. 6 (1), 98.

Aldis, O., 1975. Play fighting. Academic Press, New York.

Altmann, S.A., 1962. Social behavior of anthropoid primates: analysis of recent concepts. In: Bliss, E.L. (Ed.), Roots of behavior. Harper, New York, pp. 277–285.

Aloff, B., 2005. Canine body language. A photographic guide interpreting the native language of the domestic dog. Dogwise Publishers, Wenatebee, WA.

Bauer, E.B., Smuts, B.B., 2007. Cooperation and competition during dyadic play in domestic dogs. Anim. Behav. 73, 489–499.

Bauer, E.B., Ward, C., Smuts, B.B., 2009. Play like a puppy, play like a dog. J. Vet. Behav. 4 (2), 68–69.

Beach, F.A., 1970. Coital behaviour in dogs: VIII. Social affinity, dominance and sexual preference in the bitch. Behaviour 36 (1–2), 131–148.

Beck, A.M., 1973. The ecology of stray dogs: a study of free-ranging urban animals. York Press, Baltimore.

Bekoff, M., 1974. Social play and play-soliciting by infant canids. Am. Zoologist 14, 323–340.

Bekoff, M., 1978. Social play: structure, function, and the evolution of a cooperative social behavior. In: Burghardt, G.M., Bekoff, M. (Eds.), The development of behavior: comparative and evolutionary aspects. Garland Publishers, New York, pp. 367–383.

Bekoff, M., 1995. Play signals as punctuation: the structure of social play in canids. Behaviour 132, 419–429.

Bekoff, M., 2001. Social play behavior: cooperation, fairness, trust, and the evolution of morality. J. Consciousness Stud. 8, 81–90.

Bekoff, M., 2004. Wild justice and fair play: cooperation, forgiveness, and morality in animals. Biol. Philos. 19, 489–520.

Bekoff, M., Allen, C., 1998. Intentional communication and social play: How and why animals negotiate and agree to play. In: Bekoff, M., Byers, J.A. (Eds.), Animal play: evolutionary, comparative, and ecological perspectives. Cambridge University Press, Cambridge, pp. 97–114.

Bernstein, I.S., 1981. Dominance: the baby and the bathwater. Behav. Brain Sci. 4 (3), 419–429.

Boitani, L., Ciucci, P., 1995. Comparative social ecology of feral dogs and wolves. Ethol. Ecol. Evol. 7, 49–72.

Borchelt, P.L., 1983. Aggressive behavior of dogs kept as companion animals: classification and influence of sex, reproductive status and breed. Appl. Anim. Ethol. 10, 45–61.

Bradshaw, J.W., Lea, A.M., 1992. Dyadic interactions between domestic dogs. Anthrozoos 5 (4), 245–253.

Bradshaw, J.W., Nott, H.M.R., 1995. Social and communication behaviour of companion dogs. In: Serpell, J. (Ed.), The domestic dog. Its evolution, behaviour and interactions with people. Cambridge University Press, Cambridge, pp. 115–130.

Bradshaw, J.W.S., Blackwell, E.J., Casey, R.A., 2009. Dominance in dogs: useful construct or bad habit? J. Vet. Behav. 3, 176–177.

Burghardt, G.M., 2005. The genesis of animal play: testing the limits. MIT Press, Cambridge, MA.

Byosiere, S., Smuts, B.B. (In prep). The contexts of play bows and play bow mimicry in pet domestic dogs.

Cafazzo, S., Valsecchi, P., Bonanni, R., Natoli, E., 2010. Dominance in relation to age, sex, and competitive contexts in a group of free-ranging domestic dogs. Behav. Ecol. 21, 443–455.

Capra, A., Barnard, S., Valsecchi, A.C., 2011. Flight, foe, fight! Aggressive interactions between dogs. J. Vet. Behav. 6 (1), 62.

Carrier, L.O., Cyr, A., Anderson, R.E., Walsh, C., 2013. Exploring the dog park: relationships between social behaviours, personality and cortisol in companion animals. Appl. Anim. Behav. Sci. 146 (1), 96–106.

Casey, R.A., Loftus, B., Bolstert, C., Richards, G.J., Blackwell, E.J., 2013. Inter-dog aggression in a UK owner survey: prevalence, co-occurrence in different contexts and risk factors. Vet. Rec. 172 (5), 127–132.

Cools, A.K.A., Van Hout, A.J.M., Nelissen, M.H.J., 2008. Canine reconciliation and third-party-initiated post conflict affiliation: do peacemaking social mechanisms in dogs rival those of higher primates? Ethology 114, 53–63.

Cordoni, G., Palagi, E., 2008. Reconciliation in wolves (Canis lupus): new evidence for a comparative perspective. Ethology 114, 298–308.

Csyani, V., 2005. If dogs could talk. North Point Press, New York.

Derix, R.M., van Hooff, J., de Vries, H., Wensing, J., 1993. Male and female mating competition in wolves: female suppression vs. male intervention. Behaviour 127 (1–2), 141–174.

de Vries, H., 1998. Finding a dominance order most consistent with a linear hierarchy: a new procedure and review. Anim. Behav. 55, 827–843.

de Waal, F., 1986. The integration of dominance and social bonding in primates. Q. Rev. Biol. 61, 459–479.

de Waal, F., van Roosmalen, A., 1979. Reconciliation and consolation among chimpanzees. Behav. Ecol. Sociobiol. 5, 55–66.

de Waal, F., Harcourt, S.H., 1992. Coalitions and alliances: a history of ethological research. In: Harcourt, A.H., de Waal, F. (Eds.), Coalitions and alliances in humans and other animals. Oxford University Press, Oxford, pp. 1–19.

Drews, C., 1993. The concept and definition of dominance. Anim. Behav. 125, 283–313.

Duffy, D.L., Hsu, Y., Serpell, J.A., 2008. Breed differences in canine aggression. Appl. Anim. Behav. Sci. 114, 441–460.

Dutton, E.E., Anderson, R.E., Walsh, C.J., 2011. "Do I know you?" Does partner familiarity influence social interactions among dogs in a park setting? J. Vet. Behav. 6 (1), 65.

Fagen, R., 1981. Animal play behavior. Oxford University Press, New York.

Farago, T., Pongracz, P., Range, F., Viranyi, Z., Miklósi, A., 2010. "The bone is mine": affective and referential aspects of dog growls. Anim. Behav. 79, 917–925.

Fatjo, J., Amat, M., Mariotti, V.M., de la Torre, J.L.R., Manteca, X., 2007. Analysis of 1040 cases of canine aggression in a referral practice in Spain. J. Vet. Behav. 2, 158–165.

Feddersen-Petersen, D., 1991. The ontogeny of social play and agonistic behaviour in selected canid species. Bonn. Zool. Beitr. 42, 97–114.

Font, E., 1987. Spacing and social organization: urban stray dogs revisited. Appl. Anim. Behav. Sci. 17, 319–328.

Fox, M.W., 1969. The anatomy of aggression and its ritualization in canidae: a developmental and comparative study. Behaviour 35, 242–258.

Fox, M.W., 1970. A comparative study on the development of facial expressions in canids; wolf, coyote and foxes. Behaviour 36, 49–73.

Fox, M.W., 1971. Socio-infantile and soci-sexual signals in canids: a comparative and ontogenetic study. Zeitschrift Tierpsychologie 28, 185–210.

Goddard, M.E., Beilharz, R.G., 1985. Individual variation in agonistic behaviour in dogs. Anim. Behav. 33, 1338–1342.

Goodloe, L.P., Borchelt, P.L., 1998. Companion dog temperament traits. J. Appl. Welfare Sci. 1 (4), 303–338.

Guy, N.C., Luescher, U.A., Dohoo, S.E., Spangler, E., Miller, J.B., Dohoo, I.R., Bate, L.A., 2001. Demographic and aggressive characteristics of dogs in a general veterinary caseload. Appl. Anim. Behav. Sci. 74, 15–28.

Handleman, B., 2008. Canine behavior: a photo illustrated handbook. Dogwise Publishers, Wanatchee, WA.

Haug, L.I., 2008. Canine aggression toward unfamiliar people and dogs. Practical applications and new perspectives in veterinary behavior. Vet. Clin. N. Am., Small Anim. Pract. 38, 1023–1041.

Horowitz, A., 2009. Attention to attention in domestic dog *(Canis familiaris)* dyadic play. Anim. Cogn. 12, 107–118.

Jenks, S.M., 2011. A longitudinal study of the sociosexual dynamics in a family group of wolves: the University of Connecticut wolf project. Behav. Gen. 41, 810–829.

Kerasote, T., 2007. Merle's door. Lessons from a freethinking dog. Harcourt, San Diego, CA.

Le Boeuf, B.J., 1967. Interindividual associations in dogs. Behaviour 29, 268–295.

Lindsay, S.R., 2001. Handbook of applied dog behavior and training, Vol. 2. Etiology and assessment of behavior problems. Iowa State University Press, Ames, IA.

Lund, J.D., Vestergaard, K.S., 1998. Development of social behavior in four litters of dogs (*Canis familiaris*). Acta Veterinaria Scandinavia 39, 183–193.

Marshall Thomas, E., 2001. The social lives of dogs. The grace of canine company. Simon & Schuster, New York.

Mech, L.D., Boitani, L., 2003. Wolf social ecology. In: Mech, L.D., Boitani, L. (Eds.), Wolves. behavior, ecology and conservation. University of Chicago Press, Chicago, IL, pp. 1–34.

Miklósi, A., 2007. Dog behavior, evolution and cognition. Oxford University Press, New York.

Pal, S.K., 2010. Play behavior during early ontogeny in free-ranging dogs (*Canis familiaris*). Appl. Anim. Behav. Sci. 126, 140–153.

Palagi, E., Cordoni, G., 2009. Postconflict third-party affiliation in *Canis lupus*: do wolves share similarities with the great apes? Anim. Behav. 78 (4), 979–986.

Parker, H.G., Shearin, A.L., Ostrander, E.A., 2010. Man's best friend becomes biology's best in show: genome analysis in the domestic dog. Annu. Rev. Genet. 44, 309–336.

Pellis, S.M., Pellis, V.C., 1987. Play-fighting differs from serious fighting in both target of attack and tactics of fighting in the laboratory rat *Rattus norvegicus*. Aggressive Behav. 13, 227–242.

Pellis, S.M., Pellis, V.C., Foroud, A., 2005. Play fighting: aggression, affiliation and the development of nuanced social skills. In: Tremblay, R., Hartup, W.W., Archer, J. (Eds.), Developmental origins of aggression. Guilford Press, New York, pp. 47–62.

Petak, I., 2013. Communication patterns within a group of shelter dogs and implications for their welfare. J. Appl. Anim. Welfare Sci. 16, 118–139.

Ritter, C., Viranyi, Z., Range, F., 2012. Who is more tolerant? Cofeeding in pairs of pack-living dogs (*Canis familiaris*) and wolves (*Canis lupus*). Third International Canine Science Forum. http://www.csf2012.com/Pages/abstracts.html (accessed 27.06.13.).

Rugbjerg, H., Proschowsky, H.F., Ersboll, A.K., Lund, J.D., 2003. Risk factors associated with interdog aggression and shooting phobias among purebred dogs in Denmark. Prev. Vet. Med. 58, 85–100.

Sampson, J., 2006. The kennel club and the early history of dog shows and breed clubs. *The dog and its genome*. Cold Spring Harb. Monogr. Ser. 44, 19–30.

Sands, J., Creel, S., 2004. Social dominance, aggression and faecal glucocorticoid levels in a wild population of wolves. Anim. Behav. 67, 387–396.

Santos, O., Polo, G., Garcia, R., Oliveira, E., Vieira, A., Calderon, N., de Meester, R., 2013. Grouping protocols in shelters. J. Vet. Behav. 8, 3–8.

Schenkel, R., 1947. Ausdrucks-Studien an Wolfen: Gefangenschafts-Beobachtungen. Behaviour 1, 81–129.

Schenkel, R., 1967. Submission: Its features and functions in wolf and dog. Am. Zoologist 7, 319–329.

Scott, J.P., Fuller, J.L., 1965. Genetics and the social behavior of the dog. University of Chicago Press, Chicago, IL.

Serpell, J., Jagoe, J.A., 1995. Early experience and the development of behaviour. In: Serpell, J. (Ed.), The domestic dog. Its evolution, behaviour and interactions with people. Cambridge University Press, Cambridge, pp. 79–102.

Sherman, C.K., Reisner, I.R., Taliaferro, L.A., Houpt, K.A., 1996. Characteristics, treatment, and outcome of 99 cases of aggression between dogs. Appl. Anim. Behav. Sci. 47, 91–108.

Shyan, M.R., Fortune, K.A., King, C., 2003. "Bark Parks." A study of interdog aggression in a limited-control environment. J. Appl. Anim. Welfare Sci. 6 (1), 25–32.

Smuts, B.B. (In prep). Communication and negotiation of relationships among domestic dogs.

Smuts, B.B., Ward, C., 2010. Does your dog need a BFF? Bark 60, 68–71.

Spinka, M., Newberry, R.C., Bekoff, M., 2001. Mammalian play: training for the unexpected. Q. Rev. Biol. 76, 141–168.

Symons, D., 1978. Play and aggression: a study of rhesus monkeys. Columbia University Press, New York.

Trisko, R., 2011. Dominance, egalitarianism and friendship at a dog daycare facility. Ph.D. Thesis. Department of Psychology, University of Michigan.

Van der Borg, J.A.M., Schilder, M.B.H., Vinke, C., 2012. Dominance and its behavioural measures in a pack of domestic dogs. Third International Canine Science Forum. http://www.csf2012 .com/Pages/abstracts.html (accessed 27.6.13.).

van Hooff, J.A.R.A.M., Wensing, J.A.B., 1987. Dominance and its behavioral measures in a captive wolf pack. In: Frank, H. (Ed.), Man and wolf: advances, issues, and problems in captive wolf research. Dr. W. Junk Publishers, Dordrecht, The Netherlands, pp. 219–252.

Vila, C., Savolainen, P., Maldonado, J.E., Amorim, I.R., Rice, J.E., Honeycutt, R.L., Crandall, K.A., Lundeberg, J., Wayne, R.K., 1997. Multiple and ancient origins of the domestic dog. Science 276, 1687–1689.

Ward, C., Bauer, E.B., Smuts, B.B., 2008. Partner preferences and asymmetries in social play among domestic dog, *Canis lupus familiaris*, littermates. Anim. Behav. 76, 1187–1199.

Ward, C., Trisko, R., Smuts, B.B., 2009. Third-party interventions in dyadic play between littermates of domestic dogs, *Canis lupus familiaris*. Anim. Behav. 78, 1153–1160.

Watson, D.M., Croft, D.B., 1996. Age-related differences in play fighting strategies of captive male red kangaroos (*Macropus rufogriseus banksiansus*). Ethology 102, 336–346.

Wright, J.C., 1980. The development of social structure during the primary socialization period in German shepherds. Dev. Psychobiol. 13 (1), 17–24.

Zimen, E., 1981. The wolf. A species in danger. Delacorte Press, New York.

# Chapter 5

# Auditory Communication in Domestic Dogs: Vocal Signalling in the Extended Social Environment of a Companion Animal

Anna Magdalena Taylor, Victoria Frances Ratcliffe, Karen McComb, and David Reby
*School of Psychology, University of Sussex, Falmer, UK*

## 5.1 INTRODUCTION

Vocal signals play a key role in the communication systems of most mammalian species. Crucially, vocalisations can transmit information about the signaller that receivers may use to mediate their responses during sexual and social interactions. Amongst mammals, the vocal repertoire and perceptual abilities of the domestic dog *(Canis familiaris)* are particularly interesting because their communication system seems to have undergone significant changes as they have adapted to the human environment. Domestication is characterised by selective breeding, and whilst natural selection pressures are relaxed, human-controlled selection typically leads to the development of new traits (Price, 1999). An important aspect of human-driven selection in many domesticated species is a requirement for individuals to cooperate with people, which can lead to a differentiation of morphological and behavioural traits as compared to the wild ancestor. Because domestication tends to favour individuals that are able to exploit human perceptual abilities and biases (e.g., domestic cats: McComb et al., 2009) as well as those best able to perceive and make functional assessments of human vocal signals, this chapter considers not only dog–dog vocal communication, but also the use of vocal signalling in dog–human interactions.

In the first section of this chapter, we review existing knowledge on the vocalisations of domestic dogs with a view to understanding their communicative function. The different calls made by dogs are familiar to most humans; barks,

The Social Dog. http://dx.doi.org/10.1016/B978-0-12-407818-5.00005-X

**131**

growls, and whimpers are ubiquitous in our everyday environment and even non-dog owners are likely to form an opinion of what a dog might be signalling. Whilst the contextual use of these calls may provide us with a broad under-standing of their functional role, in order to fully investigate the communicative potential of signals also, we need to look at their acoustic structure, which we characterise in this chapter using the source–filter framework of vocal produc-tion. We emphasise how the relationship between anatomical characteristics and acoustic output directly influences the type and reliability of information contained within a call. In this light, we discuss in detail how dog vocalisations are able to broadcast specific information about static characteristics of the sig-naller, such as body size, and dynamic attributes, such as motivational state. We also review evidence that receivers, including dogs, humans, and potentially other species, can perceive these acoustic cues to make functional assessments and adapt their own behavioural responses.

Throughout this chapter, it is evident that whilst domestic dog vocalisations retain socially relevant information that can be identified by other dogs, func-tional changes in their vocal communication system may also have occurred to facilitate their interaction with humans. Finally, because vocal communica-tion between dogs and humans is not unilateral, we should consider how dogs perceive human voices in order to fully appreciate their ability to assess vocal signals. Just as humans can extract information from dog vocalisations, so dogs can extract information from human vocal signals. In the final section of this chapter, we discuss evidence that dogs appear to respond to both verbal (syntax and semantics) and subverbal information (vocal identity and emotional pros-ody) in human speech, as well as to human non-verbal vocalisations (such as cries and laughter).

## 5.2 HOW DOGS PRODUCE VOCAL SIGNALS

To explore the communicative function of domestic dog vocalisations, we need to first understand more generally how specific acoustic parameters can encode relevant information about the signaller. An efficient way to do this is to consider the structure of vocalisations in the light of their mechanism of production. In most mammals, including dogs and humans, the anatomy of the vocal apparatus is fundamentally similar, so that theories of vocal production can be applied across different species. The source–filter model of human speech production (Fant, 1970; Titze, 1994) has successfully enabled bioacousticians to interpret the acoustic structure of many vertebrate vocal signals within the context of production, enabling researchers to determine how the structure of signals is influenced by the physical and physiological attributes of the caller. According to this model, the production of vocal signals involves independent contribu-tions from two different parts of the vocal apparatus, the *source* and the *filter*.

The *source* of mammalian vocal production is the larynx, a mostly cartilagi-nous organ that is situated low in the throat where the oesophagus and trachea join (see Figure 5-1). At the superior border of the larynx, protected by the elastic cartilage of the epiglottis, the glottis consists of soft tissue layers of muscle and vocal ligament (known as the vocal folds) and the spacing between them. The production of vocalisations begins here: air from the lungs forces its way through the closed glottis and the vocal folds are pushed apart. Biomechanical forces then cause the vocal folds to snap shut again, and this sequence of opening and closing of the glottis causes a cyclic and self-sustaining variation in air pressure across the larynx (see Titze, 1994). The resulting waveform is known as the source signal or glottal wave. The rate of oscillation of the vocal folds determines the fundamental frequency (henceforth F0), and associated harmonics of the source signal, and is perceived as pitch by human listeners. F0 is primarily determined by the length and mass of the vocal folds: longer and heavier vocal folds vibrate at a slower rate than smaller vocal folds. In humans, these properties can, to a certain extent, be manipulated by flexion/relaxation of the muscles controlling the lengthening/shortening and tension of the vocal folds. Other characteristics of the source signal include tempo, duration, and amplitude contour, all of which are controlled by a variety of muscular interactions and changes in airflow or subglottal pressure (Titze, 1994). Source characteristics can thus vary between and within vocalisations from the same caller either on a volitional (intonation in human speech: Ohala, 1984; Banse & Scherer, 1996) or on an involuntarily basis (emotional expression in humans: Ohala, 1996; Aubergé & Cathiard, 2003; affective state in baboons: Rendall, 2003; stress in pigs: Düpjan et al., 2008). The implications of this are discussed in more detail later in the chapter.

The source signal then travels through the vocal tract *filter*, which is composed of the caller's pharyngeal, oral, and nasal cavities, before it is radiated into

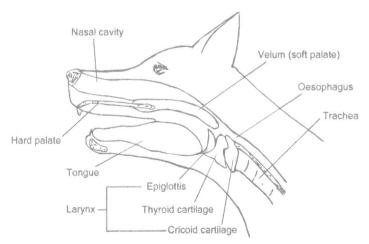

**FIGURE 5-1**   Major components of the vocal apparatus of the domestic dog. Note that only the outer structures of the larynx are represented in this diagram. *(Based on work by Piérard, 1963.)*

the environment through the mouth and/or nostrils. The vocal tract is referred to as the 'filter' because its resonant properties (determined primarily by its length and shape) selectively enhance or dampen some of the harmonic frequencies of the source signal. Those frequencies that resonate well form spectral peaks, or formants (Fant, 1970). Their perceptual quality is best described as the 'depth' or 'timbre' of a vocalisation, which is not to be confused with its pitch (Titze, 1994). The degree to which formant positioning can be actively controlled is dependent on the species. Humans, for example, demonstrate highly sophisticated neuromotor control of their vocal apparatus: subtle changes in the position of the pharynx, velum (soft palate), tongue, and lips influence the resonant properties of the vocal tract, affecting the relative frequency position of formants, which is necessary for speech production (see Lieberman & Blumstein, 1988; Titze, 2000). In contrast, most mammals, including dogs, do not appear to have fine motor control over their vocal tract characteristics and so the resonant properties of the vocal tract are usually more predictable. Formant frequencies in most non-human calls therefore appear as evenly distributed horizontal bands, often directly reflecting the length of the vocal tract (Fitch, 1997). This distribution of formants is quantified under the term 'formant dispersion' (Fitch, 1997; Reby & McComb, 2003) and constitutes one of the key factors affecting the perceived quality of the vocalisation.

The source- and filter-related acoustic components in mammalian vocalisations are therefore constrained by different production characteristics. By applying this framework to the dog's vocal repertoire, we can explore how dogs produce the acoustic features which characterise their different call types. Moreover, we can see how this influences the functional content of different calls and discuss their evolutionary origins (see Taylor & Reby, 2010, for a detailed review of how the source–filter theory can be applied as a generalised conceptual and methodological framework to investigate vocal communication in non-human mammals).

## 5.3 DESCRIPTION OF DOG VOCAL REPERTOIRE (AND COMPARISON WITH WOLVES, *CANIS LUPUS*)

Domestic dogs produce a range of vocalisations, many of which are likely to be familiar to non-expert human listeners. We will now provide an overview of the dog's most common vocal signals, focussing both on production aspects and on their evolutionary functionality. In this regard, it is useful to draw comparisons between dogs and equivalent calls in the ancestral wolf. At a first glance, the acoustic structure of dog vocalisations appears to be very similar to that of wolves; several studies present acoustic analyses of calls from both species, and on a structural level, one would be hard-pressed to distinguish between the barks or growls of dogs and wolves (Feddersen-Petersen, 2000; Schassburger, 1993). However, when we look at the context of emission, it appears that the vocalisations of these species may not be the same on a functional level. Indeed, it is likely that the domestication of dogs has had an important impact on how their vocalisations are used. A broad comparison between the vocal behaviour of domestic dogs, wolves, and other canids is presented in Table 5-1.

The four most common characteristic calls produced by dogs, namely barks, growls, whimpers, and howls (illustrated in Figure 5-2), appear to be shared across breeds (Cohen & Fox, 1976). While we primarily focus on these four call types, we also briefly discuss anecdotal reports of other vocalisations, including both breed-specific calls and those generally occurring across breeds, and highlight where future investigation would be particularly relevant.

## 5.3.1 Barks

One of the most stereotypical vocalisations of the domestic dog is the bark. Barks are short, plosive signals that can be produced as part of a sequence or in isolation, with an F0 range that can vary considerably across breeds, individuals, and contexts (Cohen & Fox, 1976; Feddersen-Petersen, 2004). The acoustic output is achieved by the dog lowering the larynx and raising the velum, thereby closing off the nasal passage. The sound is then emitted from the mouth, with an open jaw, as shown in Figure 5-3a. Barks are often described as noisy or chaotic because the harmonic-to-noise ratio varies due to the irregular oscillation of the vocal folds (Riede & Fitch, 1999; also see Figure 5-2a). Harrington and Mech (1978) proposed that wolves bark as an aggressive signal, advertising their willingness to defend themselves and their companions or territory. This instinctive response to draw attention to potential intruders remains common in domestic dogs today and is highly likely to have been one of the factors facilitating the association between early hominids and canids. Yet despite similarities, the barking behaviour of dogs and wolves has clearly undergone different selection pressures. While adult wolves bark in specific aggressive contexts, adult dogs appear to bark readily across many different contexts, including playful and positive situations (Cohen & Fox, 1976). This vocal distinction remains apparent even when individuals of both species have been raised together in similar environments, suggesting that the ontogeny of barks in adult domestic dogs has radically changed from its ancestral, context-specific form (Frank & Frank, 1982; Feddersen-Petersen, 2000).

There also appear to be breed-typical differences in the production of barks. In an analysis of vocalisations from nine different breeds of dog, Feddersen-Petersen (2000) found a high level of variability in barking between breeds, with each breed showing between 2 and 12 subtypes of barking based both on their spectrographic features and on behavioural correlates. This suggests that domestication may have affected the vocal behaviour of different breeds in different ways, potentially due to selection by humans for different behavioural roles. Certainly, the observation that dogs bark in many different situations led earlier researchers to speculate that dog barks were a 'hypertrophy' of a previously functional behaviour that

**TABLE 5-1** Comparative Overview of Dog and Wolf Common Vocalisations

| Signal | Acoustic Features | | Context | Other Canids Known to Produce this Call Type |
|---|---|---|---|---|
| Bark | Plosive Variable pitch Long distance | Domestic dog | Non-specific All contexts | Dingo, *Canis lupus dingo* (Corbett, 1995) New Guinea singing dog (Koler-Matznick et al., 2003) Coyote, *Canis latrans* (Cohen & Fox, 1976) |
| | | Grey wolf | Specific Aggressive | Ethiopian wolf, *Canis simensis* (Sillero-Zubiri & Gottelli, 1994) Golden jackal, *Canis aureus* (Estes, 1991) Side-striped jackal, *Canis adustus* (Estes, 1991) Black-backed jackal, *Canis mesomelas* (Moehlman, 1983; Estes, 1991) Dhole, *Canis alpinus* (Volodin et al., 2001) African wild dog, *Lycaon pictus* (Robbins, 2000) Crab-eating fox, *Cerdocyon thous* (Brady, 1981) Hoary fox, *Lycalopex vetulus* (Sillero-Zubiri et al., 2004) Maned wolf, *Chrysocyon brachyurus* (Brady, 1981) Bush dog, *Speothos venaticus* (Brady, 1981) |
| Growl | Harsh/ broadband Low pitch Short range | Domestic dog | Non-specific Agonistic Playful | Dingo, *Canis lupus dingo* (Déaux & Clarke, 2013) New Guinea singing dog (Koler-Matznick et al., 2003) Coyote, *Canis latrans* (Cohen & Fox, 1976) |
| | | Grey wolf | Specific Agonistic | Ethiopian wolf, *Canis simensis* (Sillero-Zubiri & Gottelli, 1994) Golden jackal, *Canis aureus* (Estes, 1991) Side-striped jackal, *Canis adustus* (Estes, 1991) African wild dog, *Lycaon pictus* (Robbins, 2000) Short-eared dog, *Atelocynus microti* (Sillero-Zubiri et al., 2004) Crab-eating fox, *Cerdocyon thous* (Brady, 1981) Maned wolf, *Chrysocyon brachyurus* (Brady, 1981) Bush dog, *Speothos venaticus* (Brady, 1981) |

| Vocalization | Acoustic properties | Species | Function | Examples |
|---|---|---|---|---|
| Whimper/ Whine | Tonal High pitch Short range | Domestic dog | Specific Social distance reducing | Dingo, *Canis lupus dingo* (Déaux & Clarke, 2013) New Guinea singing dog (Koler-Matznick et al., 2003) Coyote, *Canis latrans* (Cohen & Fox, 1976) Ethiopian wolf, *Canis simensis* (Sillero-Zubiri & Gottelli, 1994) Golden jackal, *Canis aureus* (Estes, 1991) Side-striped jackal, *Canis adustus* (Estes, 1991) Black-backed jackal, *Canis mesomeias* (Moehlman, 1983; Estes, 1991) Dhole, *Canis alpinus* (Volodin et al., 2001) African wild dog, *Lycaon pictus* (Robbins, 2000) Crab-eating fox, *Cerdocyon thous* (Brady, 1981) Maned wolf, *Chrysocyon brachyurus* (Brady, 1981) Bush dog, *Speothos venaticus* (Brady, 1981) |
|  |  | Grey wolf | Specific Social distance reducing |  |
| Howl | Harmonic Frequency modulated Long distance | Domestic dog | Non-specific Function unclear, possibly vestigial response to high-pitched sounds and social isolation | Dingo, *Canis lupus dingo* (Corbett, 1995) New Guinea singing dog (Koler-Matznick et al., 2003) Coyote, *Canis latrans* (Cohen & Fox, 1976) Ethiopian wolf, *Canis simensis* (Sillero-Zubiri & Gottelli, 1994) Golden jackal, *Canis aureus* (Estes, 1991; Jaeger et al., 1996) Side-striped jackal, *Canis adustus* (Estes, 1991) Black-backed jackal, *Canis mesomeias* (Moehlman, 1983; Estes, 1991) Dhole, *Canis alpinus* (Volodin et al., 2001) Crab-eating fox, *Cerdocyon thous* (Brady, 1981) Maned wolf, *Chrysocyon brachyurus* (Brady, 1981) |
|  |  | Grey wolf | Specific Social cohesion |  |

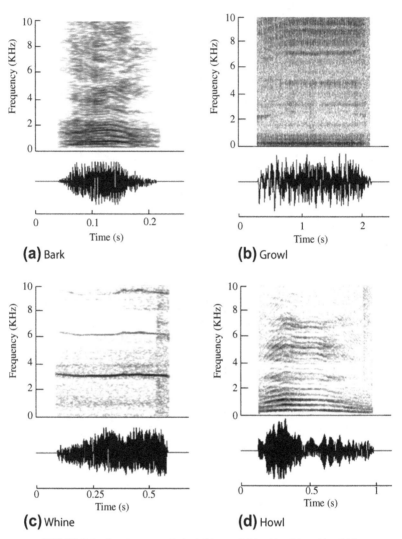

**FIGURE 5-2**   Spectrograms of a bark (a), growl (b), whine (c), and howl (d).

may have developed due to relaxed selection pressures (Cohen & Fox, 1976). For a long time, it was believed that dog barks were a functionally insignificant, 'non-communicative' by-product of domestication (Coppinger & Feinstein, 1991; Bradshaw & Nott, 1995). However, recent studies have demonstrated that barks do, in fact, broadcast reliable information to receivers, which we discuss in more detail in the following sections.

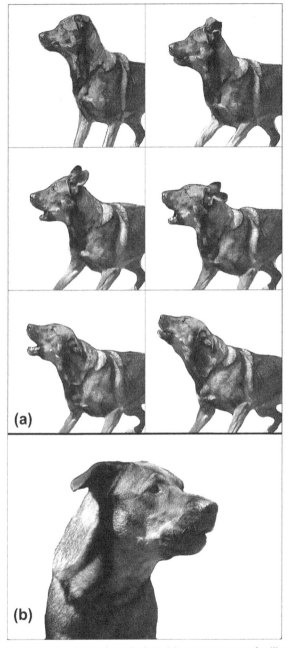

**FIGURE 5-3**   Tess, a crossbreed, emitting a single bark in response to an unfamiliar person (a) and a growl in response to an unfamiliar dog (b).

## 5.3.2 Growls

The next most common vocalisation, likely to be just as familiar to most human listeners, is the growl. The growls of dogs and wolves are structurally identical: harsh, low-frequency broadband vocalisations, with an F0 range between 80 and 300 Hz (Cohen & Fox, 1976; Harrington & Mech, 1978; see Figure 5-2b). In contrast to the plosive barks that can propagate over longer distances, it seems likely that growls evolved for the purpose of close-range communication (Cohen & Fox, 1976). In adult wolves, growls commonly occur during confrontations and dominance interactions and near the site of a fresh kill (Fox, 1984), and they may also be used to warn off subordinate individuals or scavengers. Growls accompanied by threatening body posture are more likely to be followed by an actual attack both on other wolves and on humans (Rutter & Pimlott, 1968; Harrington & Mech, 1978). In subadult wolves, growls are also commonly emitted during play fights. Fox (1984) observed that dominant cubs play growled more than subordinate cubs, whereas subordinate cubs whimpered more, indicating that play growling could be part of the maturation process as juvenile wolves practise behaviours that will be needed in adulthood. In contrast, play growls are not heard during interactions between adult wolves (Fox, 1984). This represents a clear distinction between the vocal repertoires of wolves and dogs. Indeed, while adult dogs clearly growl in aggressive contexts (Figure 5-3b shows Tess, a crossbreed, emitting an aggressive growl with closed jaw and drawn-back flews in response to another dog), they also frequently growl in playful situations, mirroring behaviour observed in juvenile wolves (Cohen & Fox, 1976). In adult dogs such play growling can occur during positive interactions with humans, dogs, other animals, or inanimate objects (e.g., toys), and so here too it seems likely that the relaxation of natural selection triggered by domestication has affected the function of the vocalisation.

## 5.3.3 Whimpers and Whines

The terms 'whimper' and 'whine' are taken here to refer to high-pitched, tonal vocalisations that can be emitted through the nasal cavity (Fitch, 2000b; Figure 5-2c). In both dogs and wolves, they are produced in a greeting context and as a solicitation for food or care (Fox, 1984; Cohen & Fox, 1976). For the production of whimpers, the larynx remains inserted in the nasopharynx, and the velum position remains open (Fitch, 2000b). In wolves, greeting whimpers might be exchanged between adults or between cubs and their dam. In dogs, similar greeting whimpers are mostly likely to occur when a familiar human is approaching (Cohen & Fox, 1976). It is as yet unknown whether this mirrors cub–dam behaviour in the same way as face licking and play bowing (Bekoff, 1974), or whether it is a greeting from a subordinate adult to a dominant individual. Importantly, whimpers and whines are produced exclusively by a non-aggressive caller and often serve to appease a dominant individual and decrease social distance. Based on this observation, it can be proposed that whimpers serve as the functional opposites of growls.

## 5.3.4 Howls

Howls are frequency-modulated harmonic vocalisations that are perhaps more commonly associated with wolves than with domestic dogs. Indeed, wolves are well known for their group howling, a harmonious chorus containing both howls and barks and able to propagate across wide distances up to two kilometres (Harrington & Mech, 1978). Recent vocal analyses in Italian wolves found that the mean F0 of howls emitted by the same pack could range from 274 to 908 Hz (Zaccaroni et al., 2012) and there is some speculation that the frequency range broadcast by a chorus advertises the presence and location of a pack, thereby enabling different packs to minimise the risk of meeting each other (Joslin, 1967). In addition to chorus howling for territory maintenance, wolf howls may enable separated individuals to regroup, possibly because they are individually recognisable by other group members (Harrington & Mech, 1978; Theberge & Falls, 1967). A recent study found that captive wolf pack howling rates were higher when dominant individuals or close social partners were absent from the group. Whilst the remaining pack members' stress hormone (cortisol) levels were higher when the dominant individuals were missing, there was no difference in cortisol levels when preferred pack mates were absent. This suggests that howl production is not entirely driven by the physiological or emotional state of the signaller, but rather is to some extent under voluntary cognitive control and can be flexibly emitted depending on the social context (Mazzini et al., 2013). Coyotes, jackals, and domestic dogs also howl (see Figure 5-2d), although the informational content and function of howls in these species and specifically domestic dogs remains largely unstudied.

## 5.3.5 Other Vocalisations

In addition to the four main call types described in the preceding sections, dogs produce more unusual, sometimes breed-specific vocalisations. The work of Feddersen-Petersen (2000) and others suggests that domestication may have affected the vocal behaviour of different breeds in different ways. Indeed, selective breeding for specific behavioural roles appears to have involved the selection of vocal behaviour. In some breed groups, this may have influenced the rate of occurrence of shared call types. For example, guarding breeds were selected for protective instinct and thus display strong defensive behaviours, including exaggerated barking and growling. Terriers were likewise bred to bark frequently in order to alert hunters to their location whilst pursuing prey animals in underground burrows, but gundogs must remain silent whilst hunting so as not to frighten away any game (Coren, 2005). Moreover, distinct vocalisations may have developed in specific breeds. Scent hounds, for example, appear to remain in constant communication with each other while hunting using a very distinct 'baying' vocalisation, a melodious call containing many tonal variations (Coren, 2005). The propensity to produce unusual call types appears to be higher in breeds descended from the Indian plains wolf (*Canis lupus pallipes*) such as the dingo, New Guinea singing dog, and basenji dog, which all show a number of unique attributes that do not occur in other breeds. Examination of the vocal tract of the New Guinea singing dog revealed a rudimentary two-lobed uvula (a mass of tissue that is suspended from the velum and hangs above the throat), which has not been observed in any other canid. It is thought that the vibration of the uvula may allow these dogs to produce their distinctive 'trill', a high-frequency pulsed vocalisation (Koler-Matznick et al., 2003). Basenjis, an African breed selected for pack hunting large game, are known to produce a high-pitched, harmonic call known as a 'yodel' or 'baroo'. A study of two basenji dogs revealed that their laryngeal anatomy was significantly different from that of other breeds (Ashdown & Lea, 1979). Similarly, the dingo also produces a distinctive stereotyped composite 'bark-howl' (Deaux & Clarke, 2013). More research is needed to fully understand these breed-specific vocalisations.

Overall, despite some functional differences, it is clear that the vocalisations of domestic dogs and wolves follow a broad, predictable pattern of acoustic output, with lower-pitched sounds occurring more commonly in aggressive contexts and higher-pitched sounds occurring in social and playful contexts. In the previous section, we noted that the relaxed selection pressures of domestication may have contributed to a change in the context-specificity in which barks and growls are produced. On a structural level, these vocalisations are indicative of agonistic displays, yet they are heard across many different contexts. In the next section, we review investigations of the acoustic structure of vocal signals aimed at understanding their information content. We focus primarily on barks and growls because these two call types are the most well-documented and researched vocalisations of the dog.

## 5.4 THE COMMUNICATIVE FUNCTION OF ACOUSTIC VARIATION

### 5.4.1 Indexical Information

Within the methodological framework of the source–filter model, researchers interested in mammal communication can explain why different calls contain particular acoustic features. Any vocalisation in which the acoustic properties are predictable as a function of the physical attributes of the caller has the potential to broadcast accurate information about that physical attribute (Fitch, 1997; Fitch & Reby, 2001). It is well documented that larger-sized individuals are at an advantage over smaller individuals during agonistic encounters, and individuals benefit from avoiding costly fights that they are unlikely to win (Schmidt-Nielsen, 1975; Peters, 1986). Acoustic advertising and assessment of body size are thereby central to social interactions in some species (e.g., red deer: Reby & McComb, 2003), and identifying the relevant acoustic correlates to body size is thus an important task in vocal communication research.

Because of the perceptual salience of pitch to human listeners (Ohala, 1984), some earlier studies of mammal vocal communication focussed on F0 as a potential source of size-related information (see August & Anderson, 1987). As we saw earlier, F0 is primarily dependent on the length and mass of the vocal folds. Generally speaking, the soft tissue anatomy of the vocal folds means that they are not stringently constrained by the body size of the individual (Fitch, 2000c), but it has been observed that across age and sex categories (possibly due to age-related vocal-fold growth and sexual dimorphism) F0 can sometimes be broadly correlated with caller body size. For example, in both baboons and humans, males arc larger than females and have a lower F0 (Rendall et al., 2005; Puts et al., 2006). In relation to other mammals, canid species have a comparably low level of sexual dimorphism, which is related to their largely monogamous breeding systems (Bekoff et al., 1981). The grey wolf is the most sexually dimorphic canid species; males have around 18% greater body mass than females (Hillis & Mallory, 1996). Within dog breeds, those of comparable size to wolves show similar levels of body mass dimorphism, which was found to relate to a difference of 5%–7% in vocal tract length (VTL) between adult males and females for Portuguese water dogs (Plotsky et al., 2013). However, the size difference between sexes decreases in smaller breeds (Frynta et al., 2012), and no sex-related differences in vocal anatomy or in F0 have been identified across breeds when controlling for body weight (Riede & Fitch, 1999; Taylor et al., 2008). However, artificial selection by humans has led to an exceptionally large level of size variation between breeds, giving dogs the highest level of morphological variation of any mammal (Wayne, 1993). Consequently, across breeds, F0 can provide a broad indication of body size, with smaller breeds producing growls with a higher F0 than larger breeds (Taylor et al., 2008). On an individual level, F0 is influenced by changes to respiration and muscular control of the tension of the vocal folds. Therefore, F0 is highly variable both within and between calls, across individuals, and across breeds (Feddersen-Petersen, 2000; Yin, 2002; Taylor et al., 2008).

In contrast, the vocal tract cannot grow independently of the rest of the body, as its development is anatomically constrained by skeletal structures (Fitch, 2000a, 2000c). Vocal tract length is thus more directly dependent on body size, and investigations have confirmed a strong positive correlation between anatomical VTL and body size in many species, including domestic dogs (Riede & Fitch, 1999). As formant dispersion is controlled by VTL, this means that unlike F0, formants have the potential to provide accurate or 'honest' information about the caller (Fitch, 1997; Reby & McComb, 2003). A direct negative correlation between formant dispersion and body size has been confirmed in the signals of many species, including dog growls (Riede & Fitch, 1999; Taylor et al., 2008). When we analyse formant dispersion within any growl, it is thus possible to predict the body size of the caller. It is also noteworthy that amongst dog vocalisations, the growl is particularly suited to transmitting size-related information. The harsh, broad bandwidth structure of growls transmits acoustic energy across a wide range of frequencies, which increases the perceptual salience of the formants (also see Ryalls & Lieberman, 1982). As growls evolved, this aspect of their acoustic structure may have originally been selected for in the context of agonistic interactions because it most effectively advertised body size to competitors.

In terms of perceptual relevance, several experiments show that domestic dogs are sensitive to size-related acoustic variation in conspecific vocalisations. Taylor et al. (2010a) observed that dogs responded to variations in the formant dispersion of growls in a manner consistent with size assessment. In the absence of visual size cues, dogs' behavioural responses to playbacks simulating small dogs were significantly different than their responses to playbacks simulating large dogs. Responses were additionally influenced by the subjects' own body size. Indeed, large dogs showed more motivation to investigate when growls simulating a smaller intruder were played than when growls simulating a larger intruder were played. In contrast, small dogs did not respond differentially to small versus large intruders and reacted in a more consistent manner to all playbacks (Taylor et al., 2010a). Taylor and colleagues' observations of a potential effect of the relative size of signallers and receivers are consistent with the results of an earlier study of visual communication in dogs (Leaver & Reimchen, 2008). These authors found that only dogs larger than a Labrador retriever model responded differentially depending on modified tail positions in the latter, while dogs smaller than the model did not respond differently depending on tail position and, in fact, showed lower motivation to approach or interact with the model overall. The two studies strongly suggest that such responses may not be universal across dogs but may be influenced by their own body size. In fact, small dogs (under 10 kg) are the most likely victims of bites and injuries inflicted by other dogs (Shamir et al., 2002) and may therefore benefit the most from avoiding interactions with unfamiliar individuals. Because domestication is likely to have eliminated most competition for resources, we can speculate that small dogs may have little benefit to gain from a confrontation with a potentially aggressive intruder. As such, assessing caller body size on the basis of auditory information may well be a functionally relevant ability even within the context of a domestic animal.

Integrating auditory information with related visual information is an ability most humans take for granted, and there is increasing evidence that many species are able to process some types of information across modalities (Adachi & Fujita, 2007; Adachi et al., 2007; Ghazanfar et al., 2007; Proops et al., 2009). In two separate experiments using cross-modal designs, dogs demonstrated that they were able to reliably match acoustic size information in growls with corresponding visual information [for the visual stimuli, Faragó et al. (2010) used images of a small or large variant of a dog projected onto a white screen, whereas Taylor et al. (2011) used taxidermies of a Jack Russell terrier and a German shepherd dog]. In both studies, dogs preferentially looked towards the visual stimulus that matched the apparent acoustic size of the caller, showing not only that they had the ability to perform this association, but also that they were motivated to do so, suggesting that this task was functionally relevant to dogs.

Size-related acoustic variation may moreover be functionally relevant across species. In a psychoacoustic investigation, Taylor et al. (2008) demonstrated that human listeners are able to accurately attribute associated body size to growls that had been resynthesised to vary only in formant dispersion, with all other acoustic parameters standardised across stimuli. Interspecific size assessment is also interesting to consider in the context of the predatory potential of domestic dogs. As larger predators are more successful hunters and generally outperform smaller predators (Gittleman, 1989; Herrel & Gibb, 2006; MacNulty et al., 2009), the body size of predators is valuable information to any prey animal. In support of this, species at risk of predation by dogs such as domestic sheep or red deer appeared to respond to playbacks of domestic dog growls in a manner consistent with size assessment. In a playback experiment, the individual vigilance of groups of domestic sheep and individual red deer hinds was found to be greater in response to growls typical of large dogs than to growls typical of small dogs (Taylor & Reby, unpublished data). A confirmation of these observations would bring further support to the theory that size perception in the auditory domain is universal across mammalian species and applicable in a number of different functional contexts.

Attending to information about body size and using it to assess the physical attributes of potential competitors can thus have important implications for the reproductive opportunities and survival of receivers. As such, it is very likely that the functional decisions of receivers, based on this information, may place additional selection pressure on signals at the level of their production. In an influential paper based on a comparative study of vocalisations used in agonistic displays in a range of mammalian and avian species, Morton (1977) suggested that audible frequency differences in vocalisations reflect ritualised signalling: animals with aggressive motivation produce low-pitched, broadband vocalisations (such as growls and hisses), whereas animals with affiliative or submissive motivation produce high-pitched vocalisations (such as whimpers and whines). This theory, known as Morton's motivation-structural code, is based on the observation across several species that aggressive and dominant animals seek to project (both visually and acoustically) a larger impression of body size and greater threat, whereas friendly or submissive animals seek to project a smaller impression of body size and lesser threat (Morton, 1977; Ohala, 1984; Owings & Morton, 1998).

Acoustic variation of signals encoding information that is associated with attributes such as body size may thereby become ritualised, broadcasting more transient information pertaining to the motivational state of the caller (Ohala,

1984; Taylor & Reby, 2010). While mechanisms for visual maximisation or exaggeration of body size such as piloerection is relatively common among mammals, few acoustic equivalents have been reported (although see the retraction of the larynx in red deer; Reby & McComb, 2003). The use of lower frequency vocalisations could thus be representative of aggressive motivation not just across call types as described by Morton (1977) but within call types as a ritualised function of the acoustic size code (Ohala, 1984, 1996; Taylor & Reby, 2010).

## 5.4.2 Dynamic Information

Whilst dogs do not appear to modify the formant frequency scaling in their vocalisations, and thus only seem to provide a static indication of body size, they are able to produce an impressive range of F0 both within and between calls (Yin, 2002). Variations in the source signal may provide reliable information about the affective state of callers because it is influenced by changes to rate of respiration and/or muscular tension in the vocal folds (Hauser, 2000; Bachorowski & Owren, 2008), both of which are linked to physiological arousal. One outcome of this is that these temporal characteristics can be affected by state of arousal. Indeed, in barks and growls, inter-call interval and call duration are affected by production context: barks recorded in an aggressive context (being approached by a stranger) were found to have very short inter-bark intervals compared to barks recorded in a playful context (Yin, 2002), whereas growls recorded in an aggressive context are characterised by a low calling rate and have long call durations (Taylor et al., 2009). Temporal characteristics might thus provide a guide to motivational state, although the salience of this to listeners is yet to be established. Another, better understood, outcome is the effect of arousal on F0 in the context of broadcasting of motivational state information. Domestic dog barks occur on a graded frequency scale: lower peak, mean, and range of F0 are found in barks recorded in an aggressive situation than in those recorded in a playful situation (Feddersen-Petersen, 2000; Yin, 2002). This has been confirmed in single-breed (Pongrácz et al., 2006) and mixed-breed experiments (Taylor et al., 2009) and is sufficiently predictive that barks can be reliably categorised as aggressive or playful on the basis of statistical analysis (Yin & McCowan, 2004). Moreover, Pongrácz et al. (2005) showed that human listeners are able to discriminate between barks recorded when dogs were approached by a stranger, antagonised by a Schutzhund trainer, going on a walk, left in isolation, or playing. In addition, listeners attributed emotional content to barks from the preceding contexts (aggressiveness, fear, playfulness, or happiness; Pongrácz et al., 2006). Similarly, dogs appear to physiologically discriminate between barks from unfamiliar dogs recorded in two different situations, but the functional relevance of their discrimination is unclear because dogs did not respond with quantifiable behaviours (Marós et al., 2008).

Growls have similarly been studied within the framework of the acoustic size code hypothesis. Investigating the theory that more aggressive dogs may manipulate their formants to sound larger, Taylor et al. (2009) did not find any difference in the formant dispersion of growls recorded in an aggressive (approaching human) or playful context. However, Faragó et al. (2010) analysed growls recorded in an extremely high emotional valence situation (food guarding against an unfamiliar dog) and found that formant dispersion in these growls was, in fact, representative of size maximisation. A cineradiographic study showed that domestic dogs may be able to do this by modifying the level of nasalisation of a signal via small jaw and tongue movements (Fitch, 2000c). In terms of perceptual significance, dogs avoided food when they were presented with food-guarding growls but did not avoid food when presented with either playful growls or defensive (threatening stranger approach) growls (Faragó et al., 2010). It is noteworthy that manipulating size-related formant frequency scaling can also potentially affect the perception of motivation by humans, despite the lack of relationship between formant and motivation in dog growls. Indeed, humans attributed higher levels of aggressiveness to growls where formants indicate larger dogs than to growls where they indicate smaller dogs (Taylor et al., 2010b). This suggests that there may be a pre-existing perceptual bias in humans to confound information about body size and motivation in the acoustic domain (possibly such a bias may have developed because of the human tendency to use size-related acoustic variation in voice to rate traits such as masculinity and dominance; e.g., Puts et al., 2007) and that this bias occurs despite, or perhaps in addition to, their ability to accurately judge caller body size (Taylor et al., 2008, 2010b).

Overall, the acoustic variation in barks appears to be better suited to broadcasting information about dynamic state than the acoustic variation in growls. However, size-related information in growls may in some cases of very high emotional valence serve to broadcast a highly aggressive motivation. Similarities in signal production across different individuals allow receivers to make reliable judgements about the dynamic and static attributes of the signaller; however, differences between individuals in vocal anatomy and physiology mean that some level of individual variation within calls still occurs. If these differences are both static and perceived by receivers, it becomes possible to identify individuals from their calls.

### 5.4.3 Individual Recognition

Some mammalian vocalisations may contain a vocal signature, i.e., a unique combination of acoustic parameters that enables receivers to recognise or identify the signaller. In an early experiment, Yin and McCowan (2004) found that dog barks could be statistically attributed to individual callers (irrespective of production context) using a discriminant function analysis. Similarly, Molnár et al. (2008) developed computer-based learning software, which categorised individual dogs on the basis of their barks. Human listeners, however, struggled to reliability discriminate between barks from dogs of the same breed (Molnár et al., 2006), although the presentation of five-bark sequences did somewhat improve discrimination. Molnár et al. (2009) suggest that the interaction between harmonic-to-noise ratio and mean F0 may be individually different between dogs and that although these cues may be too subtle for human listeners, they may be perceptually accessible to other dogs. To date, however, no study has conclusively demonstrated individual recognition on the basis of auditory information in dogs.

Although it is possible to ask human listeners what information they can extract from dog vocalisations, innovative paradigms are needed to determine how this information is both assessed and used by other dogs. Whilst this leaves many areas for advancement in determining how dogs perceive information within their own species' calls, it is possible to look further into how dogs may have adapted these perceptual and cognitive abilities to extract information from human vocal signals.

## 5.5 THE PERCEPTION OF HUMAN VOCAL SIGNALS BY DOGS

So far, it is evident that dog vocalisations can broadcast relevant information about the physical and motivational state of signallers, and that listeners may make functionally useful assessments on the basis of this information. However, domestication has strongly exposed dogs to the human environment, and thus, it is interesting to consider what perceptually relevant information dogs may be able to extract from human vocal signals. It has long been known that domestic dogs are highly sensitive to the visual cues used in human social communication, learning to use pointing gestures and gaze direction to locate hidden rewards from an early age (e.g., Miklósi et al., 1998; Hare & Tomasello, 1999). Furthermore, when asked about their dogs' responsiveness to verbal utterances, owners tend to state that their dogs are good at interpreting communicative intentions in speech. This belief often leads owners to attempt to communicate verbally with their dogs, but what information do dogs actually perceive when they are spoken to?

Although the human voice contains the same anatomically related acoustic components as other mammals, providing indexical and dynamic information about the speaker, the most crucial dimension in human vocal communication is the ability to transmit semantic information through speech. Speech consists of hierarchical segments, starting at the basic level of consonants and vowels, or phonemes. These are produced through intentional movements of the vocal apparatus, which alter the relative positions of the lower formant frequencies (Fant, 1970). Phonemes are then combined to create meaningful words and structured into sentences, allowing humans to communicate complex language in a spoken format. Human vocalisations thus have the potential to transmit several levels of information: indexical, emotional, and semantic. We next discuss the ability of domestic dogs to extract both the shared and unique levels of information from the human voice.

## 5.5.1 Indexical Information

The anatomical features of the human vocal apparatus differ in size between children and adults and between adult men and women (Fitch & Giedd, 1999), generating categorical acoustic differences in F0 and formant dispersion that provide an indication of the age and sex of the speaker (Smith & Patterson, 2005). Because dogs perceive indexical information broadcast in the calls of their own species, they may also have the ability to discriminate the systematic variation between human voices derived from these static attributes. F0- and formant-related gender differences allow human listeners to readily classify adult human voices as male or female (e.g., Coleman, 1976). With discrimination training, dogs can learn to distinguish between the average F0 values of adult male and female voices and to discriminate formant-related variation in the human voice (Baru, 1975). If dogs have sufficient social experience with men and women, they spontaneously learn to categorise human voices as male or female, enabling them to associate unfamiliar voices with people of the same gender (Ratcliffe et al., submitted). This shows that dogs naturally learn to discriminate acoustic variation that relates to the biological attributes of human speakers and use this information to guide their visual search towards a matching person. Further research is needed to determine which acoustic characteristics dogs use to categorise human voices as male or female, and if they also categorise other indexical cues present in the human voice, such as age or body size, to associate voices with different speakers.

## 5.5.2 Individual Recognition

The co-variation of characteristics that are unique to the vocal apparatus of each individual speaker contributes to the existence of individually distinct voices in humans (Bachorowski & Owren, 1999). Humans are able to match voices with other unique traits of known individuals such as their facial features from infancy (Bahrick et al., 2005), making individual people highly recognisable across sensory modalities. The cognitive mechanisms of individual recognition are less well understood in animals (see Proops et al., 2009), but it has been shown that domestic dogs are able to match auditory and visual cues to human identity, indicating some level of individual recognition. Adachi et al. (2007) illustrated this by using an expectancy violation (congruent/incongruent) paradigm. In this experiment, dogs were first presented with a voice recorded from either their owner or a stranger and were then presented with an image of either their owner's face or a stranger's face. The dogs looked longer at the person's face when it did not match the preceding voice, suggesting that the dogs expected to see their owner when they heard his/her voice, and did not expect to see their owner if they heard a stranger's voice. It therefore appears that dogs are able to perceive and categorise the individually distinctive acoustic cues characterising their owner's voice and use this information to recognise individual human vocal signatures. In this study, the acoustic stimulus was the dog's own name, a highly familiar signal. To establish the extent to which this ability is independent from learnt signals, we therefore need to determine if dogs can also recognise their owner's voice when he/she is saying unfamiliar phrases. It is also unknown whether dogs are able to discriminate between two familiar people, which would provide further evidence in support of auditory individual recognition.

### 5.5.3 Emotional Information

In addition to anatomically constrained information, dynamically controlled prosody is also present in human-voiced signals. At an acoustic level, vocal prosody is produced by variation in F0, duration, and intensity, which in addition to clarifying the speaker's intentions (e.g., questions versus statements) can also provide information about his/her emotional state. Different emotions can be related to predictable variation in the relative position and modulation of F0, voice quality, and formant values across speakers, and are universally identified by human listeners (Sauter et al., 2009). It appears that dogs can also discriminate and perhaps even assess prosodic variation generated by different emotions. In one study, dogs were found to respond in a similar manner to vocal commands when spoken in a neutral or positive tone, but their responses became more variable when commands were spoken in a negative tone (Mills et al., 2005). This suggests that dogs can perceive and respond to changes in vocal prosody generated by the emotional state of the speaker. In line with this, Scheider et al. (2011) found that dogs searched longer for food when an experimenter spoke to them in a high-pitched, encouraging voice. However, when spoken to in a low-pitched, imperative voice the dogs reduced their movements and responded more often by sitting or lying down.

One possible explanation of these observations is that dogs learn to associate specific prosodic cues with different behavioural responses, as the dogs produced responses associated with obedience training after hearing prosodic cues that were likely to mimic those used to give previously learnt commands. However, an intriguing alternative is that similarities between the acoustic characteristics in the vocalisations of dogs and humans in specific contexts may facilitate the generalisation of responses across species, in this case causing the dogs to respond submissively to a more negatively valenced voice. Potential support for this explanation can be derived from findings that prosodic cues indicating anger and happiness in human speech may also be related to the acoustic size code discussed earlier in this chapter. Human listeners perceive synthetic vowels created with a dynamically lower F0 and smaller formant dispersion as being spoken in an angry voice, whilst vowels with a dynamically higher F0 and a larger formant dispersion are perceived as being spoken in a happy voice (Chuenwattanapranithi et al., 2008). This potential universality in ritualisation across mammal vocalisations may allow dogs to generalise their responses to specific prosodic cues, aiding their perception of certain emotions in the human voice. Rather than learning to associate specific responses with different emotional cues in the voice, similarities in the way that emotions are advertised across mammal vocal signals may therefore cause affect–induction (Owren & Rendall, 1997, 2001), influencing the emotional state of the dog and creating an innate empathic response.

Both dog owners and non-dog owners generally agree that dogs are empathically sensitive to human emotions (Vitulli, 2006), and there is some empirical evidence to suggest that dogs may recognise the valence of human emotions. In a preferential choice paradigm in which only one box out of two contained a reward, dogs were more likely to choose a box to which they had seen a person respond with a happy reaction than one paired with a disgusted reaction, demonstrating their ability to use human emotional expressions to guide their choice (Buttlemann & Tomasello, 2012). Interestingly, dogs are better at making such decisions based on emotional expressions of familiar rather than unfamiliar people (Merola et al., 2013). This study also found that dogs' responses are more strongly guided by happy expressions than by avoiding boxes paired with neutral or fearful expressions. This suggests that dogs may learn to associate positive expressions in familiar people with positive outcomes. Finally, it has also been observed that dogs are more likely to approach a person in a submissive manner when that person is pretending to cry than when he/she is talking or humming (Custance & Mayer, 2012), although it is difficult to determine conclusively why dogs may respond differently specifically to crying.

Whilst it remains unclear exactly how dogs process emotion in human voices, they do perceive and appear to make some form of assessment on the basis of the vocal prosody related to human emotional expression. The ability of dogs to recognise prosodic cues in the voice is also likely to be facilitated by the way many dog owners speak to their pets. Just as it is possible that dogs' vocal repertoire has adapted to facilitate interspecific communication with humans (Pongrácz et al., 2005), so humans also change their speech patterns when talking to dogs and other pets. Pet-Directed Speech (PDS) mimics the way in which parents speak to their young children (Hirsh-Pasek & Treiman, 1982). Both Infant-Directed Speech (IDS) and PDS contain a higher mean F0 and a larger F0 range than Adult-Directed Speech (ADS). The higher mean F0 in IDS appears to be an important prosodic feature for engaging an infant's attention and thus promoting and facilitating social interaction (Fernald & Kuhl, 1987). Perhaps, then, it is not surprising that dogs seem to find higher-pitched voices more encouraging in a food-seeking task (Scheider et al., 2011), although to date it has not been determined whether PDS is more effective than ADS in engaging a dog's attention.

Exaggerated prosodic variation also enables adults to recognise emotion more easily in IDS independently of language or culture (Bryant & Barrett, 2007). Although the emotional salience in PDS is not rated as strongly as in IDS, it is significantly higher than in ADS (Burnham et al., 2002). The increased emotional expression in IDS is thought to communicate intentions to the infant (Fernald, 1989), and it may be that the exaggerated emotional expression in PDS also influences responses in dogs. There is some evidence to support this: rising tones are more effective than descending tones in obtaining responses from dogs that require increased movement (McConnell, 1990). Likewise, in IDS, rising F0 contours are associated with gaining attention and encouraging a response (Fernald, 1989). Instead of being a by-product of our parental attachment to dogs, PDS may thus actually share some of the functions of IDS, such as maintaining dogs' attention and manipulating their behavioural responses. Further work is now needed to assess how different intentions are expressed in PDS and whether dogs respond to these cues.

Aside from the possible functions of PDS, the tendency of dog owners to change their speech patterns when speaking to dogs in a way that mimics speech patterns directed towards infants should probably not be unexpected; many dog owners form strong attachment to their dogs, often viewing them as family members (Cain, 1985). In fact, there is some evidence that dogs may also have been selected to have childlike facial features because they trigger parental attachment in their owners (Archer & Monton, 2011). These influences could have an additive effect, causing many owners to attempt to communicate with their dogs as though they were human infants. In support of this, dog owners have been shown to spontaneously produce PDS when interacting with their dogs during situations designed to assess attachment, namely adaptations of Ainsworth's Strange Situation Test, which stimulate caregiving and protective responses. Although both men and women used PDS, the female owners spent significantly more time speaking to their dogs and were more likely to use PDS (Prato-Previde et al., 2006). It is interesting to note, however, that owners may be inherently sensitive to the limited linguistic abilities of their dogs: vowel hyperarticulation, a critical component of IDS thought to be specific to the teaching of language, is absent from PDS (Burnham et al., 2002).

## 5.5.4 Semantic Information

While, as we just discussed, human voices are comparable with the vocalisations of other mammals in terms of broadcasting indexical and motivational information, humans alone appear to have evolved a further dimension of vocal communication: speech (see Fitch, 2000a). By intentionally manipulating the resonance frequencies within vocal signals, humans are able to produce the phonological structure necessary to create words. These words can be used to refer to abstract concepts, giving them semantic meaning. Furthermore, humans have the ability to generate and process an infinite number of combinations between words using syntax or grammar, giving speech a rule-based recursive structure. Whilst human speech is therefore very different from canid vocal signals in this respect, in the context of domestication, dogs are likely to benefit from the ability to perceive semantic information in human vocal signals. In order to understand the extent to which dogs are capable of perceiving this information, it is important to review how human speech signals are produced and how semantic information is encoded. To create the precise sounds for different words, humans make specific movements of the articulators, including the tongue and lips, to alter the shape of the vocal tract. This causes dynamic variation in the formant frequencies, particularly those at the lowest frequencies, producing the different phonemes that are categorised as vowels (see Titze, 2000). We have seen that, because dogs have limited ability to control formant positions in their own calls, their formant positions show little variation during vocalisations and are dependent on the length of the vocal tract. We have already seen that dogs are perceptually aware of formant-related information, as they are capable of using the formant scaling in growls to judge the body size of other dogs. This leads to the interesting question of whether dogs can perceive more than the static scaling between formants, and if they are capable of perceiving the dynamic formant modulation used to create the different phonetic sounds in human speech. The first demonstration that dogs are indeed able to perceive the relative positions of individual formants involved training dogs to discriminate between individual vowels (Baru, 1975). Indeed, we now know that discriminating between vowel sounds is important for dogs to learn to identify individual words. Fukuzawa et al. (2005) demonstrated that changing the phonemes embedded in previously learnt commands can lead to a significant decline in the number of correct responses, illustrating that specific phonemes are a crucial part of word recognition for dogs. Although further research is needed to better understand the extent to which dogs perceive and categorise phonemes and how their abilities relate to our own perception, they are clearly able to use phonetic information in order to identify a wide variety individual words.

Through training, dogs can learn a very large number of vocal labels relating to specific objects, responses, and events, comparable with the word acquisition abilities of apes, parrots, dolphins, and sea lions (Miles & Harper, 1994). This was first demonstrated in dogs by Kaminski et al. (2004) with a border collie named Rico. Rico could be given a verbal instruction by his owner to retrieve a specific toy (e.g., 'fetch teddy') from a different room containing an array of 10 familiar toys, and successfully responded to the verbal labels of 200 different toys. Rico also showed evidence of learning new object labels on their first presentation, as when asked to fetch a new toy included amongst an array of 7 familiar toys, Rico successfully excluded the familiar toys and retrieved the new toy. He also showed retention of the new label, as after a month without testing the new object was placed with 4 familiar and 4 unfamiliar toys, and Rico still successfully retrieved the new object. Because the acquisition of the new label was so rapid, it provided evidence that rather than just learning to associate the word with the object (which often takes many repetitions), dogs could be capable of 'fast mapping', showing the ability to learn that a new word may be used to refer to something new in the environment during a single exposure, a skill that was previously thought to be limited to humans.

However, Bloom (2004) questioned whether Rico understood that words were referential in the same way as humans, as the verbal command could have been represented as a single proposition (e.g., 'fetch the sock'). If this was the case, Rico may have merely associated the command with a specific behavioural response (retrieving the object) without understanding that the label for the object was independent from the action of retrieving it. To test this, Pilley and Reid (2011) taught a border collie named Chaser the labels of more than 1,000 objects and associated these objects with different behavioural commands. Chaser responded to novel combinations of behavioural responses and objects (e.g., 'take ball' versus 'paw ball') from the first trial, illustrating that she was able to differentiate the object label from the action directed towards it. This ability was confirmed by Ramos and Ades (2012) using a crossbreed dog named Sophia, who also learnt to respond appropriately to two-item spoken instructions including both object- and action-related words (e.g., 'stick fetch' versus 'stick point'), even when the word order was reversed. Comprehension of two-item sequences has also been shown in other language-trained species [e.g., African grey parrot (*Psittacus erithacus*): Pepperberg, 1981; California sea lion (*Zalophus californianus*): Gisiner & Schusterman, 1992)], and these sequences also appear to be used naturally in some primate signals [e.g., Diana monkeys (*Cercopithecus diana*) perceive combinatory calls produced by Campbell's monkeys (*Cercopithecus campbelli*): Zuberbuhler, 2002].

Whilst these studies demonstrated that dogs, amongst some other mammal species, can learn to respond to combinatory signals, and therefore show understanding that the label is associated with the object and not the behavioural response, Bloom (2004) also argued that dogs may not recognise that words are symbolic, referring to categories and items in the environment. Although both Rico and Chaser had shown evidence of 'fast mapping', which suggested some understanding of words as verbal referents, potential problems with this inference were identified from the experimental methodology used. In the initial exclusion test, Markman and Abelev (2004) suggested that both Rico and Chaser could have shown a base-line novelty preference for the new object, and retrieved it without learning the label used to refer to it. Obtaining a reward for choosing the new object may have influenced the subsequent retention test, as when presented with the new object plus four familiar and four unfamiliar objects, the dogs may have fetched the new object again as it had been previously rewarded, whilst the labels of the four familiar objects were already known and the four unfamiliar objects had never been rewarded. Griebel and Oller (2012) investigated this possibility using a Yorkshire terrier named Bailey. After successfully showing the ability to learn a range of word labels, Bailey was given the same exclusion and retention tests as used in the previous studies, although in this case using two novel objects, and was similarly successful. She was then given an additional two-choice identification test, where both novel objects were presented together with no other objects. Bailey failed to retrieve the correct object in this task, demonstrating that she had not learnt the labels that referred to the new objects. This suggests that her success in the retention trial was through an extended form of exclusion learning, rather than showing evidence of 'fast mapping'. Therefore, it remains unclear if dogs can understand that words are symbolically referential, rather than learning that words match specific objects through associative learning. To determine more conclusively whether dogs perceive words as representational, researchers could employ a testing paradigm used to show this ability in bottlenosed dolphins (*Tursiops truncatus*) by establishing that they could understand reference to an absent object. Herman and Forestell (1985) trained a captive dolphin named Ake to successfully give 'yes' or 'no' responses by pressing different paddles when asked if specific objects were present in her tank. If dogs also understand that labels can refer to absent objects, this would demonstrate that they perceive words as abstract symbols for objects, rather than purely associating the word and object together.

As well as dogs' potentially using different cognitive mechanisms to humans, it is interesting to note that the associations dogs form to match word labels to the correct objects may be based on different object features to those one might expect. Humans tend to generalise a word that refers to one object to new objects based on their similarity in shape. In contrast, dogs may focus on different attributes, such as the size and texture of objects. This was demonstrated by another border collie (Gable), who had previous training in learning object labels. After initial training to fetch a novel object using an arbitrary word label (e.g., 'dax'), Gable was asked to fetch the 'dax' object and was presented a choice of two new objects, both of which differed from the original in size, shape, or texture. In similar experiments, humans tend to choose the new object with the same shape as the original (Landau et al., 1988), whereas Gable initially chose objects of the same size and with further familiarisation chose objects that had the same texture as the original object (Van der Zee et al., 2012). This could, of course, be linked to the fact that dogs frequently manipulate objects in their mouths, but also brings to light intriguing possibilities of potential differences to humans in the way that dogs may perceive object features

Finally, as well as forming concepts based on individual words or simple combinations, humans meaningfully arrange words using syntactic or grammatical rules, giving a limitless generative capacity to express and combine different concepts. By exploring how dogs respond to grammatical rules, researchers gain insights into the extent to which dogs share these cognitive abilities. To explore this, Pilley (2013) recently adapted a testing paradigm originally developed to demonstrate that bottlenosed dolphins could successfully respond to different relational combinations of known words. In his study, Pilley (2013) successfully taught Chaser to respond to sentences including three elements of grammar: a prepositional object, a verb, and a direct object (e.g., 'to ball take frizbee'). In itself, this ability shows an impressive capacity in the working memory, but even more interestingly, Chaser responded correctly when the object labels were reversed, illustrating her awareness of word order and simple grammatical rules. As well as sharing this ability with bottlenosed dolphins, non-human primates have also been shown to process rule-governed sequences in acoustic signals [bonobo (*Pan paniscus*): Savage-Rumbaugh et al., 1993; cotton-top tamarins (*Saguinus oedipus*): Fitch & Hauser, 2004]. However, although dogs show comparable abilities to other language-trained species in spontaneously learning simple grammatical rules, known as finite state grammars, it is not clear whether they are able to process more complicated phrase structures. For example, although cotton-top tamarins also respond correctly to finite state grammars, they appear unable to spontaneously process phrase structures where related elements are distantly placed (e.g., 'if' followed by 'then') (Fitch & Hauser, 2004). It has yet to be established if dogs are able to learn these higher-order phrase-structure grammars.

Thus, despite being relatively constrained in their own vocal production, dogs have shown the ability to adapt their perceptual abilities in order to discriminate the phonetic elements of human speech. Whilst dogs show comparable comprehension skills to other language-trained animal species, we still have much to learn about the cognitive mechanisms underlying their speech perception, particularly how speech-related information is represented by dogs. Further to this, another interesting avenue to be explored is the extent to which dogs partition linguistic information from prosodic cues, and if this mirrors the way in which humans separate this information. Overall, it certainly seems that as well as responding exceptionally well to human spoken commands via label learning, dogs can also extract information about the speaker, allowing them to identify familiar individuals and discriminate some of their physical attributes. Dogs may furthermore recognise some level of emotion in the human voice, facilitated perhaps by dog owners' automatic use of exaggerated prosody. Much more research is needed to determine exactly how dogs learn to respond to cues in the human voice, particularly the extent to which interspecific communication can be explained by similarities across our species' vocal signals.

## 5.6 CONCLUSIONS

Throughout this chapter, we have seen that at a superficial level domestic dogs appear to possess a relatively limited vocal repertoire with fewer functional distinctions between call types than their wild ancestors. This initially appeared to indicate functional atrophy as a result of relaxed natural selection pressures during domestication. However, on closer examination of the acoustic structure of dog vocalisations, it is evident that their vocal repertoire has the ability to broadcast a range of socially relevant information about the signaller, including his/her body size, motivational state, and almost certainly some measure of individual identity. Indeed, rather than indicating a functional decline, some differences in vocal production, as we have seen, appear to have become adapted to facilitate communication with humans.

Despite ongoing research, there remain many aspects of dog vocal communication that we do not yet fully understand, from the broadcasting of dynamic attributes such as motivational state within individual call types to the encoding of static attributes such as individual identity across different calls. It is also important to note that many studies to date have focussed on a single breed (e.g., Hungarian mudi; Pongrácz et al., 2005, 2006; Marós et al., 2008) or a very small sample resulting in a limited number of breeds (e.g., ten dogs; Yin, 2002; Yin & McCowan, 2004). Because the range of morphological and behavioural variation between different dog breeds is so great, this might have led to biased representations of the occurrence and communicative content of dog vocalisations. Using a wide range of different breeds in research can ensure that such potential differences are controlled for, thus promoting greater generalisability across domestic dogs as a species.

Although breed differences are undoubtedly present, at a species level, it is clear that acoustic communication plays a major role during social interactions, allowing domestic dogs to transmit a broad range of important social information. Recent experiments have greatly advanced not only our understanding of the physiological, perceptual, and cognitive processes underlying the dog's vocal communication system, but also of how some of these mechanisms may have adapted to facilitate interaction with humans during the process of domestication. Further research remains crucial to provide deeper insights into the full communicative potential of dog vocalisations during social interactions, both within and across species.

### Future Directions

- Are motivational state and emotional valence encoded within call types?
- Do vocal signatures across call types in dogs support individual recognition?
- What is the function of the full range of acoustically diverse call types within and between breeds?
- Do dogs perceive and make use of emotional and motivational cues in human voices?
- How is acoustic information integrated with information from other modalities?
- Are dog vocal production and auditory perception adapted to facilitate communication with humans?

# REFERENCES

Adachi, I., Fujita, K., 2007. Cross-modal representation of human caretakers in squirrel monkeys. Behav. Processes 74 (1), 27–32.

Adachi, I., Kuwahata, H., Fujita, K., 2007. Dogs recall their owner's face upon hearing their owner's voice. Anim. Cogn. 10 (1), 17–21.

Archer, J., Monton, S., 2011. Preferences for infant facial features in pet dogs and cats. Ethology 117 (3), 217–226.

Ashdown, R.R., Lea, T., 1979. The larynx of the basenji dog. J. Small Anim. Pract. 20 (11), 675–679.

Aubergé, V., Cathiard, M., 2003. Can we hear the prosody of smile? Speech Comm. 40, 87–97.

August, P.V., Anderson, J.G., 1987. Mammal sounds and motivation-structural rules: a test of the hypothesis. J. Mammal., 1–9.

Bachorowski, J.A., Owren, M.J., 1999. Acoustic correlates of talker sex and individual talker identity are present in a short vowel segment produced in running speech. J. Acoustical Soc. Am. 106, 1054.

Bachorowski, J., Owren, M.J., 2008. Vocal expressions of emotion. In: Lewis, M., Haviland-Jones, J.M., Feldman, L. (Eds.), Handbook of emotions, 3rd ed. Guilford Press, New York, pp. 196–210.

Bahrick, L.E., Hernandez-Reif, M., Flom, R., 2005. The development of infant learning about specific face-voice relations. Develop. Psychol. 41 (3), 541.

Banse, R., Scherer, K.R., 1996. Acoustic profiles in vocal emotion expression. J. Pers. Soc. Psychol. 70 (3), 614.

Baru, A.V., 1975. Discrimination of synthesized vowels [a] and [i] with varying parameters (fundamental frequency, intensity, duration and number of formants) in dog. Aud. Anal. Percept. Speech, 91–101.

Bekoff, M., 1974. Social play and play-soliciting by infant canids. Am. Zoologist 14 (1), 323–340.

Bekoff, M., Diamond, J., Mitton, J.B., 1981. Life-history patterns and sociality in canids: body size, reproduction, and behavior. Oecologia 50 (3), 386–390.

Bloom, P., 2004. Can a dog learn a word? Science 304 (5677), 1605–1606.

Bradshaw, J., Nott, H., 1995. Social and communication behaviour of the dog. In: Serpell, J. (Ed.), The domestic dog: its evolution, behaviour, and interactions with people. Cambridge University Press, Cambridge, pp. 116–130.

Brady, C.A., 1981. The vocal repertoires of the bush dog (Speothos venaticus), crab-eating fox (Cerdocyon thous), and maned wolf (Chrysocyon brachyurus). Anim. Behav. 29 (3), 649–669.

Bryant, G.A., Barrett, H.C., 2007. Recognizing intentions in infant-directed speech evidence for universals. Psychol. Sci. 18 (8), 746–751.

Burnham, D., Kitamura, C., Vollmer-Conna, U., 2002. What's new, pussycat? On talking to babies and animals. Science 296 (5572), 1435.

Buttelmann, D., Tomasello, M., 2012. Can domestic dogs (Canis familiaris) use referential emotional expressions to locate hidden food? Anim. Cogn. 16 (1), 137–145.

Cain, A.O., 1985. Pets as family members. Marriage Fam. Rev. 8 (3–4), 5–10.

Chuenwattanapranithi, S., Xu, Y., Thipakorn, B., 2008. Encoding emotions in speech with the size code—a perceptual investigation. Phonetica 65, 210–230.

Cohen, J.A., Fox, M.W., 1976. Vocalizations in wild canids and possible effects of domestication. Behav. Processes 1 (1), 77–92.

Coleman, R.O., 1976. A comparison of the contributions of two voice quality characteristics to the perception of maleness and femaleness in the voice. Lang. Hear. Res. 14, 2–3.

Coppinger, R., Feinstein, M., 1991. Hark-hark, the dogs do bark and bark and bark—whether barking is an adaptive trait and has a purpose. Smithsonian 21 (10), 119.

Corbett, L.K., 1995. The dingo in Australia and Asia. Comstock/Cornell University Press, Ithaca, NY.

Coren, S., 2005. How to speak dog: the age of dog–human communication. Pocket Books, Simon and Schuster, London.

Custance, D., Mayer, J., 2012. Empathic-like responding by domestic dogs (*Canis familiaris*) to distress in humans: an exploratory study. Anim. Cogn. 15 (5), 851–859.

Deaux, E.C., Clarke, J.A., 2013. Dingo (*Canis lupus dingo*) acoustic repertoire: form and contexts. Behaviour 150 (1), 75–101.

Düpjan, S., Schön, P.C., Puppe, B., Tuchscherer, A., Manteuffel, G., 2008. Differential vocal responses to physical and mental stressors in domestic pigs (*Sus scrofa*). Appl. Anim. Behav. Sci. 114 (1), 105–115.

Estes, R.D., 1991. The behavior guide to African mammals: including hoofed mammals, carnivores, primates. University of California Press, Berkeley and Oxford.

Fant, G., 1970. Acoustic theory of speech production (No. 2). Walter de Gruyter, New York.

Faragó, T., Pongrácz, P., Miklósi, Á., Huber, L., Virányi, Z., Range, F., 2010. Dogs' expectation about signalers' body size by virtue of their growls. PLoS One 5 (12), e15175.

Fedderden-Petersen, D., 2004. Communication in wolves and dogs. In: Bekoff, M. (Ed.), Encyclopedia of animal behaviour, vol. 1. Garland SMTP Press, New York, pp. 385–394.

Feddersen-Petersen, D.U., 2000. Vocalization of European wolves (*Canis lupus lupus L.*) and various dog breeds (*Canis lupus f. fam.*). Archiv fur Tierzucht 43 (4), 387–398.

Fernald, A., 1989. Intonation and communicative intent in mothers' speech to infants: is the melody the message? Child Dev. 60 (6), 1497–1510.

Fernald, A., Kuhl, P., 1987. Acoustic determinants of infant directed preference for motherese speech. Infant Behav. Dev. 10 (3), 279–293.

Fitch, W.T., 1997. Vocal tract length and formant frequency dispersion correlate with body size in rhesus macaques. J. Acoustical Soc. Am. 102, 1213.

Fitch, W.T., 2000a. The evolution of speech: a comparative review. Trends Cogn. Sci. 4 (7), 258–267.

Fitch, W.T., 2000b. The phonetic potential of nonhuman vocal tracts: comparative cineradiographic observations of vocalizing animals. Phonetica 57 (2–4), 205–218.

Fitch, W.T., 2000c. Skull dimensions in relation to body size in nonhuman mammals: the causal bases for acoustic allometry. Zoology-Anal. Complex Syst. 103 (1–2), 40–58.

Fitch, W.T., Giedd, J., 1999. Morphology and development of the human vocal tract: a study using magnetic resonance imaging. J. Acoustical Soc. Am. 106 (3), 1511–1522.

Fitch, W.T., Hauser, M.D., 2004. Computational constraints on syntactic processing in nonhuman primate. Science 303 (5656), 377–380.

Fitch, W.T., Reby, D., 2001. The descended larynx is not uniquely human. Proc. Royal Soc. London. Series B: Biol. Sci. 268 (1477), 1669–1675.

Fox, M.W., 1984. Behaviour of wolves, dogs, and related canids. Krieger Publishing Co, Malabar, FL.

Frank, H., Frank, M.G., 1982. On the effects of domestication on canine social development and behavior. Appl. Anim. Ethol. 8 (6), 507–525.

Frynta, D., Baudyšová, J., Hradcová, P., Faltusová, K., Kratochvíl, L., 2012. Allometry of sexual size dimorphism in domestic dog. PloS One 7 (9), e46125.

Fukuzawa, M., Mills, D.S., Cooper, J.J., 2005. The effect of human command phonetic characteristics on auditory cognition in dogs (*Canis familiaris*). J. Comp. Psychol. 119 (1), 117.

Ghazanfar, A.A., Turesson, H.K., Maier, J.X., van Dinther, R., Patterson, R.D., Logothetis, N.K., 2007. Vocal-tract resonances as indexical cues in rhesus monkeys. Curr. Biol. 17 (5), 425–430.

Gisiner, R., Schusterman, R.J., 1992. Sequence, syntax, and semantics: responses of a language-trained sea lion (*Zalophus californianus*) to novel sign combinations. J. Comp. Psychol. 106 (1), 78–91.

Gittleman, J.L., 1989. Carnivore behavior, ecology and evolution. Cornell University Press, New York.

Griebel, U., Oller, D.K., 2012. Vocabulary learning in a Yorkshire terrier: slow mapping of spoken words. PloS One 7 (2), e30182.

Hare, B., Tomasello, M., 1999. Domestic dogs (*Canis familiaris*) use human and conspecific cues to locate hidden food. Comp. Psychol. 113 (2), 173–177.

Harrington, F., Mech, L., 1978. Wolf vocalization. In: Hall, R.L., Sharp, H.S. (Eds.), Wolf and man: evolution in parallel. Academic Press, New York, pp. 109–132.

Hauser, M.D., 2000. The sound and the fury: primate vocalizations as reflections of emotion and thought. In: Wallin, N.L., Merker, B., Brown, S. (Eds.), The origins of music. The MIT Press, Cambridge, MA, pp. 77–102.

Herman, L.M., Forestell, P.H., 1985. Reporting presence and absence of named objects by a language-trained dolphin. Neurosci. Behav. Rev. 9 (4), 667–681.

Herrel, A., Gibb, A.C., 2006. Ontogeny of performance in vertebrates. Physiol. Biochem. Zool. 79 (1), 1–6.

Hillis, T.L., Mallory, F.F., 1996. Sexual dimorphism in wolves (*Canis lupus*) of the Keewatin District, Northwest Territories, Canada. Can. J. Zool. 74 (4), 721–725.

Hirsh-Pasek, K., Treiman, R., 1982. Doggerel: motherese in a new context. J. Child Lang. 9 (1), 229–237.

Jaeger, M.M., Pandit, R.K., Haque, E., 1996. Seasonal differences in territorial behavior by golden jackals in Bangladesh: howling versus confrontation. J. Mammal, 768–775.

Joslin, P.W.B., 1967. Movements and home sites of timber wolves in Algonquin Park. Am. Zoologist 7 (2), 279–288.

Kaminski, J., Call, J., Fischer, J., 2004. Word learning in a domestic dog: evidence for 'fast mapping'. Science 304 (5677), 1682–1683.

Koler-Matznick, J., Brisbin, I.L., Feinstein, M., Bulmer, S., 2003. An updated description of the New Guinea singing dog (*Canis hallstromi*). J. Zool. 261 (2), 109–118.

Landau, B., Smith, L.B., Jones, S.S., 1988. The importance of shape in early lexical learning. Cogn. Dev. 3, 299–321.

Leaver, S.D.A., Reimchen, T.E., 2008. Behavioural responses of *Canis familiaris* to different tail lengths of a remotely-controlled life-size dog replica. Behaviour 145 (3), 377–390.

Lieberman, P., Blumstein, S., 1988. Speech physiology, speech perception, and acoustic phonetics. Cambridge University Press, Cambridge.

MacNulty, D.R., Smith, D.W., Mech, L.D., Eberly, L.E., 2009. Body size and predatory performance in wolves: is bigger better? J. Anim. Ecol. 78 (3), 532–539.

Markman, E.M., Abelev, M., 2004. Word learning in dogs? Trends Cogn. Sci. 8 (11), 479–481.

Marós, K., Pongrácz, P., Bárdos, G., Molnár, C., Faragó, T., Miklósi, A., 2008. Dogs can discriminate barks from different situations. Appl. Anim. Behav. Sci. 114, 159–167.

Mazzini, F., Townsend, S.W., Virányi, Z., Range, F., 2013. Wolf howling is mediated by relationship quality rather than underlying emotional stress. Curr. Biol. 23 (17), 1677–1680.

McComb, K., Taylor, A.M., Charlton, B., Wilson, C., 2009. The cry embedded within the purr. Curr. Biol. 19, 507–508.

McConnell, P.B., 1990. Acoustic structure and receiver response in domestic dogs, *Canis familiaris*. Anim. Behav. 39 (5), 897–904.

Merola, I., Prato-Previde, E., Lazzaroni, M., Marshall-Pescini, S., 2013. Dogs' comprehension of referential emotional expressions: familiar people and familiar emotions are easier. Anim. Cogn. 2013 Aug 18. [Epub ahead of print].

Miklósi, Á., Polgárdi, R., Topál, J., Csányi, V., 1998. Use of experimenter-given cues in domestic dogs. Anim. Cogn. 1, 113–121.

Miles, H.L., Harper, S.E., 1994. 'Ape language' studies and human language origins. Hominid Cult. in Primate Perspect., 253–278.

Mills, D.S., Fukuzawa, M., Cooper, J.J., 2005. The effect of emotional content of verbal commands on the response of dogs. In: Current issues and research in veterinary behavioural medicine. Purdue University Press, West Lafayette, IN (pp. 207–220). Papers presented at the 5th International Veterinary Behavior Meeting.

Moehlman, P.D., 1983. Socioecology of silverbacked and golden jackals (*Canis mesomelas* and *Canis aureus*). Adv. Study Mammal. Behav. 7, 423–453.

Molnár, C., Kaplan, F., Roy, P., Pachet, F., Pongrácz, P., Dóka, A., Miklósi, Á., 2008. Classification of dog barks: a machine learning approach. Anim. Cogn. 11, 389–400.

Molnár, C., Pongrácz, P., Dóka, A., Miklósi, A., 2006. Can humans discriminate between dogs on the base of the acoustic parameters of barks? Behav. Proces. 73, 76–83.

Molnár, C., Pongrácz, P., Faragó, T., Dóka, A., Miklósi, Á., 2009. Dogs discriminate between barks: the effect of context and identity of the caller. Behav. Processes 82, 198–201.

Morton, E.S., 1977. On the occurrence and significance of motivation-structural rules in some bird and mammal sounds. Am. Nat., 855–869.

Ohala, J., 1984. An ethological perspective on common cross-language utilization of F0 of voice. Phonetica 41, 1–16.

Ohala, J., 1996. Ethological theory and the expression of emotion in the voice. In: Proc. ICSLP 96, [4th International Conference on Spoken Language Processing, Philadelphia], 3–6 October, 1996, Vol. 3, University of Delaware, Wilmington, DE, pp. 1812–1815. Accessed online 12 June, 2008. http://www.linguistics.berkeley.edu/phonlab/users/ohala/papers/emotion_in_the_voice.pdf.

Owings, D., Morton, E., 1998. Animal vocal communication: a new approach. Cambridge University Press, Cambridge.

Owren, M.J., Rendall, D., 1997. An affect-conditioning model of nonhuman primate vocal signalling. Perspect. Ethol. 12, 299–346.

Owren, M.J., Rendall, D., 2001. Sound on the rebound: bringing form and function back to the forefront in understanding nonhuman primate vocal signalling. Evolutionary Anthropol.: Issues, News, Rev. 10 (2), 58–71.

Pepperberg, I.M., 1981. Functional vocalizations by an African grey parrot (*Psittacus erithacus*). Ethology 55 (2), 139–160.

Peters, R.H., 1986. The ecological implications of body size. Cambridge University Press, Cambridge.

Piérard, J.A.M., 1963. Comparative anatomy of the carnivore larynx: with special reference to the cartilages and muscles of the larynx in the dog. Cornell University, New York.

Pilley, J.W., 2013. Border collie comprehends sentences containing prepositional object, verb, and direct object. Learn. Motiv. 44 (4), 229–240.

Pilley, J.W., Reid, A.K., 2011. Border collie comprehends object names as verbal referents. Behav. Processes 86 (2), 184–195.

Plotsky, K., Rendall, D., Riede, T., Chase, K., 2013. Radiographic analysis of vocal tract length and its relation to overall body size in two canid species. J. Zool. 291 (1), 76–86.

Pongrácz, P., Molnár, C., Miklósi, Á., 2006. Acoustic parameters of dog barks carry emotional information for humans. Appl. Anim. Behav. Sci. 100 (3), 228–240.

Pongrácz, P., Molnár, C., Miklósi, Á., Csányi, V., 2005. Human listeners are able to classify dog (Canis familiaris) barks recorded in different situations. J. Comp. Psychol. 119 (2), 136.

Prato-Previde, E., Fallani, G., Valsecchi, P., 2006. Gender differences in owners interacting with pet dogs: an observational study. Ethology 112 (1), 64–73.

Price, E.O., 1999. Behavioral development in animals undergoing domestication. Appl. Anim. Behav. Sci. 65 (3), 245–271.

Proops, L., McComb, K., Reby, D., 2009. Cross-modal individual recognition in domestic horses (Equus caballus). Proc. Natl. Acad. Sci. 106 (3), 947–951.

Puts, D.A., Gaulin, S.J., Verdolini, K., 2006. Dominance and the evolution of sexual dimorphism in human voice pitch. Evolution Hum. Behav. 27 (4), 283–296.

Puts, D.A., Hodges, C.R., Cárdenas, R.A., Gaulin, S.J., 2007. Men's voices as dominance signals: vocal fundamental and formant frequencies influence dominance attributions among men. Evolution Hum. Behav. 28 (5), 340–344.

Ramos, D., Ades, C., 2012. Two-item sentence comprehension by a dog (Canis familiaris). Plos One 7 (2), e29689.

Ratcliffe, V.F., McComb, K., & Reby, D. (submitted). Effect of experience on domestic dogs' (Canis familiaris) ability to spontaneously associate the gender of unfamiliar humans across modalities.

Reby, D., McComb, K., 2003. Anatomical constraints generate honesty: acoustic cues to age and weight in the roars of red deer stags. Anim. Behav. 65, 519–530.

Rendall, D., 2003. The affective basis of referential grunt variants in baboons. J. Acoustical Soc. Am. 113, 3390–3402.

Rendall, D., Kollias, S., Ney, C., Lloyd, P., 2005. Pitch (F0) and formant profiles of human vowels and vowel-like baboon grunts: the role of vocalizer body size and voice-acoustic allometry. J. Acoustical Soc. Am. 117, 944–955.

Riede, T., Fitch, T., 1999. Vocal tract length and acoustics of vocalisation in the domestic dog (Canis familiaris). J. Exp. Biol. 202, 2859–2867.

Robbins, R.L., 2000. Vocal communication in free-ranging African wild dogs (Lycaon pictus). Behaviour 137 (10), 1271–1298.

Rutter, R., Pimlott, D., 1968. The world of the wolf. Lippincott, Philadelphia, PA.

Ryalls, J.H., Lieberman, P., 1982. Fundamental-frequency and vowel perception. J. Acoustical Soc. Am. 72, 1631–1634.

Sauter, D.A., Eisner, F., Ekman, P., Scott, S.K., 2009. Cross-cultural recognition of basic emotions through nonverbal emotional vocalizations. Proc. Natl Acad. Sci. 107 (6), 2408–2412.

Savage-Rumbaugh, E.S., Murphy, J., Sevcik, R.A., Brakke, K.E., Williams, S.L., Rumbaugh, D.M., Bates, E., 1993. Language comprehension in ape and child. Monogr. Soc. Res. Child Dev. i-252.

Schassburger, R.M., 1993. Vocal communication in the timber wolf, Canis lupus: structure, motivation, and ontogeny. Paul Parey Scientific Publishers, Berlin.

Scheider, L., Grassmann, S., Kaminski, J., Tomasello, M., 2011. Domestic dogs use contextual information and tone of voice when following a human pointing gesture. Plos One 6 (7), e21676.

Schmidt-Nielsen, K., 1975. Scaling in biology: the consequences of size. J. Exp. Zool. 194 (1), 287–307.

Shamir, M.H., Leisner, S., Klement, E., Gonen, E., Johnston, D.E., 2002. Dog bite wounds in dogs and cats: a retrospective study of 196 cases. J. Vet. Med. Series A. 49 (2), 107–112.

Sillero-Zubiri, C., Gottelli, D., 1994. Canis simensis. Mammalian Species 485, 1–6.

Sillero-Zubiri, C., Hoffmann, M., Macdonald, D.D.W. (Eds.), 2004. Canids: foxes, wolves, jackals and dogs: status survey and conservation action plan, vol. 62. IUCN, Oxford.

Smith, D., Patterson, R., 2005. The interaction of glottal pulse rate and vocal tract length in judgement of speaker's size, sex and age. J. Acoustical Soc. Am. 118, 3177–3186.

Taylor, A.M., Reby, D., 2010. The contribution of the source–filter theory to mammal vocal communication. J. Zool. 280 (3), 221–236.

Taylor, A.M., Reby, D., McComb, K., 2008. Human listeners attend to size information in domestic dog growls. J. Acoustical Soc. Am. 123, 2903.

Taylor, A.M., Reby, D., McComb, K., 2009. Context-related variation in the vocal growling behaviour of the domestic dog (Canis familiaris). Ethology 115 (10), 905–915.

Taylor, A.M., Reby, D., McComb, K., 2010a. Size communication in domestic dog (Canis familiaris) growls. Anim. Behav. 79, 205–210.

Taylor, A.M., Reby, D., McComb, K., 2010b. Why do large dogs sound more aggressive to human listeners: acoustic bases of motivational misattributions. Ethology 116, 1155–1162.

Taylor, A.M., Reby, D., McComb, K., 2011. Cross modal perception of body size in domestic dogs (Canis familiaris). PLoS One 6 (2), e17069.

Theberge, J.A., Falls, J.B., 1967. Howling as a means of communication in timber wolves. Am. Zoologist 7 (2), 331–338.

Titze, I.R., 1994. Principles of vocal production. Prentice Hall, Englewood Cliffs, NJ.

Titze, I.R., 2000. Principles of voice production. National Centre for Voice and Speech, Iowa City, IA.

Van der Zee, E., Zulch, H., Mills, D., 2012. Word generalization by a dog (Canis familiaris): is shape important? Plos One 7 (11), e49382.

Vitulli, W.F., 2006. Attitudes towards empathy in domestic dogs and cats 1, 2. Psychol. Rep. 99 (3), 981–991.

Volodin, I.A., Volodina, E.V., Isaeva, I.V., 2001. Vocal repertoire in the dhole Cuon alpinus (Carnivora, Canidae) in captivity. Entomol. Rev. 81 (2), S346.

Wayne, R.K., 1993. Molecular evolution of the dog family. Trends Genetics 9 (6), 218–224.

Yin, S., 2002. A new perspective on barking in dogs (Canis familiaris). J. Comp. Psychol 116, 189–193.

Yin, S., McCowan, B., 2004. Barking in domestic dogs: context specificity and individual identification. Anim. Behav. 68, 343–355.

Zaccaroni, M., Passilongo, D., Buccianti, A., Dessì-Fulgheri, F., Facchini, C., Gazzola, A., Apollonio, M., 2012. Group specific vocal signature in free-ranging wolf packs. Ethol. Ecol. Evol. 24 (4), 322–331.

Zuberbuhler, K., 2002. A syntactic rule in forest monkey communication. Anim. Behav. 63, 293–299.

# Chapter 6

# The Immaterial Cord: The Dog–Human Attachment Bond

Emanuela Prato Previde[1] and Paola Valsecchi[2]

[1]*Dipartimento di Fisiopatologia Medico-Chirurgica e dei Trapianti, Sezione di Neuroscienze, Università degli Studi di Milano, Segrate, Italy,* [2]*Dipartimento di Neuroscienze, Università degli Studi di Parma, Parma, Italy*

## 6.1 BEING SOCIAL, BEING BONDED, BEING A DOG

*'It is certain that associated animals [i.e., those living together in social groups] have a feeling of love for each other which is not felt by adult and non-social animals'.*

Charles Darwin, *The Descent of Man*

An increasing number of ethological and psychological studies investigating the nature and the development of social relationships in a variety of non-human species indicate that animals and, in particular, mammals form different types of social relationships with their conspecifics, which vary along different dimensions: functions, duration, emotional intensity, and exclusivity. Overall, these relationships facilitate reproduction, provide a sense of security, and reduce negative emotional states determined by stress and anxiety (Wilson, 1975; Hinde, 1976, 1982; Aureli & de Waal, 2000; Dunbar & Shulz, 2010; Massen et al., 2010). Intense and long-lasting emotional attachments in mammals are mainly restricted to two social contexts: the parent–offspring relationship and the relationship between an unrelated adult male and female. The most numerous social groups are bound together not only by parental care or reproductive bonds, but also by relationships such as friendship between adults that, despite being part of the same group, are not necessarily relatives. Primates, horses, cows, meerkats, and crows are just some of the animals in which the existence of friendship has been revealed in recent years (for a review on friendship, see Massen et al., 2010).

In general, under natural conditions the different forms of social relationships including friendship occur among conspecifics, but it often happens that individuals of different species develop a strong bond. One of the most classical examples of interspecific bond formation comes from Konrad Lorenz (1952) in studies on imprinting in wild geese. In his best-selling *King Solomon's Ring,*

*The Social Dog.* http://dx.doi.org/10.1016/B978-0-12-407818-5.00006-1

Lorenz tells the story of Martina, a gosling that, despite her beak, feathers, and webbed feet, chooses him as her foster mother, thereby becoming inseparable. Whenever Lorenz moved away from her for just a few minutes, Martina fell in a state of distress and despair, protesting strenuously and emitting a 'vivivi' call that Lorenz interpreted as 'I am here, where are you?' Martina, soon after hatching, learned to follow Lorenz around and seemed to consider him as her mother. This process, known as filial imprinting, involves visual and auditory stimuli from the parents that elicit a following response in the young. Thus, newly hatched birds stay close to their parents that provide them with food, care, and security.

Besides ducks and geese, among domestic animals, dogs occupy an outstanding position for their ability to develop intense and long-lasting bonds with humans. The dog is the first domesticated species, its relationship with humans has a long history (Savolainen, 2007; see Chapter 3 by Bonanni and Cafazzo in this book), and today it is one of the most popular companion animals (Hart, 1995). Its success as a domestic animal depends on a wide range of social competencies that allow it to engage in complex communicative, relational, and cooperative interactions with humans. How these abilities evolved in dogs is still a matter of debate (Hare et al., 2002; Miklósi et al., 2003; Udell & Wynne, 2008; Hare et al., 2010), but undoubtedly its social skills made it 'man's best friend' (Miklósi & Topal, 2013).

As Miklósi (2007) pointed out, the dog–human relationship and affiliative bond have been described and investigated mainly from two different perspectives: according to the first, it is a dominant-subordinate relationship derived from the model of wolf society, in which humans are, and must be, unquestioned leaders (i.e., 'lupomorph model' based on the metaphor of 'wolf in dog's clothing'). However, many researchers have called into question the appropriateness of the wolf social model, to interpret the human–dog relationship, particularly the role of the owners as individual alphas within the human–dog dyads (see McGreevy et al., 2012, for an overview of this debate, but see also Chapter 3 by Bonanni and Cafazzo in this book for updated information of intraspecific social behaviour of dogs).

According to the second point of view, the dog–human relationship is an asymmetrical one based on cognitive and behavioural similarities between dogs and children (Hare & Tomasello, 2005; Tomasello & Kaminski, 2009). In this perspective, dogs are supposed to form an infant-like bond towards humans who play the role of caregivers ('babymorph model', 'dog in infant's clothing'). In support of this view, there is evidence that pet owners, especially women, talk to their dogs using 'motherese' or baby-talk (Hirsh-Pasek & Treiman, 1982; Mitchell, 2001; Burnham et al., 2002; Prato-Previde et al., 2006) and hold and cuddle them, seeking and maintaining physical contact with them (Katcher & Beck, 1983; Serpell, 1986).

Miklósi (2007) suggested a third way to look at the human–dog relationship: friendship, which is a common, perhaps universal, feature of human societies, implying an intimate, supportive, egalitarian relationship. However, this concept is difficult to apply to animals. Thus, primatologists use the term 'friendship' in a more general way; for Tomasello and Call (1997), friendship is a synonym for close affiliative bonds. Goodall (1986) defined friendly relationships in chimpanzees as those in which 'affiliative behaviors outweigh aggressive ones, both in quantity and quality', and defined friends as dyads who have strong and enduring friendly relationships that are 'characterized by two-way affiliative, supportive interactions' (Silk, 2002). This perspective is intriguing as dogs fulfill some of the requirements for being a friend (i.e., companionship, loyalty, affection, social support, cooperation); however, the relationship with us is not egalitarian nor symmetrical, and it is strongly shaped and influenced by human needs, desires, and lifestyle (Marinelli et al., 2007).

So far the majority of experimental studies investigating the dog–human bond conformed to the 'babymorph' model, adopting the framework borrowed from the human infant literature and embedded in attachment theory. Interestingly, Bowlby's original attachment theory was inspired by work on non-human animals, and hence it was rooted in biology with a concern for the adaptive value of attachment and other related behavioural systems (Bretherton, 1992).

## 6.2 ATTACHMENT: A BRIDGE BETWEEN ETHOLOGY AND PSYCHOLOGY

*'Hier bin ich, wo bist du?'*

Konrad Lorenz, *Hier bin ich—wo bist du? Ethologie der Graugans*

The construct of attachment was used for the first time by John Bowlby (1958) to explain the nature of the bond that develops between human infants and their mother/caregiver. At that time, the two most widely accepted theories providing explanations for the child's bond to the mother were the psychoanalytic and the social learning theories (Freud, 1910/1957; Sears et al., 1957), both suggesting that the infant's bond to the mother emerged as a secondary response to the rewarding function of food. Bowlby, influenced by the ethological work on imprinting (Lorenz, 1935), on instinct in animals (Tinbergen, 1951), as well as studies on the behaviour of infant rhesus monkeys (Harlow, 1958; Harlow & Zimmerman, 1959), de-emphasised the mother's role in feeding as a basis for the development of a strong mother–child relationship. Reading these studies, he became aware of a completely new world in which outstanding scientists were investigating in non-human species the same issues he was working on in humans. Both Lorenz and Harlow showed, in very different species, that food was not necessary for an attachment bond to develop: ducklings, as precocial birds, find their own food, but they learn to follow their mother shortly after hatching, moving safely in the environment in her presence and being distressed (contact call) in her absence; infant macaques, as altricial animals, depend on their mothers' milk, but Harlow found that they have an overwhelming preference for a surrogate cloth-covered mother compared to a bare-wire surrogate mother, even when only the wire mother could provide nourishment. Infant macaques used the wire mother to obtain milk and the cloth mother as a source of comfort and security. Harlow concluded that in the mother–infant relationship, there was much more than milk, and that this 'contact comfort' was essential to the psychological development and health of infant monkeys and children.

Bowlby sought new understanding from the fields of evolutionary biology, ethology, developmental psychology, cognitive science, and control systems theory. With his co-worker Mary Ainsworth, he formulated the innovative proposition that mechanisms underlying an infant's emotional tie to the caregiver emerged as a result of evolutionary pressure (Cassidy, 1999).

Within an ethological perspective, attachment can be considered as a particular kind of affectional bond that an individual can form throughout the life span with another individual perceived as stronger and wiser (Ainsworth & Bell, 1970; Ainsworth, 1989).

According to Ainsworth (1989), affectional bonds fulfil the following criteria: (1) endure over time; (2) involve a specific individual which is not interchangeable with anyone else (the 'reference figure'); (3) are emotionally significant; (4) promote proximity- and contact-seeking behaviours towards the reference figure; and (5) are characterised by a distress reaction when involuntary separation from the reference figure occurs (Cassidy, 1999). There is, however, one additional criterion that is unique of the attachment bond: the individual seeks security and comfort in the relationship with the partner and yet 'has the ability to move off from the secure base provided by the partner, with confidence to engage in other activities' (Ainsworth, 1989, p. 711). The attachment response is activated under certain circumstances either external (danger, threat) or internal to the individual (hunger, pain, illness). It is expressed through species-specific attachment behaviours aimed at promoting and restoring proximity to, and contact with, the attachment figure, such as following, approaching, vocalising, clinging, and crying. Attachment behaviours are thought to be organised into a behavioural system (Bowlby 1969/1982) that has a complex interplay with other systems regulating the individual's interaction with the environment: exploratory and fear behavioural systems. Furthermore, the attachment behavioural system is strictly linked to the care-giving behavioural system, the latter functioning to protect the young and, ultimately, enhance one's own reproductive fitness. Situations that offspring perceive as frightening, dangerous, and stressful should activate the attachment system, while situations that parents perceive as frightening, dangerous, and stressful should activate the care-giving system (Solomon & George, 1999). This concept of an interplay between the attachment and the care-giving systems is further supported by experimental evidence showing that all mammals have emotional networks in homologous brain regions that mediate affective experience when animals are emotionally aroused (for a review, see Panksepp, 2011a, b).

Mary Ainsworth (1970) developed a standardised experimental procedure aimed at characterising attachment relationships in human infants based on Bowlby's theories: the Strange Situation. This procedure involves controlled observations of a subject's response to being placed in an unfamiliar room, introduced to an unfamiliar adult (the stranger), and subjected to three short episodes of separation from the attachment figure. This test was designed to elicit attachment and exploratory behaviour under conditions of increasing though moderate stress (Table 6-1). The Strange Situation has been widely used in the study of mother–infant attachment and has allowed a classification of different infant attachment styles: secure, avoidant, resistant, and disorganised/disoriented (Table 6-2).

Going back to the animal world, which originally inspired Bowlby's work, researchers focused their attention on the animal–human bonds: using the ethological attachment theory as a framework and the Strange Situation as a testing procedure, researchers have tested other species such as chimpanzees, cats, and dogs (Bard, 1991; Topal et al., 1998; Prato-Previde et al., 2003; Edwards et al., 2007).

**TABLE 6-1**  Description of the Original Strange Situation Procedure*

| Episode | Description |
|---|---|
| **Episode 1**: Parent & Infant (1 minute) | The dyad is introduced to the room. |
| **Episode 2**: Parent & Infant (3 minutes) | Infant settles in, explores. Parent assists only if necessary. |
| **Episode 3**: Parent, Infant, & Stranger (3 minutes) | Introduction of a stranger. Stranger plays with infant during the last minute. |
| **Episode 4**: Infant & Stranger (3 minutes) | Parent leaves infant with stranger. *First separation.* |
| **Episode 5**: Parent & Infant (3 minutes) | Parent returns; stranger leaves quietly. *First reunion.* |
| **Episode 6**: Infant (3 minutes) | Parent leaves infant alone in the room. *Second separation.* |
| **Episode 7:** Infant & Stranger (3 minutes) | Stranger enters room and stays with infant. |
| **Episode 8**: Parent & Infant (3 minutes) | Parent returns; stranger leaves quietly. *Second reunion.* |

*Based on Ainsworth et al. (1978)

**TABLE 6-2** Attachment Style Descriptions Based on the Original Works Reported in Table 6-1

| Group | Description |
|---|---|
| **Secure (B)**<br>(Ainsworth et al., 1978) | - Mother is used as a secure base for exploration.<br>- Signs of missing mother during separations and active greeting upon reunions.<br>- When distressed, seeks contact with mother; once comforted, engages in exploration and play. |
| **Avoidant (A)**<br>(Ainsworth et al., 1978) | - Little evidence of affect or secure base behaviour.<br>- Responds minimally to separation.<br>- Active avoidance of mother upon reunion; focused on toys. |
| **Ambivalent or resistant (C)**<br>(Ainsworth et al., 1978) | - Distress when entering the room; no exploration.<br>- Unsettled during separations.<br>- Ambivalent behaviour upon reunions: does not find comfort in mother. |
| **Disorganised/disoriented**<br>(Main & Solomon, 1990) | - Lack of a coherent attachment strategy: contradictory behavioural displays, incomplete movements, stereotypies, freezing, disorientation. |

## 6.3  ON THE NATURE OF THE DOG–HUMAN BOND

*'He never makes it his business to inquire whether you are in the right or wrong, never bothers as to whether you are going up or down life's ladder, never asks whether you are rich or poor, silly or wise, sinner or saint. You are his pal. That is enough for him'.*

Jerome K. Jerome, *Three Men in a Boat*

So far, the Strange Situation has been the privileged method to empirically investigate whether the close and individualised relationship between adult dogs and their human owners represents an attachment bond (e.g., Topál et al., 1998; Prato-Previde et al., 2003; Palmer & Custance, 2008). The rationale for this choice has been the hypothesis that the relationship between dogs and their owners might represent an infant-like attachment. This hypothesis is partly based on the evidence that dogs are entirely dependent on human care for welfare, protection, and affect; furthermore, they present both morphological and behavioural traits typical of young animals (neotenia), which elicit their owners' care-giving system (Askew, 1996; Archer, 1997).

All published studies based on the Strange Situation share a number of characteristics: dogs are confronted with an unfamiliar setting (the testing room) and an unfamiliar person (the stranger) entering the room; they are exposed to brief separations from the owner, who leaves the room, and periods of reunion with him/her (Figure 6-1). These elements are aimed at inducing mild stress in dogs,

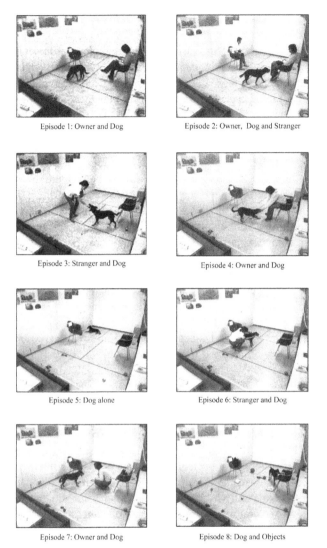

FIGURE 6-1    Strange Situation episodes from Prato Previde et al. (2003). *Reproduced with permission from Koninklijke Brill NV, Leiden, The Netherlands.*

with the general goal to assess whether they exhibit the three behavioural components characterising the attachment system: proximity seeking/maintenance, separation distress, and secure base effect.

Topal and colleagues (1998) first used an adapted version of the Strange Situation procedure to explore the relationship between adult dogs and humans. They found that the dogs explored and played more in the presence of the owner compared to the stranger, greeted their owners more enthusiastically than the stranger during reunion episodes, and stood by the door more when separated from their owner than from the stranger. Furthermore, their behavioural analysis suggested that dogs' reaction in the Strange Situation could be described in terms of three variables: response to the unfamiliar environment (Anxiety), attitude towards the stranger (Acceptance), and response to separation from the owner (Attachment). A few years later, Prato-Previde and colleagues (2003) included an additional episode at the end of the standard Ainsworth procedure, in which the owner and stranger left an article of clothing and his/her shoes in the room while the dog was left alone. This episode aimed at testing the degree to which the dogs would orient towards their owner's objects and the degree to which these objects could provide a source of comfort. As in studies on human infants, the authors examined the dogs' behavioural changes across episodes and analysed each episode separately. Results confirmed the presence of attachment behaviours: dogs searched for the owner when separated from him/her, scratching and jumping up on the door, remained oriented to the door or the owner's empty chair, and vocalised. Dogs also greeted their owner more enthusiastically and for longer durations compared to the stranger. Finally, they contacted the owner's clothing more often and longer compared to the stranger's clothing and spent more time next to the owner's chair when the owner's objects were present. A secure base effect was suggested by the fact that the dogs accepted playing with the stranger more in the presence of their owner than during his/her absence. Dogs also explored more in the presence of their owner, but exploration dropped from episode 2 onwards. Despite this detailed analysis, due to order effects inherent in the Strange Situation, clear evidence that the dog–human bond constitutes an attachment was still lacking, since the secure base effect considered by Ainsworth as the most important component of the attachment system was not clearly demonstrated. Prato-Previde and colleagues suggested that the significant decline in exploration may simply have been due to lessening curiosity, and that the decrease in social play with the stranger could possibly have been due to fatigue or reduced interest rather than the absence of the owner. Miklósi (2007) pointed out that dogs and children might differ fundamentally in their reactions to stressful situations and also in their pattern of exploratory and play behaviours, suggesting that these differences in species-specific behaviours could complicate the assessment of the secure base effect in dogs. In 2008, Palmer and Custance carried out a simple experiment in which the order of owner and stranger presence was counterbalanced. In condition A, dogs entered an unfamiliar room with their owner; a stranger entered; the owner left the dog with the stranger; the dog was left alone in the room; the owner returned; and finally the dog was left with the stranger again. In condition B, the order in which owner and stranger were present was reversed. Secure base effects were shown in that the dogs explored, remained passive (defined as sitting, standing, or lying down quietly without any specific attention to the environment), played with the stranger, and engaged in individual play more when in the presence of their owner than when left with the stranger or alone.

Overall, these studies carried out on pet dog populations support the idea that dogs form an attachment which shares many similarities at a behavioural level with that observed in human infants towards their mother and in infant chimps towards their human caregivers: dogs approach the owner during emotional distress, greet him/her after separation, and use the owner as a secure base to cope with the novel environment/situation. This infant-like attachment that bonds the dog to its owner, besides being functionally similar to the filial attachment described for humans and for other species, is peculiar because it extends beyond the period of puppyhood and persists without significant behavioural changes in aged dogs (Mongillo et al., 2013). Bowlby himself referred to the attachment as a life-span phenomenon, which is an integral part of human behaviour from 'the cradle to the grave' (Bowlby, 1979). In this sense, the dog–human attachment bond appears to be different from the human infant–mother attachment that changes, as a function of maturation, into a bond in which the primary attachment figure is substituted by the partner. A central aspect in the transition from the filial attachment to pair bond is mutuality: the asymmetrical attachment of early life, in which an infant seeks security from caregivers but does not provide security in return, becomes a reciprocal attachment in adulthood (Hazan & Zeifman, 1999). Although dogs and their owners are social partners and many owners declare that their dogs are a source of security and comfort (Kurdek, 2009), it is not surprising that dog attachment towards humans retains infant-like features since pet dogs spend all their life in a state of physical and psychological dependence on humans, who are responsible for and meet their emotional needs: whenever an adult dog has to cope with distressing or novel situations, humans represent a source of protection (Gasci et al., 2013) and information (Merola et al., 2012 a, b; Merola et al., 2013).

## 6.4 ORIGIN OF DOG–HUMAN BOND: WHAT DO WE KNOW?

*'Buy a pup and your money will buy love unflinching that cannot lie'.*
                                        Rudyard Kipling, *The Power of the Dog*

Following the pioneering work of Scott and Fuller at the Bar Harbor Laboratories on the development of behaviour in dogs (Scott & Fuller, 1965; for a review, see Serpell & Jagoe, 1995), it has been widely accepted that puppies go through four developmental periods in which they are sensitive to environmental influences and experience: the neonatal period (from birth to 12–14 days of age), the transition period (from 13 to 18–20 days of age), the socialisation period (from 20 to 56 days of age), and the juvenile period (up to 6–8 months of age). Although boundaries of the socialisation and juvenile periods are individual/breed dependent, experimental evidence supports the idea that social contacts with humans during the socialisation period (particularly between 6 and 8 weeks of age) are fundamental to develop a positive social attitude towards people (Scott & Fuller, 1965). A further step towards the understanding of the dog–human relationship is to investigate the development and the evolution of the attachment bond once puppies enter their human family. A number of questions arise: (1) At what age do puppies show attachment behaviours towards humans? (2) To what extent does intense socialisation with humans (hand-rearing) affect the formation of this bond? (3) Is the human-attachment bond unique to dogs, or do wolf pups raised in a similar way show a similar bond towards humans? (4) Is the behaviour of dogs towards their human partners derived from the offspring–parent attachment system characterising the puppy–mother relationship?

The first three questions were addressed by Tòpal and colleagues in 2005, testing in the Strange Situation Test (SST) three experimental groups of 16-week-old puppies: dog and wolf puppies extensively socialised with humans (i.e., separated from their mother 3–5 days after birth and then hand-reared) and dog puppies normally reared by their mother until 7–9 weeks of age and living with their owner when tested. Like adult dogs, puppies of both the hand-reared and the normal-reared groups as early as 16 weeks of age showed a clear preference for their owner, greeting him/her and playing with him/her more compared to the stranger, and were more oriented to the door in the absence of the owner. On the contrary, wolf puppies did not show this pattern of attachment, and the authors concluded that domestication, but not extensive socialisation, had a significant effect on the emergence of the dog–human bond.

As regards the fourth question, it is known that young dog puppies separated from their mother and littermates become distressed and emit a series of rapid sounds called whines or yelps (Scott & Marston, 1950); these vocalisations have been classified as et-epimeletic or care-soliciting behaviours, such as when a puppy is unable to adapt to a situation, it calls for help and attention. It has also been shown that whines and yelps emitted by 2-month-old isolated puppies can be significantly reduced by the presence of a littermate (Fredricson, 1952; Ross et al., 1960) and of a human being, and, to a lesser extent, by the presence of another adult female dog (Pettijohn et al., 1977). In particular, Pettijohn and colleagues (1977) suggested that a conspecific was less effective than a human in reducing distress at separation because the adult conspecific explored the environment, thus ignoring the puppy, whereas the human interacted with it.

To understand whether there is a relationship between puppy–mother attachment and the dog–human attachment, 45–55-day-old puppies were tested in the Strange Situation in the presence of the mother (Ghirardelli, 2006; Prato-Previde et al., 2009; see Box 1 for details). This study is a preliminary step on the way to understanding whether and to what extent the interspecific bond (dog-owner) stems from the intraspecific (pup-mother) one. Although puppies showed towards their mother patterns of behaviour that adult dogs direct towards their owner, strict parallels between the SST carried out with the mother and with the owner require caution for several reasons. Mothers are themselves distressed by the test, and they cannot be

## Box 1

In this study 145 puppies of 23 different breeds were tested in the Strange Situation in the presence of their mother (as the attachment figure) and of an unfamiliar woman. In episodes 1 and 2, in which, according to the classical procedure, the attachment figure was supposed to be quiet in the room, the mother was on leash, and thus could not wander around exploring the room or interact with the stranger when both were present (Figure 6-2). Factor analysis was carried out

EPISODE 1: PUP AND MOTHER

EPISODE 2: PUP, MOTHER, AND STRANGER

EPISODES 3 & 6: PUP AND STRANGER

EPISODES 4 & 7: PUP AND MOTHER

EPISODE 5: PUP ALONE

**FIGURE 6-2**    The 7 episodes of the Strange Situation carried out with puppies.

*Continued*

**Box 1—cont'd**

considering 29 puppies' behaviours and extracted 8 principal dimensions: (1) 'affiliation toward humans' (positive/playful interaction such as play, following, and greeting behaviours); (2) 'human contact seeking' (approaching, keeping contact, and licking the person); (3) 'self-confidence', including playing alone with toys in opposition to being oriented toward the mother; (4) 'distress/protest', consisting of reactions such as remaining oriented toward and scratching the door; (5) 'affiliation toward the mother' (positive/playful interaction); (6) 'mother contact seeking' (behaviours pertaining to nurturing demands: licking, following, suckling); (7) 'apprehension toward humans' (approaching the mother, avoiding and orienting away from humans); and (8) 'exploration' (visual and olfactory). Pups showed the characteristic pattern of attachment behaviour: proximity seeking toward the mother (Figure 6-3), play in the presence of the mother, and distress and protest upon separation from her. A positive attitude towards the unfamiliar human developed over the test: initially avoidant towards the woman, pups showed affiliative behaviours upon her second entrance in the room (ep. 6;

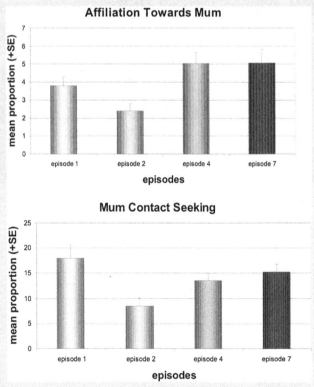

**FIGURE 6-3**   Mean proportion of sample point spent by pups in affiliative and contact-seeking behaviours toward their mothers.

**Box 1—cont'd**

lick face, greeting, contact seeking, play; Figure 6-4). Vocalizations were emitted mainly during episode 5 when puppies were left alone in the room.

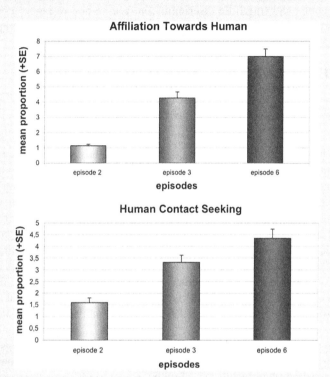

**FIGURE 6-4**   Mean proportion of sample point spent by pups in affiliative and contact-seeking behaviours toward the stranger.

instructed on how to behave; for example, mothers do not stimulate playing in pups while the owner is requested to do. So when pups engaged in play with their mothers, the latter did not necessarily reciprocate as the human stranger did. When pups approached the mothers trying to feed, they often withdrew from pups, a behaviour that owners did not show. Thus, the SST carried out in this form allowed researchers to investigate pups' behaviour towards the mother, pups' attitude towards a totally unfamiliar stranger, and, only to a lesser extent, the role of the mother as a secure base. The pups were retested after 10 months with their owner. Unfortunately, for a number of practical reasons, only 12 dogs were retested with their owners. As adults, these 12 dogs conformed to the behavioural pattern widely described in the preceding paragraph. Comparing their behaviour in the first and second SST, it emerged that, as puppies, they showed an interest towards the stranger (greeting and comfort seeking) that was redirected exclusively to their owner once adults. These results are in agreement with data showing that between 3 and 5 weeks of age pups are prone to establish social contact with unfamiliar people, showing little discrimination (see Serpell & Jagoe, 1995, for a review), and indicate a development of a clear preference towards a person within the first year of life.

## 6.5 WHEN A BOND IS NOT FOREVER

*'Quand on choisit de vivre avec un chien, c'est pour la vie. On ne l'abandonne pas. Jamais. Mettez-vous bien ça dans le cœur avant d'en adopter un'.*

Daniel Pennac, *Cabot-Caboche*

A rather dramatic life experience that can have a profound influence on the dog–human relationship is abandonment, especially if it takes place when the affectional bond is already well established and dogs are used to living within the human household. Dogs entering a rescue shelter after abandonment experience a sudden and strongly traumatic bond disruption, implying a radical change in their environmental and social conditions (novel surroundings, changes in husbandry routine, owner absence, contact with unfamiliar shelter staff; Tuber et al., 1999; Hennessy et al., 2002). Several studies indicate that dogs experience anxiety and fear almost immediately upon admittance into rescue shelters, as shown by both behavioural (Hennessy et al., 2001; Valsecchi et al., 2007) and physiological (Hennessy et al., 1997) measures in the short and long term (Stephen & Ledger, 2006).

Unfortunately, many people are still reluctant to adopt shelter dogs thinking that, as a consequence of negative experiences, they may have behavioural problems and more difficulties in reforming a new bond. These ideas were partially supported by a number of clinical and epidemiological studies reporting a high incidence of separation-related problems in dogs adopted from rescue shelters (McCrave, 1991), and suggesting that dogs which experience the loss of a primary attachment figure are more likely to develop insecure attachments towards new owners (Borchelt & Voith, 1982).

However, experimental evidence available so far suggests a more faceted scenario. Gàsci and collaborators (2001) reported that, despite bond disruption and the limited social interaction with humans that characterises life in a shelter, adult dogs maintain the capacity to form new bonds. These authors observed shelter dogs in the Strange Situation and reported that even a recent and limited social interaction with a person (being handled three times for 10 minutes) was sufficient to make this person preferred compared to a totally unfamiliar person. Similarly, it has been shown that in abandoned shelter dogs, brief positive social interactions with humans (three 10-minute periods of petting, obedience training, and habituation to a passive human carried out on consecutive days) are effective in promoting attachment behaviours (Martson et al., 2005). The rapid development of this positive attitude towards a human by shelter dogs could be mediated by a decrease in the level of stress, as demonstrated by numerous studies. For example, shelter dogs placed in a novel environment in the presence of their caretaker show lower plasma glucocorticoids and more proximity seeking compared to dogs exposed to the same situation in the presence of a familiar conspecific (Tuber et al., 1996). Recently, Shiverdecker and colleagues (2013) demonstrated that dogs receiving three forms of human interaction (exposure to a passive human, petting, or playing sessions) showed fewer behavioural signs of excitation, vocalizing, and fear-related behaviour. Furthermore, Barrera and colleagues (2010) showed that shelter dogs showed a higher frequency of physical proximity to an unfamiliar person compared to pet dogs. Overall, these observations indicate that shelter dogs, despite reduced human contact, are able and willing to rapidly initiate a novel relationship with human beings (Coppola et al., 2006). Similarly, it has been observed that orphaned children show affectionate and friendly behaviour towards adult people, including strangers, a phenomenon described as 'indiscriminate friendliness' (Tizard, 1977; Chisholm, 1998). It has been suggested that this behaviour may have an adaptive function: in an orphanage where resources are extremely limited, a very friendly child may receive more attention compared to passive children and is considered a favourite in the institution (Chisholm et al., 1995). This could be true for dogs living in rescue shelters where social contacts with humans are very limited, competition for human attention is very high, and friendliness could be a 'business card' for adoption (Wells & Hepper, 1992).

Interestingly, only one study investigated the attachment bond formed by dogs adopted from rescue shelters towards their new human partner (Prato-Previde & Valsecchi, 2007). This study investigated potential differences in attachment behaviour between adult dogs adopted from rescue shelters and pet dogs reared in the same family home from puppyhood. Adopted dogs and pets showed comparable patterns of attachment in the SST, as no differences emerged in exploration, proximity seeking, physical contact, and greeting towards their owner nor in vocal behaviour during separations. However, abandoned dogs, compared to pets, showed (1) more locomotion throughout the test and especially when left alone in the room (episode 5); (2) less play behaviour with both their owners (episodes 4 and 7) and the stranger (episodes 3 and 6) and, interestingly, when the stranger solicited play in the presence of the owner (episode 2); and (3) more visual contact with humans, especially with their owners upon reunions (episodes 4 and 7). These results also indicate that adopted dogs were more distressed and anxious compared to pet dogs: whether this difference in coping with emotional distress is caused by an anxiety state due to the negative experience of abandonment or by a less secure attachment needs further investigation.

'... y la antigua amistad,/la dicha/de ser perro y ser hombre/convertida/en un solo
anima/que camina moviendo/seis patas/y una cola/con rocío'.

Pablo Neruda, *Oda al Perro*

Guide dogs are a potentially unique example of a dog–human relationship because they establish their bond with the blind person after the formation and breakdown of two previously important affectional bonds with different attachment figures: the puppy-walker, who during the first year of life treats them exactly as a pet, and the trainer, who conversely for 6–8 months trains them at the school. The rationale for giving 2-month-old puppies to families for the first year of their lives relies on the observation that puppies that lived in a kennel for 12 weeks or more after birth, and therefore could not familiarise with different people and environments, were generally insecure even when accompanied by their attachment figure and rarely succeeded in becoming guide dogs (Pfaffenberger et al., 1976). Although giving pups in adoption became a common practice in guide dog schools in the late 1960s, until recently there were no studies assessing the potentially negative effects of this bond breakdown on the establishment of a secure attachment with the final human partner, i.e., the blind person. The only evidence was that puppies scoring very high in attachment towards their puppy-walker and his/her family failed to become guide dogs because they were judged too emotional and excitable and excessively protective and aggressive towards strangers (Serpell & Hsu, 2001). Fallani and colleagues carried out a series of studies to evaluate possible effects of multiple breakdowns of the dog–human bond, focusing their attention on guide dogs. In particular, the authors tested three groups of dogs from the National School for Guide Dogs (Scuola Nazionale Cani Guida per Ciechi, Regione Toscana) with the attachment figures they were living with during the test period: custody dogs were still living with their puppy-walker; apprentice dogs were living at the school with their trainer; guide dogs were living with their blind owner. These three groups of dogs were compared with a group of pet dogs that had lived with their owner since puppyhood.

A factor analysis carried out in the first study (Fallani et al., 2006) highlighted two different profiles of response to the testing situation: a relaxed reaction (characterised by a high play activity), which was distinctive of dogs living with the puppy-walker and under training, and an anxious reaction (characterised by a high degree of proximity-seeking behaviours) distinctive of pet dogs. Interestingly, guide dogs were intermediate between these two extremes, expressing their attachment to the owners (proximity seeking, greeting, and physical contact) but showing a more controlled emotional reaction when alone in the room (more passivity, fewer protest behaviours). Furthermore, there were differences in behaviour between golden and Labrador retrievers, the two more represented breeds in guide dogs schools (Box 2).

**Box 2**

109 dog-owner pairs were tested using the Strange Situation: dog-puppy walker (n = 34), dog-trainer (n = 26), guide dog-blind owner (n = 25), and pet dog-owner (n = 24). The sample included 42 golden retrievers, 53 Labrador retrievers, 8 golden x Labrador crosses, and 6 German shepherds. Comparisons between golden retrievers and Labrador retrievers (the two most represented breeds, Figure 6-5) revealed two different behavioural profiles: Golden retrievers showed a higher level of demands for affection while Labrador retrievers were more playful (Figure 6-6).

FIGURE 6-5    A golden retriever and a Labrador retriever tested at the Scuola Nazionale Cani Guida per Ciechi (Regione Toscana).

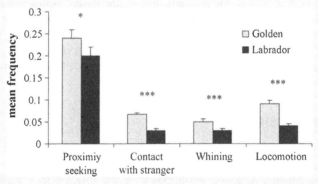

FIGURE 6-6    Mean frequency of proximity seeking, contact with stranger, whining, and locomotion expressed by golden and Labrador retrievers during Strange Situation. *P < 0.05; ***P < 0.0001 (Figure from: Fallani et al., 2006).

In a replication of the preceding studies following a longitudinal design, the same group of dogs was repeatedly tested at different stages of the training process (Valsecchi et al., 2010): the first test was carried out when the dogs were still living with their puppy-walker (11 months of age), the second test took place after 4 months of training (17 months of age), and the third one was carried out after 1 year of service as guide dogs (36 months of age). Interesting developmental changes were observed: young dogs exhibited an intense play activity and a limited discrimination of the attachment figure, because during separation from the puppy-walker, their attention was directed towards the stranger who could offer comfort rather than to the owners exiting the room. On the contrary, the same dogs tested after 1 year with the blind owner were specifically interested in regaining contact with their owner despite the presence of another friendly human (the stranger) available for support. In summary, the results of these studies indicate that guide dogs are able to form an intense bond with their blind owners, which is not compromised by repeated bond breaking.

A further aspect that emerged from the work on guide dogs is the importance of integrating physiology and behaviour to obtain a comprehensive view of dogs' attachment towards humans. In Fallani et al. (2007), besides recording behaviour of guide dogs during the SST, their cardiac activity was recorded using a Polar telemetric system, a method previously adopted in a study carried out on pet dogs' testes in the SST (Palestrini et al., 2005). As in the first study (Fallani et al., 2006), custody, apprentice, and guide dogs were tested along with a control group of pet dogs. Heart rate (HR) was recorded both in a pre-test condition, during which dogs were quietly resting next to their owner, to obtain a baseline value of heart rate for each individual, and during the whole SST procedure. Heart rate analyses revealed that in all dogs cardiac activity increased during separation from their attachment figure, as found by Sroufe and Waters (1977) in human infants. Interestingly, despite the lower baseline value of HR and a more controlled behavioural reaction, guide dogs had a strong cardiac activation comparable to that of pet dogs. Breed differences emerged between golden retrievers and Labrador retrievers, with the latter showing a higher cardiac activation across the SST despite an apparently less distressed behavioural display.

## 6.6 CONCLUSIONS

Since the pioneering work by Topal and colleagues (1998), attachment in dogs has been empirically investigated using modified versions of the original Ainsworth Strange Situation developed to assess attachment behaviour and attachment styles in young children towards their caregiver (Ainsworth & Bell, 1970). Overall, the available literature shows that dogs react to the Strange Situation carried out with their owner in a similar way as infants and hand-reared chimpanzees: this supports the existence of an infant-like attachment bond between dogs and their owners. In addition, the coherence in the behavioural patterns emerging in the test carried out on dogs with different life experiences (pets, guide dogs, shelter and abandoned dogs) demonstrates that this test is a useful tool for studying the dog–human bond.

Nevertheless, the use of the Strange Situation has also raised some methodological concerns (Prato-Previde et al., 2003; Palmer & Custance, 2008; Rehn et al., 2013). For example, it has been suggested that the standardised order of episodes compromises the possibility to isolate specific secure base effects. However, Palmer & Custance (2008), using a counterbalanced version of the test, counteracted order effects inherent in the traditional test, thereby producing more clear evidence of secure base effects. Recently, Rehn et al. (2013) showed an effect of the sequence of events in the protocol, and further stressed the importance of controlling the inherent order effects in future studies on dog–human attachment. In particular, these authors used the traditional Strange Situation procedure with a familiar person (caretaker) and an unfamiliar human and a new procedure with two unfamiliar persons to test laboratory dogs matched for breed and age, living in groups at a laboratory research facility and walked around the campus by their caretaker. Each dog was subjected to the two procedures. No differences emerged in exploration (behaviour related to the secure base effect) between the two procedures, as dogs explored as much in the presence of the first stranger entering the room as they did in the presence of the familiar person. However, dogs showed more physical contact and greeting with the familiar person than with the stranger. Based on these findings, Rehn and colleagues (2013) concluded that exploratory behaviour is not a good measure of attachment because it is affected by the order of episodes, whereas it would be profitable to focus on proximity-seeking and greeting behaviours that do not seem to be affected by the order effect. As in human infants, reunion between children and their parents is considered a key feature to look at in the assessment of the attachment bond (Ainsworth et al., 1978); a closer analysis into reunion between dog and owner could provide new insight into dog–human attachment. Finally, in recent years the secure base effect in dogs has been investigated using experimental paradigms other than the Strange Situation: the effect of the owners' presence on dogs' behaviour has been investigated, presenting a frightening stimulus (Gacsi et al., 2013) or a problem-solving task (Horn et al., 2013a). Gasci and colleagues exposed dogs to a threatening stranger either in the presence or in the absence of the owner, providing evidence of the owner's secure base effect in a situation perceived as dangerous. Horn and collaborators proposed a problem-solving task to dogs in the absence of the owner, in the presence of the owner either being silent or encouraging, or in the presence of an unfamiliar human. Dogs manipulated the food dispenser longer in the presence of the owner independently from his/her behaviour and were affected to a lesser extent by the presence of an unfamiliar human. Nevertheless, their proximity-seeking behaviours were directed mainly to the owner; thus, the authors concluded that the secure base effect was specific to the owner. These studies are interesting, as they stimulate researchers to devise new experimental and observational protocols aimed at catching new facets of the old dog–human story: the investigation of dogs' attachment styles, the influence of owner personality and of life experience on dogs' attachment, and the relationships between attachment and cognitive performance are some of the possible intriguing new directions (Box 3).

**Box 3   Future Directions**

- Attachment styles described for human infants have been so far neglected in dog research (but see Topal et al., 1998; Fallani et al., 2006). Thus, an interesting challenge for future research is to assess whether the construct of attachment styles is applicable to the dog–human bond.
- In humans, the secure base effect provided by the attachment figure is a buffer against distress, but extends to other aspects of individual life as cognitive performance. The connection between cognition and attachment in dogs deserves further attention (see Horn et al., 2013a; Marshall-Pescini et al., 2011).
- Reunion between children and their parents is considered a key feature in the assessment of the attachment bond. Following the suggestion of Rehn and colleagues (2013), a procedure focused specifically on the reunion between dog and owner could provide new insights into the dog–human bond.
- Little is known about the connection between the filial attachment of a puppy towards its mother and its subsequent attachment towards the owner (but see Prato-Previde et al., 2009), particularly the role of early experiences in modulating adult dogs' attachment bond. Does the pup's attachment to the mother and the quality of maternal care received influence future bonds with humans?
- In humans, gender differences in attachment start to emerge around 6–7 years of age (Del Giudice & Belsky, 2010). Gender differences in dog–human attachment still have to be experimentally investigated.
- There is evidence that dogs in a cognitive test attended more to those familiar humans with whom they also had a close relationship (Horn et al., 2013b). It would be interesting to investigate whether dogs form multiple attachment bonds towards members of the same family and whether these bonds have different intensity and characteristics.
- It has been suggested that a dysfunctional attachment relationship with the owner could be the underlying cause in the development of separation anxiety problems. So far, the available experimental evidence does not support this hypothesis (Parthasarathy & Crowell-Davis, 2006). Given the relevance of this topic for a successful relationship, new methodological approaches aimed at clarifying whether and how separation anxiety and attachment are related would be beneficial.

## ACKNOWLEDGEMENTS

We are grateful to the owners and their dogs that participated as volunteers in our studies. We particularly want to mention the staff of the Scuola Nazionale Cani Guida per Ciechi (Regione Toscana, Italy), who for 3 years participated in our research, providing active support, suggestions, and a warm welcome, and to the blind owners who generously agreed to be involved in our studies. Thanks to the Unione Italiana Ciechi, which provided a 2-year post doc grant to Gaia Fallani. Many thanks to the breeders who allowed us to test their puppies; in particular, a special thanks to Paola Daffunchio (Allevamento Welsea, Il Biancospino). We are grateful to Gaia Fallani and Giorgia Ghirardelli, who participated as Ph.D. students, and to all the undergraduate students who helped us in data collection. We thank Juliane Kaminski and Sarah Marshall-Pescini for inviting us to contribute to this book. Researchers were supported by grants from the Università degli Studi di Milano and the Università degli Studi di Parma. Last but not least, we thank our dogs, Tipota, Geppo, Pimpa, Laika, and the many others that shared their life with us in the past, for giving us the magical experience of a unique bond.

# REFERENCES

Ainsworth, M.D.S., 1989. Attachments beyond infancy. Am. Psychol. 44, 709–716.

Ainsworth, M.D.S., Bell, S.M., 1970. Attachment, exploration, and separation: illustrated by the behavior of one-year-olds in a strange situation. Child Devel. 41, 49–67.

Ainsworth, M.D.S., Blehar, M.C., Waters, E., Wall, S., 1978. Patterns of attachment: a psychological study of the strange situation. Erlbaum, Hillsdale.

Archer, J., 1997. Why do people love their pets? Evol. Hum. Behav. 18, 237–259.

Askew, H.R., 1996. Treatments of behaviour problems in dog and cat. A guide for the small animal veterinarian. Blackwell Science, Oxford.

Aureli, F., de Waal, F.B.M., 2000. Natural conflict resolution. University of California Press, Berkeley.

Bard, K., 1991. Distribution of attachment classifications in nursery chimpanzees. Am. J. Prim. 24, 88.

Barrera, G., Jakovcevic, A., Elgier, A.M., Mustaca, A., Bentosela, M., 2010. Responses of shelter and pet dogs to an unknown human. J. Vet. Behav. 5, 339–344.

Borchelt, P.L., Voith, V.L., 1982. Diagnosis and treatment of separation-related behaviour problems in dogs. Vet. Clin. N. Am.: Small Anim. Pract. 12, 625–636.

Bowlby, J., 1958. The nature of the child's tie to his mother. Int. J. Psycho-Anal. 39, 350–373.

Bowlby, J., 1969/1982. Attachment and loss: 1. Attachment. Basic Books, New York.

Bowlby, J., 1979. The making and breaking of affectional bonds. Tavistock, London.

Bretherton, I., 1992. The origins of attachment theory: John Bowlby and Mary Ainsworth. Devel. Psychol. 28, 759–775.

Burnham, D., Kitamura, C., Wollmer-Conna, U., 2002. What's new, pussycat? On talking to babies and animals. Science 296, 1435.

Cassidy, J., 1999. The nature of the child's ties. In: Cassidy, J., Shaver, P.R. (Eds.), Handbook of attachment: theory, research, and clinical applications. Guilford Press, New York, pp. 3–20.

Chisholm, K., 1998. A three year follow-up of attachment and indiscriminate friendliness in children adopted from Romanian orphanages. Child Develop 4, 1092–1106.

Chisholm, K., Carter, M.C., Ames, E.W., Morison, S.J., 1995. Attachment security and indiscriminately friendly behaviour in children adopted from Romanian orphanages. Develop Psychopathol. 7, 283–294.

Coppola, C.L., Grandin, T., Enns, M., 2006. Human interaction and cortisol: can human contact reduce stress for shelter dogs? Physiol. Behav. 87, 537–541.

Del Giudice, M., Belsky, J., 2010. Sex differences in attachment emerge in middle childhood: an evolutionary hypothesis. Child Develop. Persp. 4, 97–105.

Dunbar, R.I.M., Schultz, S., 2010. Bondedness and sociality. Behaviour 147, 775–803.

Edwards, C., Heiblum, M., Tejeda, A., Galindo, F., 2007. Experimental evaluation of attachment behaviors in owned cats. J. Vet. Behav. 2, 119–125.

Fallani, G., Prato-Previde, E., Valsecchi, P., 2006. Do disrupted early attachments affect the relationship between guide dogs and blind owners? App. Anim. Behav. Sci. 100, 241–257.

Fallani, G., Prato-Previde, E., Valsecchi, P., 2007. Behavioral and physiological responses of guide dogs to a situation of emotional distress. Physiol. Behav. 90, 648–655.

Fredricson, E., 1952. Perceptual homeostasis and distress vocalizations in puppies. J. Personal. 20, 427–477.

Freud, S., 1957. Five lectures on psycho-analysis. In: Strachey, J. (Ed.), The standard edition of the complete psychological works of Sigmund Freud, vol. 11. Hogarth Press, London, pp. 3–56. Original work published 1910.

Gacsi, M., Maros, K., Sernkvist, S., Farago, T., Miklósi, A., 2013. Human analogue safe haven effect of the owner: behavioural and heart rate response to stressful social stimuli in dogs. PLoS One 8 (3), e58475.

Gacsi, M., Topal, J., Miklósi, A., Doka, A., Csanyi, V., 2001. Attachment behaviour of adult dogs (Canis familiaris) living at rescue centers: Forming new bonds. J. Comp. Psychol. 115, 423–431.

Ghirardelli, G., 2006. L'attaccamento nel cane domestic (Cannis familiaris): analisi comparative del legame cucciolo-madre e cane adulto-proprietario. Ph.D. dissertation thesis. Università degli Studi di Milano, Italy.

Goodall, J., 1986. The chimpanzees of Gombe: patterns of behavior. The Belknap Press, Cambridge.

Hare, B., Brown, M., Williamson, C., Tomasello, M., 2002. The domestication of cognition in dogs. Science 298, 1634–1636.

Hare, B., Rosati, A., Kaminski, J., Brauer, J., Call, J., Tomasello, M., 2010. The domestication hypothesis for dogs' skills with human communication: a response to Udell et al., 2008 and Wynne et al., 2008. Anim. Behav. 79, e1–e6.

Hare, B., Tomasello, M., 2005. Human-like social skills in dogs? Trends Cogn. Sci. 9, 439–444.

Harlow, H.F., 1958. The nature of love. Am. Psychol. 13, 673–685.

Harlow, H.F., Zimmermann, R.R., 1959. Affectional responses in the infant monkey. Science 130, 421–432.

Hart, L.A., 1995. Dogs as human companions: a review of the relationship. In: Serpell, J. (Ed.), The domestic dog: its evolution, behavior, and interactions with people. Cambridge University Press, Cambridge.

Hazan, C., Zeifman, D., 1999. Pair bonds as attachment. In: Cassidy, J., Shaver, P.R. (Eds.), Handbook of attachment: theory, research, and clinical applications. Guilford Press, New York.

Hennessy, M.B., Davis, H.N., Williams, M.T., Mellott, C., Douglas, C.W., 1997. Plasma cortisol levels of dogs at a county animal shelter. Physiol. Behav. 62, 485–490.

Hennessy, M.B., Voith, V.L., Hawke, J.L., Young, T.L., Centrone, J., McDowell, A.L., Linden, F., Davenport, G.M., 2002. Effects of a program of human interaction and alterations in diet composition on activity of the hypothalamic-pituitary-adrenal axis in dogs housed in a public animal shelter. J. Am. Vet. Med. Ass. 221, 65–71.

Hennessy, M.B., Voith, V.L., Mazzei, S.J., Buttram, J., Miller, D.D., Linden, F., 2001. Behavior and cortisol levels of dogs in a public animal shelter, and an exploration of the ability of these measures to predict problem behavior after adoption. Appl. Anim. Behav. Sci. 73, 217–233.

Hinde, R.A., 1976. On describing relationships. J. Child Psychol. Psych. 17, 1–19.

Hinde, R.A., 1982. Attachment: some conceptual and biological issues. In: Parkes, C.M., Stevenson-Hinde, J. (Eds.), The place of attachment in human behaviour. Basic Books, New York.

Hirsh-Pasek, K., Treiman, R., 1982. Doggerel: motherese in a new context. J. Child Lang. 9, 229–237.

Horn, L., Huber, L., Range, F., 2013a. The importance of the secure base effect for domestic dogs: evidence from a manipulative problem-solving task. PLoS One 8 (5), e65296.

Horn, L., Range, F., Huber, L., 2013b. Dogs' attention towards humans depends on their relationship, not only on social familiarity. Anim. Cogn. 16, 435–443.

Katcher, A.H., Beck, A.M., 1983. New perspectives on our lives with companion animals. University of Pennsylvania Press, Philadelphia, PA.

Kurdek, L.A., 2009. Pet dogs as attachment figures for adult owners. J. Fam. Psych. 23, 439–446.

Lorenz, K.Z., 1935. Der Kumpan in der Umwelt des Vogels. In: Schiller, C.H. (Ed.), Instinctive Behavior. International University Press, New York.

Lorenz, K.Z., 1952. King Solomon's ring. Crowell, New York.

Marinelli, L., Adamelli, S., Normando, S., Bono, G., 2007. Quality of life of the pet dog: influence of owner and dog's characteristics. Appl. Anim. Behav. Sci. 108, 143–156.

Marshall-Pescini, S., Prato-Previde, E., Valsecchi, P., 2011. Are dogs (*Canis familiaris*) misled more by their owners than by strangers in a food choice task? Anim. Cog. 14, 137–142.

Martson, L.C., Bennett, P.C., Coleman, G.J., 2005. Factors affecting the formation of a canine-human bond. Int. Working Dog Breeding Assoc. Conf. Proc. P. 132.

Massen, J.J.M., Sterck, E.H.M., de Vos, H., 2010. Close social associations in animals and humans: functions and mechanisms of friendship. Behaviour 147, 1379–1412.

McCrave, E.A., 1991. Diagnostic criteria for separation anxiety in the dog. Vet. Clin. N. Am.: Small Anim. Pract. 21, 329–342.

McGreevy, P.D., Starling, M., Branson, N.J., Cobb, M.L., Calnon, D., 2012. An overview of the dog–human dyad and ethograms within it. J. Vet. Behav. 7, 103–117.

Merola, I., Prato-Previde, E., Lazzaroni, M., Marshall-Pescini, S., 2013. Dogs' comprehension of referential emotional expression: familiar people and familiar emotions are easier. Anim. Cog. http://dx.doi.org/10.1007/s10071–013–0668–1.

Merola, I., Prato-Previde, E., Marshall-Pescini, S., 2012a. Social referencing in dog-owner dyads? Anim. Cog. 15, 175–185.

Merola, I., Prato-Previde, E., Marshall-Pescini, S., 2012b. Dogs' social referencing towards owners and strangers. PLoS One 7, e47653.

Miklósi, A., 2007. Dog behaviour, evolution, and cognition. Oxford University Press, Oxford.

Miklósi, A., Kubinyi, E., Topal, J., Gasci, M., Viranyi, Z., Csanyi, V., 2003. A simple reason for a big difference: wolves do not look back at humans but dogs do. Current Biol. 13, 763–766.

Miklósi, A., Topal, J., 2013. What does it take to become 'best friends'? Evolutionary change in canine social competence. Trends Cog. Sci. in press.

Mitchell, R.W., 2001. Americans' talk to dogs: similarities and differences with talk to infants. Res. Lang. Soc. Interact. 34, 183–210.

Mongillo, P., Pitteri, E., Carnier, P., Gabai, G., Adamelli, S., Marinelli, L., 2013. Does the attachment system towards owners change in aged dogs? Physiol. Behav. 120, 64–69.

Palestrini, C., Prato-Previde, E., Spiezio, C., Verga, M., 2005. Heart rate and behavioural responses of dogs in the Ainsworth's Strange Situation: A pilot study. App. Anim. Behav. Sci. 94, 75–88.

Palmer, R., Custance, D., 2008. A counterbalanced version of Ainsworth's Strange Situation Procedure reveals secure-base effects in dog–human relationships. App. Anim. Behav. Sci. 109, 306–319.

Panksepp, J., 2011a. Cross-species affective neuroscience decoding of the primal affective experience of humans and related animals. PLoS One 6, e21236.

Panksepp, J., 2011b. The basic emotional circuits of mammalian brains: do animals have affective lives? Neurosci. Biobehav. Rev. 35, 1791–1804.

Parthasarathy, V., Crowell-Davis, S., 2006. Relationship between attachment to owners and separation anxiety in pet dogs (*Canis lupus familiaris*). J. Vet. Behav. 1, 109–120.

Pettijohn, T.F., Wont, T.W., Ebert, P.D., Scott, J.P., 1977. Alleviation of separation distress in 3 breeds of young dogs. Develop. Psychobiol. 10, 373–381.

Pfaffenberger, C.J., Scott, J.P., Fuller, J.L., Ginsburg, B.E., Bielfelt, S.W., 1976. Guide dogs for the blind: their selection, development, and training. Elsevier Scientific Publishing Company, Amsterdam.

Prato-Previde, E., Custance, D.M., Spiezio, C., Sabatini, F., 2003. Is the dog–human relationship an attachment bond? An observational study using Ainsworth's Strange Situation. Behaviour 140, 225–254.

Prato-Previde, E., Fallani, G., Valsecchi, P., 2006. Gender differences in owners interacting with pet dogs: an observational study. Ethology 111, 1–16.

Prato-Previde, E., Ghirardelli, G., Marshall-Pescini, S., Valsecchi, P., 2009. Intraspecific attachment in domestic puppies (Canis familiaris). J. Vet. Behav. 4, 89–90.

Prato-Previde, E., Valsecchi, P., 2007. Effect of abandonment on attachment behaviour of adult pet dogs. J. Vet. Behav. 2, 87–88.

Rehn, T., McGowan, R.T.S., Keeling, L.J., 2013. Evaluating the Strange Situation Procedure (SSP) to assess the bond between dogs and humans. PLoS One 8 (2), e56938.

Ross, S., Scott, J.P., Cherner, M., Dennenberg, V.H., 1960. Effects of restraint and isolation on yelping in pups. Anim. Behav. 8, 1–5.

Savolainen, P., 2007. Domestication of dogs. In: Jensen, P. (Ed.), The behavioural biology of dogs. CAB International, London.

Scott, J.P., Fuller, J.L., 1965. Genetics and the social behavior of the dog. University of Chicago Press, Chicago, IL.

Scott, J.P., Marston, M.V., 1950. Critical periods affecting the development of normal and mal-adjustive social behavior of puppies. J. Genetic Psychol. 77, 25–60.

Sears, R.R., Maccoby, E.E., Levin, H., 1957. Patterns of child rearing. Row, Peterson, Evanston, IL.

Serpell, J., 1986. In the company of animals. Blackwell, Oxford.

Serpell, J., Jagoe, J.A., 1995. Early experience and the development of behaviour. In: Serpell, J. (Ed.), The domestic dog. Its evolution, behaviour and interactions with people. Cambridge University Press, Cambridge, pp. 79–102.

Serpell, J.A., Hsu, Y., 2001. Development and validation of a novel method for evaluating behaviour and temperament in guide dogs. Appl. Anim. Behav. Sci. 72, 347–364.

Shiverdecker, M.D., Schiml, P.A., Hennessy, M.B., 2013. Human interaction moderates plasma cortisol and behavioral responses of dogs to shelter housing. Physiol. Behav. 109, 75–79.

Silk, J.B., 2002. Using the 'F'-word in primatology. Behaviour 139, 421–446.

Solomon, J., George, C., 1999. The measurement of attachment security in infancy and childhood. In: Cassidy, J., Shaver, P.R. (Eds.), Handbook of attachment: theory, research, and clinical applications. Guilford Press, New York, pp. 287–316.

Sroufe, L.A., Waters, E., 1977. Heart rate as a convergent measure in clinical and developmental research. Merrill-Palmer Q. 23, 3–27.

Stephen, J.M., Ledger, R.A., 2006. A longitudinal evaluation of urinary cortisol in kennelled dogs, Canis familiaris. Physiol. Behav. 87, 911–916.

Tinbergen, N., 1951. The study of instinct. Clarendon Press, Oxford.

Tizard, B., 1977. Adoption: a second chance. Open Books, London.

Tomasello, M., Call, J., 1997. Primate cognition. Oxford University Press, Oxford.

Tomasello, M., Kaminski, J., 2009. Like infant, like dog. Science 325, 1213–1214.

Topal, J., Gasci, M., Miklósi, A., Viranyi, S., Kubinyi, E., Csanyi, V., 2005. Attachment to humans: a comparative study on hand-reared wolves and differently socialized dog puppies. Anim. Behav. 70, 1367–1375.

Topal, J., Miklósi, A., Csanyi, V., Doka, A., 1998. Attachment behavior in dogs (Canis familiaris): a new application of Ainsworth's (1969) strange situation test. J. Comp. Psychol. 112, 219–229.

Tuber, D.S., Hennessy, M.B., Sanders, S., Miller, J.A., 1996. Behavioral and glucocorticoid responses of adult domestic dogs (Canis familiaris) to companionship and social separation. J. Comp. Psychol. 110, 103–108.

Tuber, D.S., Miller, D.D., Caris, K.A., Halter, R., Linden, F., Hennessy, M.B., 1999. Dogs in animal shelters: problems, suggestions, and needed expertise. Psychol. Sci. 10, 379–386.

Udell, M.A.R., Wynne, C.D.L., 2008. A review of domestic dogs' (*Canis familiaris*) human-like behaviours: or why behavior analysts should stop worrying and love their dogs. J. Exp. Anal. Behav. 89, 247–261.

Valsecchi, P., Prato-Previde, E., Accorsi, P.A., 2007. Quality of life assessment in dogs living in rescue shelters. Anim. Welf. 16 (S), 178.

Valsecchi, P., Prato-Previde, E., Accorsi, P.A., Fallani, G., 2010. Development of the attachment bond in guide dogs. App. Anim. Behav. Sci. 123, 43–50.

Wells, D., Hepper, P.G., 1992. The behaviour of dogs in a rescue shelter. Anim. Welf. 1, 171–186.

Wilson, E.O., 1975. Sociobiology. Cambridge University Press, New York.

# Chapter 7

# The Personality of Dogs

Ádám Miklósi, Borbála Turcsán, and Enikő Kubinyi
*Department of Ethology, Eötvös University, Budapest, Hungary*

*'Acquiring a dog may be the only opportunity a human ever has to choose a relative.'*
Mordecai Siegal

## 7.1 THE STUDY OF INDIVIDUALITY

The study of individuality has a long history in human psychology. The topic was very much neglected in the case of animals, although in this case, dogs were an exception. Pavlov studied the individual differences in dogs that participated in his experiments on associative learning (see following text). The concept of individual differences and personality found its way into ethology only very slowly. The breakthrough happened about 15–20 years ago when behavioural ecologists realised that dynamics of animal populations can be modelled much better if they include stable behavioural traits of individuals.

Although dog owners are mostly interested in the personality of their companion, the concept of personality has much wider implications, especially with regard to applied aspects. Devoted efforts of researchers working on dogs provided several tools for characterising individual behaviour (e.g., Sheppard & Mills, 2002; Hsu & Serpell, 2003; Ley et al., 2008; Mirkó et al., 2012), identified breed (genetic) differences (e.g., Wilsson & Sundgren, 1997a,b; Svartberg, 2002, 2006; Strandberg et al., 2005; Turcsán et al., 2011), uncovered some gene polymorphisms associated with personality traits (e.g., Héjjas et al., 2007b; Takeuchi et al., 2009a,b; Konno et al., 2011), and studied the effect of development or stability of the behaviour characteristics over an extended period of time (e.g., Wilsson & Sundgren, 1998; Slabbert & Odendaal, 1999; Svobodová et al., 2008). Thus, understanding function and mechanisms of dog personality has a wide range of practical applications, including significant influence on the dog–human bond, dog welfare, and even human health.

### 7.1.1 The Concept of Personality

The concept of personality has been with us at least since the ancient Greeks made scientific inquiries, and human psychology has been concerned with individual differences for a long time. In humans, personality is often defined as an individual's distinctive pattern of behaviour, feeling, and thinking that is consistent across time and situations (e.g., Pervin & John, 1997). In the case of animals, however, a more operational definition is used: behavioural consistency across time and contexts. Although this definition allows for quantifying personality, it is still too general. Several critical aspects are not included in this concise definition; for example: What time frame should be applied? What range of contexts should be involved?

Let's consider a specific case. In many models of dog personality (see following text), 'sociability' (or 'amicability' or 'friendliness') is one of the main components. However, it is not really clear whether we can regard sociability as a personality trait. Should a dog behave friendly over months or years? Is it expected that such a dog behaves friendly towards anybody, including children, other dogs, cats, or postmen? Common sense and experience tell us that an adult dog will show consistent behaviour towards strangers generally, but even this behaviour may change with age and might depend on whether the other is a dog or a human.

Thus, the answer to this question depends very much on the researcher and how the model of personality is constructed. Even puppies that are just a few weeks old show characteristic, consistent behavioural patterns, but this behaviour may change relatively rapidly in the following weeks and months (Scott & Fuller, 1965), so one general consideration is to restrict the concept of personality to adult animals in which changes happen at a lower rate (Jones & Gosling, 2005). Another reason for this distinction is that the concept of personality also assumes that it allows for predicting the behaviour of the subject in future situations. This can be effective only if our estimation is based on reliable data which originate from observing behaviour that is stable over a certain period of time. If there is not enough stability in past and present behaviour, then the prediction of future action also becomes unreliable.

Another important aspect of the operational definition of personality is that by assuming behavioural similarity across similar situations (correlation in statistical terms), we hypothesise a common underlying factor which controls the behaviour in different situations and over some specific time scale. Saying that a dog is 'friendly' refers not only to a prediction that the dog will be (probably) friendly when meeting strangers in the future, but also that it has a mental representation that will tend to elicit affiliative behaviours whenever it faces such situations. Such mental representations (or states) are referred to as 'personality traits' and are usually not defined *a priori* (although researchers may have some hypotheses about their existence) but are constructed on the basis of observed behaviour across time and contexts using specific statistical methods (e.g., principal component analysis).

This means that personality models are hierarchical and have at least two levels: behaviour traits (e.g., approaching social partners) and personality traits (e.g., sociability). If the behaviour traits were associated independently from the context, then one would expect a large number of context-specific personality traits. However, the general experience is that despite a large number of personality traits isolated in humans and animals, this number is much lower than what would be expected in the case of specific associations that may change with time and context. Moreover, it also has been shown that some of the personality traits may also have common causes ('super personality traits') at a higher level of organisation. This suggests that the stability aspect of the behaviour can be modelled by the means of relatively few background variables, that is, personality traits.

## 7.1.2 Measuring Personality: Advantages and Disadvantages

The definition of personality (consistent behavioural patterns across time and context; Pervin & John, 1997) clearly shows what the main difficulty is in measuring. The general aim is to find variation among individuals, consistency in behaviour across situations (contexts), and repeatability of the traits, and for that, we have to assess the behaviour of the same individual several times and in multiple situations.

The ethological approach prefers the behavioural coding methods that are often referred to as being 'objective' because they rely on external reference systems (e.g., frequency, time) as measures. Personality test batteries are used to gather data from different test situations and evaluate them using well-defined and directly observable behaviour units (in terms of duration, frequency, latencies). Although advantages of these behaviour tests include the objectivity, quantitative data sets, and use of adept observers, there are disadvantages as well. For example, breed-specific differences in some behaviour traits (e.g., the likeliness of barking in a specific situation or of jumping up when greeting a stranger) may result in less accurate personality assessment. The number of behavioural test situations is also limited; therefore, the behaviour of the dogs in the tests could be highly context specific. Some aspects of behaviour are difficult to quantify, e.g., intensity (Svartberg, 2005). Moreover, some personality traits, such as aggression or fearfulness, are difficult to measure in behaviour tests because the dogs may not to show these behaviours in controlled situations, and the tests that can elicit such behaviours usually raise ethical and welfare concerns.

The same behaviour test also could be evaluated by using behavioural ratings (Wilsson & Sinn, 2012), which rely more heavily on internal reference systems. In this case, a human observer estimates the behaviour by the means of nominative scales (less than 10% of the time . . . more than 90% of the time) or scoring (1 = no sign of aggression . . . 3 = snap but not attack, etc.), which also may be biased by the subjective judgments of the observer.

Interviewing observers who are familiar with a dog's behaviour (by using a subjective rating method) seems to be an optimal solution to obtain measures on personality, just as one would provide similar estimates about his/her character or about a close friend (Gosling & Vazire, 2002). Owners observe their dogs' behaviour continuously; thus, one might assume that their judgements may correspond to the dogs' behaviour in everyday life. However, owners or family members who know the dog well are usually not experts and might be biased in their reporting. We should also note that there is no objective questionnaire. So, for example, how the owner interprets the scale (such as 'seldom' or 'severe sign of aggression') or how he/she interprets aggression itself may vary from owner to owner.

Despite some disadvantages, collecting behaviour data via questionnaires filled in by owners has become a popular method in dog personality research because one can investigate larger samples and more diverse populations, and the number of 'contexts' (represented by questions) is less limited. Some questionnaires are developed specifically for dogs (e.g., Hsu & Serpell, 2003) or by psychologists in order to measure various aspects of human character. Thus, there exists a dog version of the human Big Five Inventory (Gosling et al., 2003). Another questionnaire that was developed for children, measuring attention deficit hyperactivity disorder (ADHD; DuPaul, 1998), was also modified for use in dogs (e.g., Vas et al., 2007).

As we saw, personality has a hierarchical structure: each personality trait summarises several different aspects of the same background characteristics, so each personality trait can be measured by a set of correlated behavioural variables or questionnaire items. Personality traits are usually identified using factor analysis or principal component analysis (or other data reduction method) by examining the correlation pattern between behaviour variables (e.g., Svartberg & Forkman, 2002; Wright et al., 2011).

Given the assumption that similar personality traits can be estimated by using different types of data, it would be interesting to know to what degree they agree. So far, human studies do not provide strong support in this regard. Correlation coefficients among similar personality traits measured by questionnaires or by behavioural tests are usually low, falling between 0.2 and 0.3 (Gosling et al., 2003). In a recent study on family dogs, Carrier et al. (2013) reported similarly weak relationships among corresponding personality traits estimated by different methods. They observed the behaviour of many dogs while being walked in a park, and at the same time owners filled in a questionnaire on the dogs' personality (Monash Canine Personality Questionnaire–Revised; Ley et al., 2009). According to the study, dogs characterised as 'extrovert' by their owners were observed to be more active during walks ($r = 0.41$), while dogs described by the owner as being more amicable displayed more play acts during the walk ($r = 0.32$). In another study, dogs assessed by owners as more active-impulsive and inattentive showed also more activity in terms of behaviours (moving, vocalising, rapid approach of owner, etc.) in four behaviour test situations ($r = 0.53$ and $r = 0.25$, respectively) (Kubinyi et al., 2012).

In sum, different methods can be applied to estimate personality traits, and researchers should try to balance between the more objective but troublesome behaviour test batteries and the subjective but efficient questionnaires. One benefit to using both 'objective' and 'subjective' methods is that it allows better assessment of reliability and validity. This way, we can obtain information about the dogs' specific behaviour related to the context of testing (based on behavioural coding) as well as about more general tendencies reflecting broader aspects of personality (based on subjective rating).

## 7.2 THE CONCEPT OF PERSONALITY IN DOGS

Whether conscious or unconscious, selective breeding of dogs (including early times of domestication) must have relied on some sort of notion of systematic individual differences, that is, personality. Dog breeders must have based their choices for favourable dogs on stable individual characteristics. Dogs had to fulfil several functions (e.g., herding, hunting) and had to fit into the human communities, and thus, usually only the most appropriate individuals who showed these traits in a reliable manner were bred to future generations.

Scientific investigations started only with Pavlov, at the beginning of the 20th century. During his studies in connection with conditioned reflex learning, he found that dogs react differently to some stimuli, such as handling (Teplov, 1964). Utilising the categories of Hippocrates, Pavlov divided dogs into two main groups: one with a 'strong' and the other with a 'weak' nervous system. The latter, *melancholic* dogs, are sensitive, nervous, inhibited, and shy, and they struggle when restrained. *Choleric* dogs with a 'strong' but unbalanced nervous system are active and tend to be aggressive. *Phlegmatic* dogs are 'strong' and balanced but 'slow': they are quiet, restrained, and persistent. *Sanguine* dogs also have a 'strong and balanced' nervous system and also are very agile. They are active, reactive to novel stimuli, and sleepy from monotony. These categories were adapted by many dog trainers across the world in the last century. However, current personality research in dogs does not favour such typology and has adopted methods used in human personality investigations that are based on establishing personality traits as continuous variables (see previous text).

### 7.2.1 'Wesen', Temperament, Personality, and Behaviour Syndrome in Dogs

The recent renewed interest in individual differences in animals has revealed that a wide range of concepts have been used for covering more or less the same phenomenon. As mentioned previously, Pavlov and his followers referred to individual differences as personality types. At the same time, German dog breeders and trainers referred to the same concept as 'Wesen', which is closer to the English 'character', that is, being an individual. Interestingly, Cloninger (2002) modelled personality as involving two types of components. Temperament was defined as biological inheritable traits, whereas character included traits that were acquired through social interactions and experiencing culture.

Goldsmith et al. (1987) emphasised the difference between temperament and personality from a developmental point of view. They also regarded temperament as indicating early emerging behavioural tendencies that are highly heritable. Character, however, is regarded as an outcome of the developmental process, during which genetic effects are modulated by the environment.

More recently, the animal behaviour community introduced 'behaviour syndrome' into the terminology (Sih et al., 2004). This term is defined as a set of correlated behaviours that are displayed within or across behavioural context(s). Although the idea of personality and behaviour syndrome are very similar, the two concepts are used in different contexts. In the case of the former, researchers aim to find out in what way a specific behaviour syndrome supports the survival of the individual in a specific environment. Behaviour syndromes that are advantageous in one context may be disadvantageous in another one. In contrast, the concept of personality is more often referred to in studies examining the underlying mechanisms, including physiological and genetic basis. However, despite these different conceptual approaches to the problem of individuality, the most parsimonious solution is to refer to animal personality in this case.

## 7.2.2 The General Structure of Canine Personality

It should be emphasised that the modelling of dogs' personality depends on the quality and quantity of the data collected. Thus, it is not surprising that depending on the method used, personality models of dogs differ, despite the fact that some overlaps are also expected. Here, we review three holistic approaches that aim to model the 'whole' personality of dogs. Importantly, however, all these approaches necessarily run into problems because it is practically impossible to collect data from all possible situations that may occur in the life of a dog. It follows that personality models capture only a specific aspect of the whole structure. Here are the three examples:

(1) Gosling et al. (2003) based their study on a human personality inventory. The Five Factor Model (FFM or 'Big Five'; Goldberg, 1992) classifies human personality into five broad personality traits. FFM is a hierarchical model; each factor summarises several more specific traits: (1) *Neuroticism* (including traits such as nervousness, jealousy, or anxiety); (2) *Extroversion* (including traits such as energetic, talkative, bold); (3) *Openness* (or intellect) (including traits such as imaginative, artistic, or uncreative); (4) *Agreeableness* (including traits such as altruism, kindness, or warmth); (5) *Conscientiousness* (including traits such as systematic or sloppy) (Gosling & Bonnenburg, 1998). Accordingly, dog personality was moulded into these five personality traits by asking dog owners to fill in the same FFM questionnaire adapted to dog behaviour (Gosling et al., 2003). Interestingly, Gosling and John (1999) failed to identify conscientiousness as a personality trait in dogs, but this trait has shown acceptable internal consistency in the Hungarian version (Turcsán et al., 2012).

**(2)** The most prevalent questionnaire (C-BARQ: Hsu & Serpell, 2003) iden- tifies 11 dimensions (stranger-directed aggression, owner-directed aggres- sion, stranger-directed fear, non-social fear, dog-directed fear or aggression, separation-related behaviour, attachment- or attention-seeking behaviour, trainability, chasing, excitability, and pain sensitivity). Originally, 152 ques- tions about dogs' responses to specific events and situations were posed to owners, and 68 questions were retained for the final questionnaire after the factor analysis (Hsu & Serpell, 2003).

**(3)** The behaviour test battery with the highest number of subjects (Dog Mentality Assessment, or DMA: Svartberg & Forkman, 2002) provided five dimensions (playfulness, curiosity/fearlessness, chase-proneness, socia- bility, and aggressiveness). More than 15,000 dogs were tested in the ten subtests of a standardised behavioural test of the Swedish Working Dog Association.

After reviewing 51 studies and conducting a meta-analysis, Jones and Gos- ling (2005) identified seven main personality traits, which were merged to six in a subsequent publication and Dominance was renamed Submissiveness (Fratkin et al., 2013):

- *Fearfulness* (with Reactivity) is related to the approach/avoidance of novel objects, raised hackles, and activity in novel situations. This trait is frequently labelled 'excitability', 'nerve stability', 'courage', 'self-confidence', or 'bold- ness'.
- *Sociability* is indexed by such behaviours as initiating friendly interactions with people or other dogs; this trait is sometimes labelled 'extroversion'.
- *Responsiveness to training* is related to such behaviours as working with people, learning quickly in new situations, and playfulness. This trait is also labelled 'problem solving', 'willingness to work', 'trainability', and 'coopera- tive'.
- *Aggression* is characterised by biting, growling, and snapping at people or other dogs. Aggressive behaviour is sometimes divided into subcategories on the basis of the cause of the aggression or of the target of aggressive behaviour.
- *Submissiveness* (the opposite of dominance) is reflected in such behaviours as refusing to move out of a person's path, or 'self-right'.
- *Activity* is often assessed by the locomotor activity in open-field or open- field-like tests. There is some debate about whether Submissiveness and Activity should be considered independent personality traits (Gosling & John, 1999).

It remains to be seen whether this general structure of dog personality becomes a standard for investigations. We should note that the development of such complex traits is not the goal of the investigations. Instead, researchers are looking for tools to capture significant aspects of behavioural organisation. The usefulness of these personality traits could be that they help to explain the mechanisms controlling behaviour including genetic and environmental effects.

## 7.3 NATURE AND NURTURE: THE ROLE OF GENETIC AND ENVIRONMENTAL FACTORS IN DOGS' PERSONALITY

Personality traits, as any other phenotypes, are influenced by both genetic and environmental factors. It is important to remember that breeds do not have 'personality' but personality traits, just as other phenotypic measures, can be used to compare different breeds. Actually, it would be useful to provide an estimate of these phenotypic traits for each dog breed that, however, should be specific to the geographic region (breeding population) and updated regularly in order to follow possible changes in the population.

### 7.3.1 Dog Breeds and the Concept of Personality

Human artificial selection for certain morphological and behavioural features has resulted in more than 400 breeds being recognised today, which makes the dog the most phenotypically diverse mammal (Parker et al., 2004). During early domestication, dogs underwent a variety of behavioural changes as a result of infiltrating the anthropogenic environment. This may have included increased affiliation with humans and reduced aggressive tendencies (Hare & Tomasello, 2005; Miklósi & Topál, 2013). Later, dogs were selectively bred to perform a variety of practical tasks (e.g., herding sheep, hunting, guarding livestock, pulling sleds) with functions that required different morphological and breed-specific behavioural features (e.g., pointing, showing eye) and also specific changes in personality traits. Breeds with certain behavioural skills are generally suitable for a specific function, whereas representatives of other breeds are not. For example, the presence of predatory motor patterns is necessary for hunting and herding dogs, whereas the predatory behaviour should be suppressed in livestock-guarding dogs (Coppinger & Schneider, 1995). Breeds working under close visual guidance by humans are found to better utilise human communicative gestures, such as pointing, than breeds working independently out of humans' view (Gácsi et al., 2009). Herding dogs use more eye contact and vocalisation to communicate with others, whereas retrievers display preferentially more body contact towards humans (Lit et al., 2010). Christiansen et al. (2001) found differences in the hunting style of cooperative and non-cooperative hunting breeds. According to Wilsson and Sundgren (1997a), German shepherd police dogs scored higher for courage, hardness, prey drive, and defence drive, and showed better nerve stability than average for the breed. German shepherd guide dogs were more eager to cooperate and showed more courage and stronger nerve stability than family dogs of the same breed (Wilsson & Sundgren, 1997a). Compared to police dogs, these guide dogs obtained lower scores for sharpness, defence drive, and prey drive, but a higher partial index value for ability to cooperate. Dogs rejected as service dogs showed the greatest deficiencies in nerve stability, defence drive, affability, and ability to cooperate. Rooney and Bradshaw (2004) reported that dog experts in England preferred different type breeds for specific search work. The English springer spaniel was the most popular choice as a search dog for proactive drugs, whereas handlers of explosives dogs expressed a preference for the border collie. Labrador retrievers were the most commonly used as passive drug search dogs.

Breed-related differences have been reported for many personality traits. Although there are examples of direct breed comparisons (e.g., Duffy et al., 2008; van den Berg et al., 2010), most research investigates the differences between various breed groups. Traditional breed groupings were created by major kennel clubs (e.g., FCI, AKC), and are based on morphological similarity and anecdotal information about the relationship among the breeds. Many previous and recent studies suggest behavioural differences between these breed groups; for example, herding dogs and cooperative hunting dogs (pointers, retrievers) were found to be highly trainable (Seksel et al., 1999; Serpell & Hsu, 2005; Ley et al., 2009; Turcsán et al., 2011), while toy dogs and hounds reported being less trainable (Ley et al., 2009; Turcsán et al., 2011). Terriers were found to be bolder than hounds and herding dogs (Turcsán et al., 2011), and other studies characterised them also as energetic, excitable, and reactive (Scott & Fuller, 1965; Hart, 1995; Ley et al., 2009; Turcsán et al., 2011). In contrast, Svartberg (2006) failed to find any relationship between breed-typical behaviour and the historical use of the breeds. He suggested that recent selection criteria are probably more important factors in determining the behaviours of different dog breeds.

As some behavioural traits have a considerable genetic background, breed-typical behaviours are at least partly genetically determined. By analysing the genetic relatedness between breeds, Parker et al. (2007) created five distinctive breed groups: (1) 'Ancient (basal) breed' (e.g., basenij, Afghan hound, Siberian husky); (2) 'Mastiffs-Terriers' (e.g., German shepherd, boxer, bulldog); (3) 'Herding dogs and sight hounds' (e.g., border collie, Shetland sheepdog); (4) 'Mountain-types' (e.g., St. Bernard, Rotweiler); and (5) 'Modern breeds' (e.g., American cocker spaniel, standard poodle).

It is expected that genetically more related breeds behave more similarly than genetically more distant breeds. In line with this, for example, Turcsán et al. (2011) found that dogs belonging to breeds with closer genetic relationship to the wolf (Ancient/basal breeds) were less trainable than the purebred modern herding dogs, and they were also less bold than the mastiff-type dogs and terriers.

Most of the currently acknowledged breeds are relatively recent (100–200 years old; Parker et al., 2004) and were frequently bred by crossing members of other breeds, sometimes independently from their original function. Crossing individuals within the same functional types could improve the breeds' behavioural skills (e.g., herding), while selection for specific morphological traits (e.g., small size) involves crossing across functional types (vonHoldt et al., 2010). Although the functional and genetic categorisations of breeds are not independent from each other (Parker et al., 2004), genetic relatedness provides an alternative (however not mutually exclusive) explanation for the behavioural similarity among breeds' typical behaviour.

Breed groups were also generated on the basis of various personality traits (e.g., Bradshaw & Goodwin, 1998; Svartberg, 2006; Turcsán et al., 2011). For example, using a cluster analysis, Turcsán et al. (2011) classified dog breeds on the basis of trainability, boldness, calmness, and dog sociability. These groupings probably reflect actual behaviour conformation of the breed, their present uses, and selection but are not linked to the traditional or genetic classifications (e.g., Svartberg, 2006; Turcsán et al., 2011) (Figure 7-1).

The discriminative potential of personality traits among breeds is variable (Hart, 1995; Lit et al., 2010). Breeds represent genetically distinct populations; thus, behavioural traits with higher heritability (characterised by value between 0 and 1 that shows how much of the individual variability can be attributed to genetic differences) should differ more between breeds than behavioural traits that are more affected by environmental influence (low heritability). The heritability of some dog personality traits is similar to that of human traits. For example, the heritability of activity was estimated as being 0.53 (which means that 53% of the individuals' activity is determined by the variability in the genes) (Wilsson & Sundgren, 1998), the heritability of aggression ranges between 0.2 and 0.8 (Liinamo et al., 2007), and the heritability of fearfulness was found to be 0.46 (Goddard & Beilharz, 1982). It means that activity, aggressive, or fearful behaviour is, at least partly, inherited, and these traits can be the targets of selection.

Morphological features associated with a specific function may also influence the behaviour indirectly. For example, the typical weight of a livestock-guarding dog is 30–40 kg in order to scare off predators (Coppinger & Schneider, 1995). Because of physiological reasons, one should not expect the same level of activity and speed from these dogs as, for example, from small-sized hunting dogs. As a comparison, the optimal weight of sled dogs is 20–25 kg because larger dogs cannot dissipate heat quickly enough during running (Phillips et al., 1981).

Taken together, domestication, breed selection, and recent living style of dogs have had differential effects on personality traits. Breed formation in connection with specific working function (e.g., guarding dogs) has increased divergence among breeds because some types of tasks require specific behavioural attitude. Crossing individuals within or across functional groups could also alter the behavioural skills characteristic to breeds. Finally, the utilisation of many dog breeds as family pets (and show dogs), partly independently from their original function, may have led to converging personality traits supporting engagement in family life (Svartberg, 2006).

## 7.3.2  Gene-Behaviour Associations

In spite of selective breeding, acquiring a purebred dog does not perfectly guarantee breed-characteristic behaviour. Except for the environmental influence, the interplay of several hundred or thousand polymorph genes results in

| Breed | N | rank | mean | rank | mean | rank | mean | sociability rank | sociability mean |
|---|---|---|---|---|---|---|---|---|---|
| 6;2 French Bulldog | 60 | 76 | 1.403 | 19 | 1.661 | 10 | 1.387 | 34 | 1.488 |
| 5;4 Shih Tzu | 30 | 79 | 1.380 | 26 | 1.622 | 13 | 1.373 | 36 | 1.483 |
| 6;2 Bulldog | 92 | 73 | 1.430 | 29 | 1.609 | 16 | 1.363 | 35 | 1.486 |
| 3;1 Siberian Husky | 51 | 86 | 1.294 | 25 | 1.628 | 5 | 1.455 | 54 | 1.402 |
| 3;5 Saint Bernard | 21 | 95 | 1.200 | 23 | 1.635 | 30 | 1.305 | 32 | 1.500 |
| 1;4 American Cocker Spaniel | 15 | 84 | 1.307 | 27 | 1.622 | 21 | 1.347 | 79 | 1.250 |
| 5;4 Pekingese | 16 | 89 | 1.288 | 44 | 1.542 | 21 | 1.350 | 67 | 1.328 |
| 5;_ Havanese | 75 | 50 | 1.568 | 65 | 1.409 | 18 | 1.352 | 27 | 1.513 |
| 2;3 Whippet | 26 | 58 | 1.546 | 67 | 1.397 | 22 | 1.346 | 17 | 1.596 |
| 4;4 Miniature Schnauzer | 15 | 32 | 1.613 | 75 | 1.333 | 25 | 1.333 | 29 | 1.500 |
| 3;5 Greater Swiss Mo. Dog | 72 | 36 | 1.608 | 66 | 1.403 | 61 | 1.131 | 26 | 1.535 |
| 7;3 Shetland Sheepdog | 21 | 18 | 1.695 | 72 | 1.349 | 51 | 1.171 | 15 | 1.607 |
| 7;3 Bearded Collie | 46 | 38 | 1.604 | 49 | 1.515 | 43 | 1.222 | 6 | 1.685 |
| 1;4 Gordon Setter | 11 | 25 | 1.655 | 64 | 1.424 | 42 | 1.236 | 2 | 1.727 |
| 5;4 Cavalier King Ch. Spaniel | 20 | 40 | 1.600 | 79 | 1.317 | 27 | 1.320 | 3 | 1.713 |
| 1;4 Irish Setter | 30 | 17 | 1.700 | 78 | 1.322 | 33 | 1.287 | 1 | 1.767 |
| 4;2 Airedale Terrier | 24 | 53 | 1.567 | 32 | 1.597 | 7 | 1.425 | 25 | 1.552 |
| 4;2 Staffordshire Bull Terrier | 14 | 45 | 1.586 | 33 | 1.595 | 6 | 1.443 | 45 | 1.446 |
| 1;2 Labrador Retriever | 517 | 24 | 1.667 | 9 | 1.712 | 26 | 1.323 | 10 | 1.642 |
| 5;4 Pug | 82 | 62 | 1.520 | 7 | 1.724 | 8 | 1.402 | 12 | 1.613 |
| 2;4 Beagle | 200 | 52 | 1.568 | 42 | 1.544 | 46 | 1.202 | 14 | 1.609 |
| 1;4 Golden Retriever | 364 | 55 | 1.563 | 38 | 1.579 | 36 | 1.257 | 21 | 1.573 |
| 1;_ Small Munsterlander | 57 | 30 | 1.625 | 40 | 1.561 | 70 | 1.053 | 28 | 1.504 |
| 4;2 Soft Coated Wheaten T. | 19 | 42 | 1.600 | 43 | 1.544 | 59 | 1.137 | 22 | 1.566 |
| 1;4 Pointer | 10 | 22 | 1.680 | 24 | 1.633 | 57 | 1.140 | 8 | 1.650 |
| 7;3 Old English Sheepdog | 10 | 56 | 1.560 | 11 | 1.700 | 48 | 1.200 | 39 | 1.475 |
| 3;5 Bernese Mountain Dog | 94 | 71 | 1.434 | 47 | 1.525 | 14 | 1.372 | 7 | 1.673 |
| 3;5 Leonberger | 16 | 77 | 1.400 | 36 | 1.583 | 55 | 1.163 | 13 | 1.609 |
| 4;2 Border Terrier | 11 | 28 | 1.636 | 12 | 1.697 | 38 | 1.255 | 69 | 1.318 |
| 3;_ Hovawart | 96 | 19 | 1.690 | 20 | 1.649 | 39 | 1.246 | 77 | 1.297 |
| 3;2 Boxer | 135 | 29 | 1.628 | 18 | 1.664 | 50 | 1.185 | 61 | 1.348 |

FIGURE 7-1   Dendrogram illustrating the similarity between dog breeds based on the average trait scores for all individuals in each breed (based on Turcsán et al., 2011). Average of trait scores and rankings by each trait are presented. The numbers preceding the breed names represent the conventional breed groups [first number, (AKC): 1 = Sporting dogs; 2 = Hounds; 3 = Working dogs; 4 = Terriers; 5 = Toy dogs; 6 = Non-sporting dogs; 7 = Herding dogs] and genetic clusters [second number, 1 = Ancient breeds; 2 = Mastiff/Terrier cluster; 3 = Herding/Sighthound cluster; 4 = Hunting breeds; 5 = Mountain cluster (Parker et al., 2007); breeds without genetic classification are marked with underscore]. N is the number of individuals representing the breed.

| Breed | | | | | | | | | |
|---|---|---|---|---|---|---|---|---|---|
| 1;_ German Wireh. Pointer | 31 | 10 | 1.768 | 14 | 1.677 | 15 | 1.368 | 78 | 1.250 |
| 3;4 Giant Schnauzer | 63 | 15 | 1.733 | 5 | 1.757 | 31 | 1.302 | 63 | 1.337 |
| 4;2 American Staffordshire T. | 76 | 44 | 1.597 | 28 | 1.610 | 11 | 1.379 | 76 | 1.299 |
| 3;5 Rottweiler | 137 | 37 | 1.604 | 39 | 1.577 | 23 | 1.346 | 73 | 1.310 |
| 6;4 Wolfspitz | 11 | 54 | 1.564 | 56 | 1.485 | 29 | 1.309 | 70 | 1.318 |
| 4;2 Irish Terrier | 15 | 49 | 1.573 | 10 | 1.711 | 28 | 1.320 | 90 | 1.167 |
| 2;_ Wirehaired Dachshund | 50 | 41 | 1.600 | 30 | 1.607 | 44 | 1.216 | 88 | 1.185 |
| 2;_ Bavarian Mo. Hound | 12 | 46 | 1.583 | 46 | 1.528 | 52 | 1.167 | 64 | 1.333 |
| 2;4 Dachshund | 74 | 61 | 1.530 | 52 | 1.496 | 58 | 1.138 | 71 | 1.311 |
| 6;5 Poodle | 47 | 31 | 1.617 | 57 | 1.475 | 49 | 1.192 | 75 | 1.303 |
| 4;4 Cairn Terrier | 27 | 78 | 1.393 | 53 | 1.494 | 40 | 1.244 | 56 | 1.389 |
| 3;3 Great Dane | 76 | 70 | 1.437 | 59 | 1.469 | 45 | 1.208 | 57 | 1.372 |
| 4;4 West Highland White T. | 146 | 68 | 1.460 | 45 | 1.537 | 56 | 1.153 | 50 | 1.408 |
| 6;1 Tibetan Terrier | 46 | 75 | 1.422 | 34 | 1.594 | 34 | 1.274 | 53 | 1.402 |
| 3;4 Doberman Pinscher | 124 | 34 | 1.613 | 50 | 1.503 | 88 | 0.929 | 72 | 1.311 |
| 1;5 English Cocker Spaniel | 20 | 35 | 1.610 | 48 | 1.517 | 74 | 1.040 | 55 | 1.400 |
| 5;5 Miniature Pinscher | 25 | 48 | 1.576 | 21 | 1.640 | 91 | 0.896 | 66 | 1.330 |
| 6;4 Dalmatian | 123 | 67 | 1.472 | 35 | 1.594 | 76 | 1.033 | 47 | 1.435 |
| 7;3 Australian Shepherd | 167 | 3 | 1.817 | 51 | 1.501 | 65 | 1.072 | 44 | 1.448 |
| 7;3 Border Collie | 193 | 4 | 1.805 | 60 | 1.468 | 67 | 1.072 | 68 | 1.325 |
| 1;4 Vizsla | 44 | 8 | 1.777 | 55 | 1.485 | 77 | 1.018 | 19 | 1.580 |
| 4;2 Welsh Terrier | 13 | 1 | 1.877 | 8 | 1.718 | 78 | 1.015 | 51 | 1.404 |
| 7;5 German Shepherd Dog | 413 | 33 | 1.613 | 22 | 1.637 | 64 | 1.103 | 92 | 1.139 |
| 4;4 Jack Russell Terrier | 327 | 27 | 1.643 | 16 | 1.672 | 82 | 0.998 | 85 | 1.191 |
| 7;_ Entlebucher Mo. Dog | 17 | 6 | 1.800 | 17 | 1.667 | 68 | 1.071 | 81 | 1.235 |
| 4;_ Parson Russell Terrier | 86 | 11 | 1.763 | 15 | 1.674 | 75 | 1.037 | 80 | 1.236 |
| 7;3 Belgian Malinois | 66 | 14 | 1.736 | 31 | 1.606 | 89 | 0.912 | 91 | 1.152 |
| 3;4 Standard Schnauzer | 15 | 16 | 1.720 | 41 | 1.556 | 84 | 0.987 | 89 | 1.183 |
| 4;_ German Hunting Terrier | 12 | 60 | 1.533 | 4 | 1.778 | 97 | 0.817 | 87 | 1.188 |
| 1;4 Flat-Coated Retriever | 18 | 7 | 1.789 | 1 | 1.833 | 12 | 1.378 | 4 | 1.694 |
| 1;4 German Shorth. Pointer | 20 | 5 | 1.800 | 2 | 1.833 | 35 | 1.260 | 23 | 1.563 |
| 5;_ Kromfohrländer | 54 | 9 | 1.774 | 89 | 1.241 | 81 | 1.000 | 93 | 1.116 |
| 7;_ Polish Lowland Sheepdog | 20 | 2 | 1.830 | 61 | 1.433 | 87 | 0.940 | 88 | 1.188 |
| 7;_ Appenzeller Sennenhund | 27 | 26 | 1.652 | 77 | 1.333 | 96 | 0.844 | 59 | 1.352 |
| 2;4 Ibizan Hound | 30 | 65 | 1.487 | 92 | 1.200 | 92 | 0.893 | 43 | 1.450 |

FIGURE 7-1—Cont'd

| Breed | | | | | | | | | |
| --- | --- | --- | --- | --- | --- | --- | --- | --- | --- |
| 7;_ Briard | 33 | 43 | 1.600 | 85 | 1.263 | 79 | 1.012 | 83 | 1.227 |
| 5;4 Miniature Poodle | 26 | 59 | 1.546 | 80 | 1.295 | 69 | 1.069 | 82 | 1.231 |
| 7;_ Pyrenean Shepherd | 23 | 20 | 1.687 | 93 | 1.188 | 72 | 1.052 | 49 | 1.413 |
| 7;_ White Swiss Shepherd | 20 | 23 | 1.680 | 87 | 1.250 | 73 | 1.050 | 60 | 1.350 |
| 7;_ Beauceron | 10 | 13 | 1.740 | 96 | 1.167 | 80 | 1.000 | 38 | 1.475 |
| 1;4 Weimaraner | 39 | 12 | 1.759 | 86 | 1.256 | 66 | 1.072 | 41 | 1.474 |
| 2;_ Miniature Dachshund | 18 | 57 | 1.556 | 88 | 1.241 | 53 | 1.167 | 52 | 1.403 |
| 6;1 Shiba Inu | 35 | 47 | 1.577 | 95 | 1.171 | 60 | 1.131 | 48 | 1.414 |
| 7;3 Collie | 56 | 51 | 1.568 | 82 | 1.280 | 63 | 1.121 | 33 | 1.496 |
| 2;5 Rhodesian Ridgeback | 64 | 66 | 1.472 | 83 | 1.276 | 37 | 1.256 | 42 | 1.453 |
| 5;4 Chihuahua | 73 | 64 | 1.493 | 70 | 1.356 | 90 | 0.901 | 95 | 1.072 |
| 3;2 Perro de Presa Canario | 12 | 72 | 1.433 | 58 | 1.472 | 95 | 0.850 | 96 | 1.063 |
| 5;5 Maltese | 49 | 82 | 1.314 | 73 | 1.347 | 83 | 0.988 | 94 | 1.112 |
| 6;_ German Spitz | 12 | 90 | 1.283 | 37 | 1.583 | 58 | 0.883 | 58 | 1.354 |
| 5;_ Yorkshire Terrier | 111 | 81 | 1.342 | 54 | 1.487 | 86 | 0.942 | 65 | 1.331 |
| 4;_ Bull Terrier | 23 | 83 | 1.313 | 6 | 1.725 | 71 | 1.052 | 74 | 1.304 |
| 1;4 Brittany | 10 | 91 | 1.260 | 76 | 1.333 | 85 | 0.980 | 40 | 1.475 |
| 3;_ Dogue de Bordeaux | 14 | 92 | 1.229 | 63 | 1.429 | 9 | 1.400 | 37 | 1.482 |
| 6;1 Lhasa Apso | 16 | 97 | 1.125 | 71 | 1.354 | 19 | 1.350 | 46 | 1.438 |
| 6;1 Chinese Shar-Pei | 13 | 94 | 1.215 | 97 | 1.128 | 47 | 1.200 | 31 | 1.500 |
| 1;4 English Setter | 11 | 88 | 1.291 | 90 | 1.212 | 54 | 1.164 | 30 | 1.500 |
| 6;_ Coton de Tulear | 16 | 80 | 1.363 | 74 | 1.333 | 41 | 1.238 | 20 | 1.578 |
| 6;_ Eurasier | 21 | 85 | 1.295 | 84 | 1.270 | 24 | 1.343 | 24 | 1.560 |
| 3;_ Anatolian Shepherd Dog | 13 | 93 | 1.215 | 94 | 1.180 | 17 | 1.354 | 84 | 1.212 |
| 3;1 Alaskan Malamute | 11 | 96 | 1.127 | 68 | 1.394 | 2 | 1.655 | 62 | 1.341 |
| 2;3 Irish Wolfhound | 13 | 98 | 1.031 | 81 | 1.282 | 4 | 1.631 | 5 | 1.692 |
| 3;_ German Pinscher | 12 | 21 | 1.683 | 69 | 1.389 | 98 | 0.667 | 18 | 1.583 |
| 2;_ Spanish Greyhound | 30 | 87 | 1.293 | 98 | 0.933 | 94 | 0.880 | 9 | 1.650 |
| 3;1 Akita | 20 | 69 | 1.460 | 62 | 1.433 | 32 | 1.300 | 98 | 0.738 |
| 2;_ German Bracke | 11 | 63 | 1.509 | 91 | 1.212 | 62 | 1.127 | 97 | 0.841 |
| 3;_ Landseer | 12 | 39 | 1.600 | 3 | 1.806 | 1 | 1.850 | 16 | 1.604 |
| 3;2 Newfoundland | 31 | 74 | 1.426 | 13 | 1.677 | 3 | 1.632 | 11 | 1.621 |
| **MEAN** | | | **1.537** | | **1.487** | | **1.184** | | **1.406** |
| **SD** | | | **0.180** | | **0.183** | | **0.200** | | **0.187** |

FIGURE 7-1—Cont'd

considerable individual variation. Due to this interplay, mapping the effect of a single gene is not easy. However, reduced genetic heterogeneity within a breed helps to detect the usually small effect of a single gene (and/or linked genes). Thus, in contrast to humans, in which only a few isolated human populations exist, dog breeds with partially fixed genetic background enable us to map genes underlying complex diseases and personality traits (see, for example, Miyadera et al., 2012).

Decades after investigating genetic contribution to behaviour at the breed level (Scott & Fuller, 1965), molecular geneticists started to identify genotype–phenotype associations. So far, numerous genes have been mapped for morphological traits and diseases (Parker et al., 2010), but the identification of genes underlying behavioural phenotype is still lagging behind. However, several single nucleotide polymorphisms (SNPs), variable number of tandem repeats (VNTRs), and copy number variants of candidate genes relating to androgen, serotonin, dopamine, and other systems already have been sequenced (e.g., Héjjas et al., 2007b; van den Berg et al., 2008; Konno et al., 2011; for a review, see Hall & Wynne, 2012). Specific genes have been identified either by candidate gene approach (Héjjas et al., 2007a; Takeuchi et al., 2009a; Våge et al., 2010) or by genome-wide association studies (GWAS; Dodman et al., 2010; Tiira et al., 2012). Both methods look for association between genetic polymorphisms and a phenotype. The former method is hypothesis driven and assumes an association between specific gene variants (alleles) and some phenotypic trait. The second method is data driven; that is, it looks at possible associations among several phenotypic traits and random chosen genetic variability. Using the former method, researchers were able to detect association, for example, between glutamate transporter (SLC1A2) and stranger-directed aggression in Shiba Inu dogs (Takeuchi et al., 2009b); dopamine receptor (DRD1), serotonin receptor (HTR1D, HTR2C), glutamate transporter (SLC6A1), and activity level in Labrador retrievers (Takeuchi et al., 2009a); and dopamine receptor (DRD1), serotonin receptor (HTR1D, HTR2C), glutamate transporter (SLC6A1), and human-directed aggression in English cocker spaniels (Våge et al., 2010). Dodman et al. (2010) reported association between the neuronal adhesion protein (CDH2) and obsessive-compulsive disorder in Doberman pinschers, and Konno et al. (2011) suggested a relationship between the androgen receptor (AR) and aggression in male Japanese Akita Inu dogs. In the Belgian Malinois, the dopamine transporter (DAT) seemed to be affecting attention, aggression, and loss of responsiveness to environmental stimuli (Lit et al., 2013).

We have also mapped several candidate genes, including novel, previously not described markers, which might influence personality traits, and reported gene-behaviour associations in dogs (Héjjas et al., 2007a,b). Most of these efforts centred around the possible effects of the dopaminergic system, focusing on two components of the system: dopamine receptor D4 (DRD4) and tyrosine-hydroxylase gene (TH). DRD4 encodes the D4 subtype of the dopaminergic receptor, whereas TH encodes the enzyme tyrosine hydroxylase, which is involved in the synthesis of the dopamine precursor (L-DOPA). Dopamine itself is the precursor of the catecholamines norepinephrine and epinephrine. Dopamine is involved in the brain's reward system, as well as in cognition, movement control, and attention (Nieoullon, 2002). For example, the relationship between DRD4 and activity-impulsivity traits in humans is well established (Gizer et al., 2009; Varga et al., 2012). In the case of TH, direct links with activity and impulsivity traits have not been reported, but neuroticism and extroversion, which are associated with impulsivity (Whiteside & Lynam, 2001), have been linked to TH polymorphisms (Persson et al., 2000; Tochigi et al., 2006).

Generally, the results showing associations between behaviour and gene polymorphisms in dogs support relevant findings in humans. German shepherd police dogs that possessed at least one longer allele of the D4 dopamine receptor gene (DRD4) polymorphism were reported to show significantly higher activity-impulsivity than dogs lacking this allele (Héjjas et al., 2007b). Additionally, DRD4 polymorphism was found to be associated with possessiveness and gazing at humans in Belgian shepherd dogs (Kubinyi et al., 2006). In German shepherd dogs, the presence of the short allele also decreased social impulsivity when greeting a stranger (Héjjas et al., 2009). Importantly, this later study was complemented by specific *in vitro* characterisation of the polymorph DNA region. The analysis detected that there are significant functional differences in the transcriptional activity between the two allelic variants that strengthen the hypothesis on the link between DRD4 and behaviour. Further work on Siberian Huskies also showed that allelic variation influences activity in a behaviour test (Wan et al., 2013).

In parallel, the tyrosine-hydroxylase gene (TH) polymorphism is also related to activity-impulsivity assessed by both a behavioural test battery and a questionnaire in German shepherds and Siberian huskies (Kubinyi et al., 2012; Wan et al., 2013), and in the latter case, some additive effects on activity-impulsivity between the DRD4 and TH alleles cannot be excluded.

Association studies on working dogs, such as racing Siberian huskies (Wan et al., 2013), may lead to earlier and more predictive selection of suitable individuals. By understanding the genetics affecting working dog behaviour, breeders may be able to target more effectively the specific traits that working dogs need. For example, desired traits in sled dogs include endurance, speed, and willingness to pull (Huson et al., 2010), whereas guide dogs must display reduced fearfulness, aggressiveness, and distractibility (Goddard & Beilharz, 1983).

We have to emphasise that especially in the case of polygenetic behavioural phenotypes specific alleles explain only a small part of the phenotypic variation. The other, usually much larger part is related to other non-measured genetic effects and environmental influences. Accordingly, some genetic effects may be found under specific environmental conditions, whereas other environmental influences may mask genetic differences. Studying gene-by-environment interactions has specific importance because dog personality is usually studied in natural contexts that may be different across breeds and study sites (e.g., countries) and also affected by culture. Héjjas et al. (2007b) assessed the activity-impulsivity personality trait in German shepherd dogs by the means of questionnaires filled in by the owner. They found that the allele polymorphism of the DRD4 receptor had an effect on this trait only in the police dog population but not among family dogs. Researchers argued that the more diverse family environment was responsible for the absence of a genetic effect in the pet dog population, or from a different perspective, the controlled environment of police dogs amplified the minor genetic influences. Given the growing popularity of genetic association studies in dogs, it is unfortunate that little attention has been given to investigate the effect of the environment on the phenotypic manifestation of a genetic difference.

Although breeders can currently conduct genetic testing for a wide range of physical disorders—for example, progressive retinal atrophy (Mellersh, 2012)—screening for alleles affecting behaviour would be more problematic and unrealistic. Apart from the fact that selecting for a genetic factor with small effect results in small selective response, (1) the same allele might function differently on different genetic background (breed), (2) some breeds may lack one or many types of alleles, and (3) the effect of an allele may be environment-dependent; that is, selection for or against one allele type may change the behaviour in different directions.

Dogs and humans share both the physical and social environment in many respects; thus, such studies in dogs could actually inform researchers working on humans and reveal the generality of the specific genetic effect. So far, most genetic research regarding behaviour in dogs has been based on human findings; that is, candidate genes pinpointed by investigations on humans formed the basis of association studies in dogs. Convergent results, such as the effect of DRD4 receptor polymorphism on activity and impulsivity both in humans and dogs, make a stronger case for a specific association; thus, comparative studies have a special role to play. However, in the future, one may expect that candidate genes detected in dogs also could be tested for in human populations.

Consequently, more and more evidence shows that the dog is a useful model species for the study of genetic influence on human behaviour and personality. Although genetic association studies are still in their early stages, it is apparent that these methods can offer outstanding possibilities for those who are looking for the genetic mechanisms underlying typical behavioural variation, as well as behavioural disorders.

### 7.3.3 Environmental Effects on Dog Personality

In several studies, researchers aim to explain the variation in behaviour by the means of demographic and environmental factors. However, it is not always apparent what is the reason behind this work because the potential number of such factors is endless. Even for exploratory investigations, a huge sample size is needed, and it can be achieved only by using the questionnaire method and web-based collection of data. Thus, the effects found should be regarded as hypotheses for further experimental investigations in which the specific factor is manipulated systematically. It should be noted, however, that the effect of such factors is often so small that there is little chance of finding similar effects in a convenient sample of experimental subjects.

Next, we summarise a few environmental factors that presumably have a significant effect on dogs' behaviour and possibly personality traits. To begin with, we look at probably the most complex variable: cultural differences. Even in the most 'dog-loving' societies, a considerable part of the human population does not develop individual social relations with dogs, but people cannot really avoid getting in contact with them regularly. Although at most places dogs are more or less part of human society, there are large dog populations that live outside the boundaries of a human-dominated environment (Miklósi, 2007). For example, Fielding (2008) found that the percentage of owners reporting that they felt attached to their pets, considered them to be members of the family, and permitted them to live indoors was much higher in the United States than in the Bahamas. Compared to Japanese students, British students have significantly more positive attitudes about pets (Miura et al., 2002). White Americans have more experience with pets and are more attached to pets than African Americans (Friedmann et al., 1983; Katcher & Beck, 1983; Siegel, 1995; Brown, 2002). Although behavioural comparison of dogs living in these cultures is lacking, we can assume that cultural differences (manifested in public attitude towards dogs, dog-keeping practices, and owner attitude) affect dogs' behaviour. Wan et al. (2009) showed that owners from the United States were more likely to keep their dogs indoors, to report that their dogs were kept as pets, and to engage their dogs in a greater number of training varieties (e.g., conformation training, agility training) than Hungarian dog owners. Although the relationship with behaviour is not quite evident, American owners reported that their dogs are more confident and aggressive than dogs kept by Hungarian owners. Future cross-cultural studies on family dogs should combine the use of surveys with observational methods because the perception of owners living in different cultures may be biased.

Even within the same culture, owners' attitudes towards their dogs differ, which may affect dogs' behaviour. In the United Kingdom, dogs chosen for companionship—approximately 80% of the sample—showed lower rates of competitive aggression than dogs acquired for protection, breeding, or exercise (Jagoe & Serpell, 1996). In contrast, Kobelt et al. (2003) did not find such associations in an Australian pet dog population, where 52% of owners reported that companionship was the reason for getting the dog.

A more anthropomorphic attitude is to consider the dog as a family member. In a study by Kubinyi et al. (2009), 93% of the 14,004 respondents marked the 'family member' category as the function of their dogs. 'Family member' dogs were reported to be less trainable but more sociable than dogs for working, guarding, etc. Mirkó et al. (2012) asked owners, 'To what extent do you think your dog understands human speech?' taking the answer as a measure of anthropomorphism. Dogs considered to understand words only were rated more aggressive than the ones that were believed to understand sentences. Owners who considered that their dogs understand quite well what people are talking about rated their dogs the least aggressive (Mirkó et al., 2012). Similarly, owners who are highly attached to their dogs reported more separation problems (Konok et al., 2011) and scored their dogs higher on trainability (Hoffman et al., 2013).

Gender of the owner may be also associated with personality traits in dogs. Bennett and Rohlf (2007) found that men reported having more disobedient dogs. By comparing the opinions of 2,146 men and 8,372 women on the behaviour of their adult dogs, Kubinyi et al. (2009) found that women's dogs were more trainable, more sociable, and less bold than men's dogs. In contrast, Kotrschal et al. (2009) observed that dogs of male owners behaved more socially and were more active than dogs of female owners.

Age of owners also could be associated in several ways with dogs' behaviour. Older participants in Bennett and Rohlf's study (2007) reported that their dogs were more likely to appear anxious. In Kubinyi et al.'s (2009) sample, people 19–30 years old reported having the least calm dogs. The most trainable and sociable dogs could be found in the 31–60-year-old owner group. However, the boldness scores of these dogs were lower than those of dogs living with 19–30-year-old owners.

Number of people and other dogs in the household can also influence the behaviour of dogs, but this factor has received little attention yet. In an Australian sample, dogs from larger families were rated as more disobedient and more unfriendly/aggressive (Bennett & Rohlf, 2007). Kubinyi et al. (2009) reported also that dogs in larger families are found to be less social towards conspecifics. The presence of other dogs in the household is linked with higher calmness and trainability but decreased boldness (Kubinyi et al., 2009; but see Kobelt et al., 2003, for contrasting results).

Owners who engaged in training activities report that their dogs are more obedient, less nervous, and friendlier towards people and dogs (Bennett & Rohlf, 2007; Jagoe & Serpell; 1996; Kobelt et al., 2003; Kubinyi et al., 2009). However, the use of some training techniques (e.g., positive punishment) might be associated with aggression (Casey et al., 2013).

Traumatic experiences or improper socialisation probably also affects dogs' personality. Dogs acquired after 3 months of age are less calm (confident) than dogs acquired sooner (Kubinyi et al., 2009) and show more behaviour problems, such as increased fear of other dogs (Serpell & Jagoe, 1995). In line with this, dogs bought in pet stores compared to dogs obtained from non-commercial breeders show significantly greater aggression towards family members, unfamiliar people, and other dogs; greater fear of other dogs and non-social stimuli; and greater separation-related problems and house soiling according to the owners (McMillan et al., 2013).

As we have seen, many extrinsic environmental factors can influence the behaviour of dogs. However, looking at such effects is hampered by several problems. First, most studies have a relatively small sample size to account for the main effects and interactions of several potential variables included in an exploratory statistical analysis. Second, effects found could still be non-causal ones or could be based on undetected background factors. Third, such effects may vary from country to country, so the generality may be questionable, and finally, most of these data are provided by dog owners, whose attitude may also defy objectivity.

## 7.3.4 Owners and Their Dogs: Is There Personality Matching?

For pet dogs, the owner can be regarded as a special 'environmental factor'. As we saw previously, many characteristics of the owner, such as gender, age, education, and previous experience with dogs, were found to be associated with the personality of the dogs. Moreover, dog-keeping characteristics in the household also depend on the owner—for example, how much time he/she spends with the dog, how much training the dog receives, and whether or not the dog is allowed inside the house. Importantly, some of these factors may be (partly) influenced by the owner's personality. For example, extroverted (active, outgoing) owners may prefer longer or more frequent walks with their dogs, while lazy owners prefer to stay at home. Because personality affects how we interact with others, the owner's personality may also influence how he/she behaves towards the dog. For example, it was found that neurotic owners give more commands to their dogs, whereas extroverted owners praise more often during a simple obedience task (Kis et al., 2012).

Not surprisingly, in the past few years, there has been a growing research interest in the characteristics of dog owners. Studies have focused on four main topics: (1) personality differences between owners and non-owners (see Podberscek & Gosling, 2000, for a review); (2) personality differences between owners of different species (e.g., self-defined dog people are more extroverted, agreeable, conscientious, but less neurotic and open than cat people: Gosling et al., 2010); (3) the association between the owners' characteristics and some behaviour problems of the dogs (e.g., owners of high aggression dogs were more likely to be tense, emotionally less stable, shy, and undisciplined than owners of low aggression dogs: Podberscek & Serpell, 1997); and (4) direct comparisons of the owner and dog personality traits and/or behaviour.

The fourth topic leads us to an interesting question: is there a personality match between owners and dogs? One may suppose that given the overlap of traits in human and animal personality (e.g., Gosling & John, 1999), there is a positive association between corresponding personality traits in owners and their dogs. This question requires a cross-species approach; the respective personality traits should be compared between the two species (note that the manifestation of these traits may be different). Zeigler-Hill and Highfill (2010) adapted the interpersonal circumplex model to dogs (a human personality model assessing social behaviour) and showed that the owners were more satisfied with their pets if the pets' perceived level of 'warmth' was similar to their own. The human Five Factor Model could also provide a common tool for cross-species personality comparisons because these five traits show some generality across species (Gosling & John, 1999). Applying the Big Five framework, Cavanaugh et al. (2008) and Turcsán et al. (2012) found positive correlations between the owners' self-reported personality traits and their personality trait assessment of their dogs. Neurotic owners were more likely to assess their dogs as nervous, anxious, and emotionally less stable, whereas extroverted owners assessed their dogs as energetic, enthusiastic, and sociable. Although the direct similarity was only at a medium level—ranging from $r = 0.252$ (agreeableness) to $r = 0.458$ (neuroticism)—it was much higher than that between random-created dog–owner pairs—ranging from $r = -0.051$ (conscientiousness) to $r = 0.041$ (agreeableness) (Turcsán et al., 2012).

Moreover, both Kwan et al. (2008) and Turcsán et al. (2012) found that peer persons also rate the owner and dog to be similar to each other, suggesting that the correlations found between the owners' self ratings and dog ratings cannot be attributed solely to the owners' subjective judgement. The length of ownership did not influence the degree of similarity (Turcsán et al., 2012). Thus, it seems that owner and dog do not become more similar to each other over time.

But if this matching is not the result of a convergent process, what may be the cause? In personality psychology, there is an extended literature concerning the personality similarity in human social relationships. People's choice for social partners is influenced by their similarity in physical attractiveness (Feingold, 1988) and psychological traits (e.g., Luo & Klohnen, 2005). Similarity in personality also relates to the relationship functioning: it predicts greater relationship satisfaction (Karney & Bradbury, 1995). Because family dogs are usually considered to be companions, friends, or family members, the cultural, cognitive, and psychological factors that affect one's choice of social partners could be the same when choosing a dog companion. People may choose an individual dog or a breed because they find some aspects of their behaviour attractive and/or similar to themselves. Of course, different people find different traits attractive: self-dependent owners may place more value on the independence in a dog, whereas other owners regard their dogs as social support and desire dogs that show more affiliative behaviour. Accordingly, the owners' breed choice could consciously or unconsciously reflect on the owners' personality. Owners of 'vicious breeds' scored themselves higher in sensation seeking and primary psychopathy (Ragatz et al., 2009); persons with low agreeableness, high neuroticism, and high conscientiousness also preferred more aggressive breeds (Egan & MacKenzie, 2012).

Culture may strongly influence which personality traits match between dogs and owners: people expect similarity in different traits in different countries, similarly to other human social relationships (McCrae et al., 2008). Owner–dog similarity was found in extroversion and neuroticism in Austria (Turcsán et al., 2012) and in the United States (Cavanaugh et al., 2008), while all the Big Five traits were found in Hungary (Turcsán et al., 2012).

The presence of another dog in the household may also affect the similarity between the owner and dog in the household. Single dogs show the highest similarity to the owner, whereas in the case of multidog households, the dogs' similarity patterns complement each other (Turcsán et al., 2012). In this case, dogs may play different social and practical roles in the family (e.g., one is emotionally more close to the owner, while the other is more suitable for outdoor activities, such as dog sports) that lead to differences in the similarity pattern to the owner. Alternatively, owners may also need time to experience some degree of similarity; thus, the owners' choice of the second dog may reflect their need to fill the similarity gap left by their first dog.

Taken together, personality similarity is an important factor in dog–owner relationship. As in other human social relationships, owners also require some degree of similarity from their dogs, and this need for similarity may play a role in the choice of breed and individual dog.

## 7.4 CONSISTENCY OF PERSONALITY TRAITS

Although adult personality is conceptualised as being stable over a considerable length of time, it is not necessarily resistant to aging or other major influences such as hormonal changes caused by neutering. The fact that personality changes with age is well known for humans (McAdams & Olson, 2010; Soto et al., 2011), and especially old age could affect central traits of personality such as neuroticism (Teachman, 2006). Family dogs are among the few animals that are allowed to age in a relatively buffered environment. This offers an interesting possibility to investigate how old age changes some personality traits in dogs.

Neutering of dogs has become part of 'responsible ownership' in the United States, where the majority of owned dogs are neutered (Trevejo et al., 2011). Also, dogs are often neutered to bring about behavioural changes, for example, to reduce aggression in both sexes. Thus, it is expected that personality traits may be affected through the modified actions of hormones.

### 7.4.1 Factors Affecting Adult Personality

Although old dogs make up an increasing part of the dog population, specific studies on them are quite rare. Older dogs are generally calmer (less anxious), less trainable, less social (extrovert), and less bold than younger dogs (Bennett & Rohlf, 2007; Kubinyi et al., 2009; Ley & Bennett, 2008; but Strandberg et al., 2005, observed higher boldness in older dogs). Partly in harmony, partly in contrast with these results, Salvin et al. (2011) reported that activity and play levels, response to commands, and fears and phobias show considerable deterioration with growing age. However, Seksel et al. (1999) did not find any associations with age in a behavioural test battery.

Neutering may affect dogs' personality in two ways. First, changes in hormone levels may directly change attitude to specific situations; thus, reduction of testosterone may decrease the probability of initialising an aggressive interaction. Second, the subject's behaviour may be affected also because of the changed reaction of others towards it. Thus, neutered males may be perceived as females by other dogs, which could release intolerant behaviour on their side. Despite many general effects of neutering, the effect on behaviour is often individual specific.

Generally, neutered dogs are reported to be more nervous (Bennett & Rohlf, 2007; Kubinyi et al., 2009) and less bold (Goddard & Beilharz, 1983; Wilsson & Sundgren, 1997a; Ruefenacht et al., 2002). Neutered females and intact males are reported to be more trainable compared to neutered males and intact females (Hart & Miller, 1985; see Bradshaw et al., 1996; Ruefenacht et al., 2002; Notari & Goodwin, 2007; and for similar results, Bennett & Rohlf, 2007; and Seksel et al., 1999, who did not find differences between the trainability of males and females). Neutering seems to decrease boldness in females (Kubinyi et al., 2009). Although some females may show increased aggression, behavioural problems are generally reduced or disappear after neutering (74% in males, 59% in females; see Heidenberger & Unshelm, 1990). Activity is also decreased in 20%–40% of neutered dogs (Heidenberger & Unshelm, 1990).

However, we must emphasise again that the associations do not imply causal relationships. Neutering could well be the consequence of having experienced a behavioural problem, not the reason for showing a particular trait (see also Guy et al., 2001a,b).

## 7.4.2 Emergence of Personality in Development

As discussed previously, personality emerges as a result of an epigenetic process. Theoretically, this means that personality changes continuously, but in practice, there are phases of life when these changes slow down or speed up. Personality of a developing dog is subject to transient changes; that is, shorter-longer stable periods are interrupted by fast diversifications. Using data from the Dog Mentality Assessment (DMA), Svartberg (2007) estimated that dogs' personality stabilises around the age of 1–2 years (the more exact timing is probably breed-dependent), but it continues to change at a slower pace during the mature years, and very likely also in old age. For example, he observed that sociability and aggressiveness decreased slightly, whereas fearfulness increased within a range of 5 years after maturation.

Puppyhood would be a very good time to judge the personality traits of dogs, especially because such information could help in providing the most appropriate social environment for it. However, especially during these early times, puppies experience a wide range of significant environmental events for the first time in their life, which include interactions with other family members, especially littermates. Scott and Fuller (1965) provided an extensive description of the early socialisation period in dogs. They demonstrated that dogs have a sensitive period between 5 and 12 weeks of age that is critical for the development of typical social interaction with both conspecifics and heterospecifics (humans). In the same period, it is also important to collect experiences about the non-living environment. This means that puppies deprived of such experiences are more likely to become shy and fearful later in life. Early, more extensive perinatal tactile stimulation by humans affects the social behaviour of 8-week-old puppies raised in a human social environment, and such dogs may become emotionally more stable later in life (Gazzano et al., 2008).

Scott and Fuller (1965) also showed that during the same time (and somewhat later) social interactions can also affect the puppies' behaviour. In addition to the social effect of the mother, playful (or more serious) interactions with littermates can affect the behavioural tendencies of a puppy from one day to the other. Access to milk or solid food and winning or losing fights over resources may have an effect on the personality traits of the puppy later.

## 7.4.3 The Predictive Significance of Puppy Testing

After reviewing the genetic basis of personality, no one would deny the importance of puppy testing. Generally, dogs are expected to fit well in their future anthropogenic environment, readily join the owners' activities, interact with other dogs and humans in an acceptable manner, and acquire the rules of the household and the broader community. Training a dog requires considerable time and effort; thus, no training center would like to invest in a dog that is not competent in a given task. A cowardly police dog has little chance of stopping an avenging soccer hooligan, and a hyperactive service dog will be of no help to its disabled owner. Only a low percentage of dogs are able to qualify in special tasks. For example, at the Japan Guide Dog Association, more than half of all—already considerately selected—candidate dogs fail (Kobayashi et al., 2013).

The development of reliable and valid puppy tests is equally important from welfare, human health, and economical aspects. In these tests, individuals are repeatedly tested in early puppyhood, at a juvenile age, and later in adulthood with the aim of evaluating the predictability of certain early behavioural characteristics. The devoted efforts of working dog trainers and researchers offer a range of puppy tests for predicting qualification-related traits (e.g., Wilsson & Sundgren, 1998; Slabbert & Odendaal, 1999). Generally, these tests demonstrate some predictive power that depends very much on the trait investigated.

A recent meta-analysis of available puppy tests (Fratkin et al., 2013) showed that temperament tests are much better in predicting submission and aggression than fearfulness and trainability. Thus, the latter traits could be influenced more efficiently by socialisation. However, the same traits are not required for all work; thus, working dog facilities must necessarily develop their own evaluating methods.

For example, Arata et al. (2010) used a questionnaire method to assess 1-year-old dogs' behaviour during early training. They found that dogs concentrating on the handler, showing stable behaviour, obeying commands, and showing little interest in other dogs were more likely to pass the exam. These correlated traits (named 'Distraction') could be predicted at the age of 5 months by evaluating exploration, excitability towards strangers, and similar items (Kobayashi et al., 2013). The predictive power of the puppy test applied by the Czech Law Enforcement Canine Breeding Facility is also reliable for predicting future service ability and is used for selecting dogs at the age of 7 weeks. Trainers look for relatively heavy dogs with good retrieving skills and those that like to chase a rug but display no noise phobia (Svobodová et al., 2008).

Not surprisingly, predictability increases with the age of the dog at testing (e.g., in the case of fearfulness in Goddard & Beilharz, 1986; aggression in Slabbert & Odendaal, 1999). One should consider also that temperament traits may not develop in synchrony. Testing puppies for sociability, neophobia, or activity at 8 weeks of age failed to be predictive (Wilsson & Sundgren, 1998), whereas tests on startle behaviour provided more reliable results (Slabbert & Odendaal, 1999). Moreover, certain traits, e.g., fearfulness, may change (more) during development (Goddard & Beilharz, 1984). Before 12 weeks of age, dogs reduced their activity in fearful situations, but adult dogs in similar situations became either passive or overtly active.

Other factors, such as litter size and season of birth, could also predict some elements of adult behaviour (Foyer et al., 2013). Physical engagement, i.e., interest in objects and better retrieval skills, are more pronounced in 1–2-year-old dogs born during the harsh winter time. Dogs from a smaller litter are more confident, and early growth rate is linked to higher social engagement.

In summary, in the case of puppy tests, the jury is still out. More systematic tests, at different training facilities and on different dog breeds, need to be done before their usefulness for selecting dogs can be estimated.

## 7.5 CONCLUSIONS

In Western cultures and industrial societies, most dogs are considered to be members of the family. Thus, dogs offer the possibility of prospective owners to choose at least one of their family members. As we have shown here, owners try to find companion dogs that are similar to themselves, and they are usually successful. However, learning about the genetic and environmental factors affecting dogs' personality and knowing some tools for characterising behaviour enable owners and dog trainers to choose more efficiently an appropriate individual as a member of the family or for work. As a result, conflicts between pet dogs and people can be largely avoided, enhancing the well-being of dogs, reducing the health consequences of dog–human conflicts (bitings), and reducing dog-keeping costs, especially in working facilities. Uncovering the genetic risks may enable future owners and dog trainers to predict the behavioural tendencies of a chosen dog even at puppyhood. The investigation of gene–environment interactions could pave the way for animal welfare, because genetically suitable individuals could be chosen for an environment with well-known challenges. However, one should not forget that due to polygenic genetic effects, gene–environment interaction, and limited genetic penetrance, the predictive value of any genetic marker or allele will always be limited and/or restricted to particular circumstances.

From a more academic perspective, research should continue to apply dogs as the model of human social behaviour and genetics. Importantly, this animal model can be regarded as 'natural' (in contrast to standard laboratory animal species) because dogs share the anthropogenic environment with us and are exposed to similar environmental challenges (e.g., pollution) and social influences as humans. We expect that natural animal models have a better validity in many respects than the corresponding laboratory animal models (Overall, 2000), in which the actual 'condition' is often achieved by direct induction of the malformation (e.g., in knock-out animals) that emerges in dogs spontaneously.

Domestication may have facilitated the emergence of different socio-cognitive skills in dogs, enhancing their chances of survival in human social groups (Miklósi, 2007). Understanding the underlying genetic and developmental mechanisms of dog behaviour broadens our view of evolutionary processes, especially regarding the mechanism that determines how relatively small genetic effects could lead to major phenotypic changes. Dogs are also optimal models for the mapping of complex traits (Karlsson & Lindblad-Toh, 2008). The genetic sequence information of dogs is considered as a standard for comparison to the human genome system (Wayne & Ostrander, 2004). Dogs' phenotypic diversity and the large number of genetic diseases common to humans (Wayne & Ostrander, 2007) provide a unique possibility to model biologically relevant questions of basic human interests.

## ACKNOWLEDGEMENTS

This work was supported by the Hungarian Scientific Research Fund (K 84036, T049692), the Bolyai Foundation of the Hungarian Academy of Sciences, the MTA-ELTE Comparative Ethological Research Group (01 031), and the ESF Research Networking Programme 'Comp-Cog': The Evolution of Social Cognition (www.compcog.org) (06-RNP-020).

## REFERENCES

Arata, S., Momozawa, Y., Takeuchi, Y., Mori, Y., 2010. Important behavioural traits for predicting guide dog qualification. J. Vet. Med. Sci. 72, 539–545.

Bennett, P.C., Rohlf, V.I., 2007. Owner–companion dog interactions: relationships between demographic variables, potentially problematic behaviours, training engagement and shared activities. Appl. Anim. Behav. Sci. 102, 65–84.

Bradshaw, J.W.S., Goodwin, D., 1998. Determination of behavioural traits of pure-bred dogs using factor analysis and cluster analysis; a comparison of studies in the USA and UK. Res. Vet. Sci. 66, 73–76.

Bradshaw, J.W.S., Goodwin, D., Lea, A.M., Whitehead, S.L., 1996. A survey of the behavioural characteristics of pure-bred dogs in the United Kingdom. Vet. Rec. 138, 465–468.

Brown, S.-E., 2002. Ethnic variations in pet attachment among students at an American school of veterinary medicine. Soc. Anim. 10, 249–266.

Carrier, L.O., Cyr, A., Anderson, R.E., Walsh, C.J., 2013. Exploring the dog park: relationships between social behaviours, personality and cortisol in companion dogs. Appl. Anim. Behav. Sci. 146, 96–106.

Casey, R.A., Loftus, B., Bolster, C., Richards, G.J., Blackwell, E.J., 2013. Inter-dog aggression in a UK owner survey: prevalence, co-occurrence in different contexts and risk factors. Vet. Rec. 172, 127.

Cavanaugh, L.A., Leonard, H.A., Scammon, D.L., 2008. A tail of two personalities: how canine companions shape relationships and well-being. J. Bus. Res. 61, 469–479.

Christiansen, F.O., Bakken, M., Braastad, B.O., 2001. Behavioural differences between three breed groups of hunting dogs confronted with sheep. Appl. Anim. Behav. Sci. 72, 115–129.

Cloninger, C.R., 2002. Relevance of normal personality for psychiatrists. In: Ebstein, B.J., Belmaker, R. (Eds.), Molecular genetics and the human personality. American Psychiatric Publishing, Washington, DC, pp. 33–42.

Coppinger, R., Schneider, R., 1995. Evolution of working dogs. In: Serpell, J.A. (Ed.), The domestic dog: its evolution, behaviour and interactions with people. Cambridge University Press, Cambridge, pp. 21–47.

Dodman, N.H., Karlsson, E.K., Moon-Fanelli, A., Galdzicka, M., Perloski, M., Shuster, L., Lindblad-Toh, K., Ginns, E.I., 2010. A canine chromosome 7 locus confers compulsive disorder susceptibility. Mol. Psychiatry 15, 8–10.

Duffy, D.L., Hsu, Y., Serpell, J.A., 2008. Breed differences in canine aggression. Appl Anim. Behav. Sci. 114, 441–460.

DuPaul, G.J., 1998. ADHD Rating Scale-IV: checklist, norms and clinical interpretations. Guilford Press, New York.

Egan, V., MacKenzie, J., 2012. Does personality, delinquency, or mating effort necessarily dictate a preference for an aggressive dog? Anthrozoos 25, 161–170.

Feingold, A., 1988. Matching for attractiveness in romantic partners and same-sex friends: a meta-analysis and theoretical critique. Psychol. Bull. 104, 226–235.

Fielding, W.J., 2008. Attitudes and actions of pet caregivers in New Providence, The Bahamas, in the context of those of their American counterparts. Anthrozoos 21, 351–361.

Foyer, P., Wilsson, E., Wright, D., Jensen, P., 2013. Early experiences modulate stress coping in a population of German shepherd dogs. Appl. Anim. Behav. Sci. 146, 79–87.

Fratkin, J.L., Sinn, D.L., Patall, E.A., Gosling, S.D., 2013. Personality consistency in dogs: a meta-analysis. PloS One 8, e54907.

Friedmann, E., Katcher, A., Meislich, D., 1983. When pet owners are hospitalized: significance of companion animals during hospitalization. In: Katcher, A., Beck, A. (Eds.), New perspectives on our lives with companion animals. University of Pennsylvania Press, Philadelphia, pp. 346–350.

Gácsi, M., McGreevy, P., Kara, E., Miklósi, A., 2009. Effects of selection for cooperation and attention in dogs. Behav. Brain Funct. 5, 31.

Gazzano, A., Mariti, C., Notari, L., Sighieri, C., McBride, E.A., 2008. Effects of early gentling and early environment on emotional development of puppies. Appl. Anim. Behav. Sci. 110, 294–304.

Gizer, I.R., Ficks, C., Waldman, I.D., 2009. Candidate gene studies of ADHD: a meta-analytic review. Hum. Genet. 126, 51–90.

Goddard, M.E., Beilharz, R.G., 1982. Genetic and environmental factors affecting the suitability of dogs as guide dogs for the blind. Theor. Appl. Genet. 62, 97–102.

Goddard, M.E., Beilharz, R.G., 1983. Genetics of traits which determine the suitability of dogs as guide-dogs for the blind. Appl. Anim. Ethol. 9, 299–315.

Goddard, M.E., Beilharz, R.G., 1984. A factor analysis of fearfulness in potential guide dogs. Appl. Anim. Behav. Sci. 12, 253–265.

Goddard, M.E., Beilharz, R.G., 1986. Early prediction of adult behaviour in potential guide dogs. Appl. Anim. Behav. Sci. 15, 247–260.

Goldberg, L.R., 1992. The development of markers for the Big-Five factor structure. Psychol. Assess. 4, 26–42.

Goldsmith, H., Buss, A., Plomin, R., Rothbart, M., Thomas, A., Chess, S., Hinde, R., McCall, R., 1987. Roundtable: what is temperament? Four approaches. Child Dev. 58, 505–529.

Gosling, S.D., Bonnenburg, A.V., 1998. An integrative approach to personality research in anthrozoology: ratings of six species of pets ant their owners. Anthrozoos 11, 148–156.

Gosling, S.D., John, O.J., 1999. Personality dimensions in non-human animals: a cross-species review. Curr. Dir. Psychol. Sci. 8, 69–75.

Gosling, S.D., Kwan, V.S.Y., John, O.P., 2003. A dog's got personality: a cross-species comparative approach to personality judgments in dogs and humans. J. Pers. Soc. Psychol. 85, 1161–1169.

Gosling, S.D., Sandy, C.J., Potter, J., 2010. Personalities of self-identified 'dog people' and 'cat people'. Anthrozoos 23, 213–222.

Gosling, S.D., Vazire, S., 2002. Are we barking up the right tree? Evaluating a comparative approach to personality. J. Res. Pers. 36, 607–614.

Guy, N.C., Luescher, U.A., Dohoo, S.E., Spangler, E., Miller, J.B., Dohoo, I.R., Bate, L.A., 2001a. A case series of biting dogs—characteristics of the dogs, their behaviour, and their victims. Appl. Anim. Behav. Sci. 74, 43–57.

Guy, N.C., Luescher, U.A., Dohoo, S.E., Spangler, E., Miller, J.B., Dohoo, I.R., Bate, L.A., 2001b. Risk factors for dog bites to owners in a general veterinary caseload. Appl. Anim. Behav. Sci. 74, 29–42.

Hall, N.J., Wynne, C.D.L., 2012. The canid genome: behavioural geneticists' best friend? Genes. Brain Behav. 11, 889–902.

Hare, B., Tomasello, M., 2005. Human-like social skills in dogs? Trends Cogn. Sci. 9, 439–444.

Hart, B.L., 1995. Analysing breed and gender differences in behaviour. In: Serpell, J. (Ed.), The domestic dog: its evolution, behaviour and interactions with people. Cambridge University Press, Cambridge, pp. 65–77.

Hart, B.L., Miller, M.F., 1985. Behavioural profiles of dog breeds. J. Am. Vet. Med. Assoc. 186, 1175–1180.

Heidenberger, E., Unshelm, J., 1990. Changes of behaviour in dogs after castration. Tierärztliche Praxis 13, 69–75.

Héjjas, K., Kubinyi, E., Rónai, Z., Székely, A., Vas, J., Miklósi, Á., Sasvári-Székely, M., Kereszturi, E., 2009. Molecular and behavioural analysis of the intron 2 repeat polymorphism in the canine dopamine D4 receptor gene. Genes. Brain Behav. 8, 330–336.

Héjjas, K., Vas, J., Kubinyi, E., Sasvári-Székely, M., Miklósi, Á., Rónai, Z., 2007a. Novel repeat polymorphisms of the dopaminergic neurotransmitter genes among dogs and wolves. Mamm. Genome 18, 871–879.

Héjjas, K., Vas, J., Topál, J., Szántai, E., Rónai, Z.S., Székely, A., Kubinyi, E., Horváth, Z.S., Sasvári-Székely, M., Miklósi, Á., 2007b. Association of polymorphisms in the dopamine D4 receptor gene and the activity-impulsivity endophenotype in dogs. Anim. Genet. 38, 629–633.

Hoffman, C.L., Chen, P., Serpell, J.A., Jacobson, K., 2013. Do dog behavioural characteristics predict the quality of the relationship between dogs and their owners? J. Human-Anim. Interact. 1, 20–37.

Hsu, Y., Serpell, J.A., 2003. Development and validation of a questionnaire for measuring behaviour and temperament traits in pet dogs. J. Am. Vet. Med. Assoc. 223, 1293–1300.

Huson, H.J., Parker, H.G., Runstadler, J., Ostrander, E.A., 2010. A genetic dissection of breed composition and performance enhancement in the Alaskan sled dog. BMC Genet. 11, 71.

Jagoe, A., Serpell, J., 1996. Owner characteristics and interactions and the prevalence of canine behaviour problems. Appl. Anim. Behav. Sci. 47, 31–42.

Jones, A.C., Gosling, S.D., 2005. Temperament and personality in dogs (*Canis familiaris*): a review and evaluation of past research. Appl. Anim. Behav. Sci. 95, 1–53.

Karlsson, E.K., Lindblad-Toh, K., 2008. Leader of the pack: gene mapping in dogs and other model organisms. Nat. Rev. Genet. 9, 713–725.

Karney, B.R., Bradbury, T.N., 1995. The longitudinal course of marital quality and stability: a review of theory, method, and research. Psychol. Bull. 118, 3–34.

Katcher, A.H., Beck, A.M., 1983. New perspective on our lives with companion animals. University of Pennsylvania Press, Philadelphia.

Kis, A., Turcsán, B., Miklósi, Á., Gácsi, M., 2012. The effect of the owner's personality on the behaviour of owner–dog dyads. Interact. Stud. 13, 371–383.

Kobayashi, N., Arata, S., Hattori, A., Kohara, Y., Kiyokawa, Y., Takeuchi, Y., Mori, Y., 2013. Association of puppies' behavioural reaction at five months of age assessed by questionnaire with their later 'distraction' at 15 months of age, an important behavioural trait for guide dog qualification. J. Vet. Med. Sci. 75, 63–67.

Kobelt, A.J., Hemsworth, P.H., Barnett, J.L., Coleman, G.J., 2003. A survey of dog ownership in suburban Australia—conditions and behaviour problems. Appl. Anim. Behav. Sci. 82, 137–148.

Konno, A., Inoue-Murayama, M., Hasegawa, T., 2011. Androgen receptor gene polymorphisms are associated with aggression in Japanese Akita Inu. Biol. Letters 7, 658–660.

Konok, V., Dóka, A., Miklósi, Á., 2011. The behaviour of the domestic dog (*Canis familiaris*) during separation from and reunion with the owner: a questionnaire and an experimental study. Appl. Anim. Behav. Sci. 135, 300–308.

Kotrschal, K., Schöberl, I., Bauer, B., Thibeaut, A.-M., Wedl, M., 2009. Dyadic relationships and operational performance of male and female owners and their male dogs. Behav. Processes 81, 383–391.

Kubinyi, E., Tóth, L., Héjjas, K., Sasvári-Székely, M., Topál, J., Gácsi, M., Miklósi, Á., 2006. Dopamine D4 receptor gene and individual differences in dogs. European Conference on Behavioural Biology (ECBB) 4–6 September, Belfast.

Kubinyi, E., Turcsán, B., Miklósi, Á., 2009. Dog and owner demographic characteristics and dog personality trait associations. Behav. Processes 81, 392–401.

Kubinyi, E., Vas, J., Héjjas, K., Rónai, Z., Brúder, I., Turcsán, B., Sasvári-Székely, M., Miklósi, Á., 2012. Polymorphism in the tyrosine hydroxylase (TH) gene is associated with activity-impulsivity in German shepherd dogs. PloS One 7, e30271.

Kwan, V.S.Y., Gosling, S.D., John, O.P., 2008. Anthropomorphism as a special case of social perception: a cross-species social relations model analysis of humans and dogs. Soc. Cogn. 26, 129–142.

Ley, J., Bennett, P., 2008. Measuring personality in dogs. J. Vet. Behav. 3, 182.

Ley, J., Bennett, P., Coleman, G., 2008. Personality dimensions that emerge in companion canines. Appl. Anim. Behav. Sci. 110, 305–317.

Ley, J.M., Bennett, P.C., Coleman, G.J., 2009. A refinement and validation of the Monash Canine Personality Questionnaire (MCPQ). Appl. Anim. Behav. Sci. 116, 220–227.

Liinamo, A.E., van den Berg, L., Leegwater, P.A., Schilder, M.B., van Arendonk, J.A., van Oost, B.A., 2007. Genetic variation in aggression-related traits in golden retriever dogs. Appl. Anim. Behav. Sci. 104, 95–106.

Lit, L., Belanger, J.M., Boehm, D., Lybarger, N., Haverbeke, A., Diederich, C., Oberbauer, A.M., 2013. Characterization of a dopamine transporter polymorphism and behaviour in Belgian Malinois. BMC Genet. 14, 45.

Lit, L., Schweitzer, J.B., Oberbauer, A.M., 2010. Characterization of human–dog social interaction using owner report. Behav. Processes 84, 721–725.

Luo, S., Klohnen, E.C., 2005. Assortative mating and marital quality in newlyweds: a couple-centered approach. J. Pers. Soc. Psychol. 88, 304–326.

McAdams, D.P., Olson, B.D., 2010. Personality development: continuity and change over the life course. Annu. Rev. Psychol. 61, 517–542.

McCrae, R.R., Martin, T.A., Hrebickova, M., Urbanek, T., Boomsma, D.I., Willemsen, G., Costa, P.T., 2008. Personality trait similarity between spouses in four cultures. J. Pers. 76, 1137–1164.

McMillan, F.D., Serpell, J.A., Duffy, D.L., Masaoud, E., Dohoo, I.R., 2013. Differences in behavioural characteristics between dogs obtained as puppies from pet stores and those obtained from non-commercial breeders. J. Am. Vet. Med. Assoc. 242, 1359–1363.

Mellersh, C., 2012. DNA testing and domestic dogs. Mamm. Genome 23, 109–123.

Miklósi, Á., 2007. Dog behaviour, evolution and cognition. Oxford University Press, New York.

Miklósi, Á., Topál, J., 2013. What does it take to become 'best friends'? Evolutionary changes in canine social competence. Trends Cogn. Sci. 17, 287–294.

Mirkó, E., Kubinyi, E., Gácsi, M., Miklósi, Á., 2012. Preliminary analysis of an adjective-based dog personality questionnaire developed to measure some aspects of personality in the domestic dog (Canis familiaris). Appl. Anim. Behav. Sci. 138, 88–98.

Miura, A., Bradshaw, J.W.S., Tanida, H., 2002. Childhood experiences and attitudes towards animal issues: a comparison of young adults in Japan and the UK. Anim. Welfare 11, 437–448.

Miyadera, K., Acland, G.M., Aguirre, G.D., 2012. Genetic and phenotypic variations of inherited retinal diseases in dogs: the power of within- and across-breed studies. Mamm. Genome 23, 40–61.

Nieoullon, A., 2002. Dopamine and the regulation of cognition and attention. Prog. Neurobiol. 67, 53–83.

Notari, L., Goodwin, D., 2007. A survey of behavioural characteristics of pure-bred dogs in Italy. Appl. Anim. Behav. Sci. 103, 118–130.

Overall, K.L., 2000. Natural animal models of human psychiatric conditions: assessment of mechanism and validity. Prog. Neuro-Psychopharmacol. Biol. Psychiat. 24, 727–776.

Parker, H.G., Kim, L.V., Sutter, N.B., Carlson, S., Lorentzen, T.D., Malek, T.B., Johnson, G.S., DeFrance, H.B., Ostrander, E.A., Kruglyak, L., 2004. Genetic structure of the purebred domestic dog. Science 304, 1160–1164.

Parker, H.G., Kukekova, A.V., Akey, D.T., Goldstein, O., Kirkness, E.F., Baysac, K.C., Mosher, D.S., Aguirre, G.D., Acland, G.M., Ostrander, E.A., 2007. Breed relationships facilitate fine-mapping studies: a 7.8-kb deletion co-segregates with Collie eye anomaly across multiple dog breeds. Genome Res. 17, 1562–1571.

Parker, H.G., Shearin, A.L., Ostrander, E.A., 2010. Man's best friend becomes biology's best in show: genome analyses in the domestic dog. Annu. Rev. Psychol. 44, 309–336.

Persson, M.L., Wasserman, D., Jönsson, E.G., Bergman, H., Terenius, L., Gyllander, A., Neiman, J., Geijer, T., 2000. Search for the influence of the tyrosine hydroxylase (TCAT)(n) repeat polymorphism on personality traits. Psychiat. Res. 95, 1–8.

Pervin, L., John, O.P., 1997. Personality: theory and research, Seventh ed. Wiley, New York.

Phillips, C., Coppinger, R.P., Schimel, D.S., 1981. Hyperthermia in running sled dogs. J Appl. Physiol. 51, 135–142.

Podberscek, A.L., Gosling, S.D., 2000. Personality research on pets and their owners. In: Podberscek, A.L., Paul, E.S., Serpell, J.A. (Eds.), Animals and us: exploring the relationships between people and pets. Cambridge University Press, Cambridge, pp. 143–167.

Podberscek, A.L., Serpell, J.A., 1997. Aggressive behaviour in English cocker spaniels and the personality of their owners. Vet. Rec. 141, 73–76.

Ragatz, L., Fremouw, W., Thomas, T., McCoy, K., 2009. Vicious dogs: the antisocial behaviours and psychological characteristics of owners. J. Forensic Sci. 54, 699–703.

Rooney, N., Bradshaw, J.W.S., 2004. Breed and sex differences in the behavioural attributes of specialist search dogs—a questionnaire survey of trainers and handlers. Appl. Anim. Behav. Sci. 86, 123–135.

Ruefenacht, S., Gebhardt-Henrich, S., Miyake, T., Gaillard, C., 2002. A behaviour test on German shepherd dogs: heritability of seven different traits. Appl. Anim. Behav. Sci. 79, 113–132.

Salvin, H.E., McGreevy, P.D., Sachdev, P.S., Valenzuela, M.J., 2011. Growing old gracefully—Behavioural changes associated with "successful aging" in the dog, *Canis familiaris*. J. Vet. Behav. 6, 313–320.

Scott, J.P., Fuller, J.L., 1965. Genetics and the social behaviour of the dog. University of Chicago Press, Chicago, IL.

Seksel, K., Mazurski, E.J., Taylor, A., 1999. Puppy socialisation programs: short and long term behavioural effects. Appl. Anim. Behav. Sci. 62, 335–349.

Serpell, J.A., Hsu, Y., 2005. Effects of breed, sex, and neuter status on trainability in dogs. Anthrozoos 18, 196–207.

Serpell, J.A., Jagoe, J.A., 1995. Early experience and the development of behaviour. In: Serpell, J. (Ed.), The domestic dog: its evolution, behaviour and interactions with people. Cambridge University Press, Cambridge, pp. 79–102.

Sheppard, G., Mills, D.S., 2002. The development of a psychometric scale for the evaluation of the emotional predispositions of pet dogs. J. Comp. Psychol. 15, 201–222.

Siegel, J.M., 1995. Pet ownership and the importance of pets among adolescents. Anthrozoos 8, 217–223.

Sih, A., Bell, A., Johnson, J.C., 2004. Behavioural syndromes: an ecological and evolutionary overview. Trends Ecol. Evol. 19, 372–378.

Slabbert, J.M., Odendaal, J.S.J., 1999. Early prediction of adult police dog efficiency—a longitudinal study. Appl. Anim. Behav. Sci. 64, 269–288.

Soto, C.J., John, O.P., Gosling, S.D., Potter, J., 2011. Age differences in personality traits from 10 to 65: Big-Five domains and facets in a large cross-sectional sample. J. Pers. Soc. Psychol. 100, 330–348.

Strandberg, E., Jacobsson, J., Saetre, P., 2005. Direct and maternal effects on behaviour in German shepherd dogs in Sweden. Livest Prod. Sci. 93, 33–42.

Svartberg, K., 2002. Shyness-boldness predicts performance in working dogs. Appl. Anim. Behav. Sci. 79, 157–174.

Svartberg, K., 2005. A comparison of behaviour in test and in everyday life: evidence of three consistent boldness related personality traits in dogs. Appl. Anim. Behav. Sci. 91, 103–128.

Svartberg, K., 2006. Breed-typical behaviour in dogs—historical remnants or recent constructs? Appl. Anim. Behav. Sci. 96, 293–313.

Svartberg, K., 2007. Individual differences in behaviour—dog personality. In: Jensen, P. (Ed.), The behavioural biology of dogs. CAB International, Cambridge, pp. 182–206.

Svartberg, K., Forkman, B., 2002. Personality traits in the domestic dog (*Canis familiaris*). Appl. Anim. Behav. Sci. 79, 133–155.

Svobodová, I., Vápenik, P., Pinc, L., Bartos, L., 2008. Testing German shepherd puppies to assess their chances of certification. Appl. Anim. Behav. Sci. 113, 139–149.

Takeuchi, Y., Hashizume, C., Arata, S., Inoue-Murayama, M., Maki, T., Hart, B.L., Mori, Y., 2009a. An approach to canine behavioural genetics employing guide dogs for the blind. Anim. Genet. 40, 217–224.

Takeuchi, Y., Kaneko, F., Hashizume, C., Masuda, K., Ogata, N., Maki, T., Inoue-Murayama, M., Hart, B.L., Mori, Y., 2009b. Association analysis between canine behavioural traits and genetic polymorphisms in the Shiba Inu breed. Anim. Genet. 40, 616–622.

Teachman, B.A., 2006. Aging and negative affect: the rise and fall and rise of anxiety and depressive symptoms. Psychol. Aging 21, 201–207.

Teplov, B.M., 1964. Problems in the study of general types of higher nervous activity in man and animals. In: Gray, J.A. (Ed.), Pavlov's typology: recent theoretical and experimental developments from the laboratory of B.M. Teplov. Pergamon Press, Oxford, pp. 3–153.

Tiira, K., Hakosalo, O., Kareinen, L., Thomas, A., Hielm-Björkman, A., et al., 2012. Environmental effects on compulsive tail chasing in dogs. PLoS One 7, e41684. http://dx.doi.org/10.1371/journal.pone.0041684.

Tochigi, M., Otowa, T., Hibino, H., Kato, C., Otani, T., Umekage, T., Utsumi, T., Kato, N., Sasaki, T., 2006. Combined analysis of association between personality traits and three functional polymorphisms in the tyrosine hydroxylase, monoamine oxidase A, and catechol-O-methyltransferase genes. Neurosci. Res. 54, 180–185.

Trevejo, R., Yang, M., Lund, E.M., 2011. Epidemiology of surgical castration of dogs and cats in the United States. J. Am. Vet. Med. Assoc. 238, 898–904.

Turcsán, B., Kubinyi, E., Miklósi, Á., 2011. Trainability and boldness traits differ between dog breed clusters based on conventional breed categories and genetic relatedness. Appl. Anim. Behav. Sci. 132, 61–70.

Turcsán, B., Range, F., Virányi, Z., Miklósi, Á., Kubinyi, E., 2012. Birds of a feather flock together? Perceived personality matching in owner–dog dyads. Appl. Anim. Behav. Sci. 140, 154–160.

Våge, J., Wade, C., Biagi, T., Fatjó, J., Amat, M., Lindblad-Toh, K., Lingaas, F., 2010. Association of dopamine- and serotonin-related genes with canine aggression. Genes. Brain Behav. 9, 372–378.

van den Berg, L., Vos-Loohuis, M., Schilder, M.B.H., van Oost, B.A., Hazewinkel, H.A.W., Wade, C.M., Karlsson, E.K., Lindblad-Toh, K., Liinamo, A.E., Leegwater, P.A.J., 2008. Evaluation of the serotonergic genes htr1A, htr1B, htr2A, and slc6A4 in aggressive behaviour of golden retriever dogs. Behav. Genet. 38, 55–66.

van den Berg, S.M., Heuven, H.C.M., van den Berg, L., Duffy, D.L., Serpell, J.A., 2010. Evaluation of the C-BARQ as a measure of stranger-directed aggression in three common dog breeds. Appl. Anim. Behav. Sci. 124, 136–141.

Varga, G., Székely, A., Antal, P., Sárközy, P., Nemoda, Z., Demetrovics, Z., Sasvári-Székely, M., 2012. Additive effects of serotonergic and dopaminergic polymorphisms on trait impulsivity. Am. J. Med. Genet. Part B. 159B, 281–288.

Vas, J., Topál, J., Péch, É., Miklósi, Á, 2007. Measuring attention deficit and activity in dogs: a new application and validation of a human ADHD questionnaire. Appl. Anim. Behav. Sci. 103, 105–117.

vonHoldt, B.M., Pollinger, J.P., Lohmueller, K.E., Han, E., Parker, H.G., Quignon, P., Degenhardt, J.D., Boyko, A.R., Earl, D.A., Auton, A., et al., 2010. Genome-wide SNP and haplotype analyses reveal a rich history underlying dog domestication. Nature 464, 898–902.

Wan, M., Héjjas, K., Rónai, Z.S., Elek, Z.S., Sasvári-Székely, M., Champagne, F.A., Miklósi, Á., Kubinyi, E., 2013. DRD4 and TH gene polymorphisms are associated with activity, impulsivity and inattention in Siberian husky dogs. Anim. Genet. in press.

Wan, M., Kubinyi, E., Miklósi, Á., Champagne, F., 2009. A cross-cultural comparison of reports by German shepherd owners in Hungary and the United States of America. Appl. Anim. Behav. Sci. 121, 206–213.

Wayne, R.K., Ostrander, E.A., 2004. Out of the dog house: the emergence of the canine genome. Heredity 92, 273–274.

Wayne, R.K., Ostrander, E.A., 2007. Lessons learned from the dog genome. Trends Genet. 23, 557–567.

Whiteside, S., Lynam, D., 2001. The Five Factor Model and impulsivity: using a structural model of personality to understand impulsivity. Pers. Indiv. Differ. 30, 669–689.

Wilsson, E., Sinn, D., 2012. Are there differences between behavioural measurement methods? A comparison of the predictive validity of two rating methods in a working dog program. Appl. Anim. Behav. Sci. 141, 158–172.

Wilsson, E., Sundgren, P.-E., 1997a. The use of a behaviour test for selection of dogs for service and breeding. I. Method of testing and evaluating test results in the adult dog, demands on different kinds of service dogs, sex and breed differences. Appl. Anim. Behav. Sci. 53, 279–295.

Wilsson, E., Sundgren, P.-E., 1997b. The use of a behaviour test for selection of dogs for service and breeding. II. Heritability for tested parameters and effect of selection based on service dog characteristics. Appl. Anim. Behav. Sci. 54, 235–241.

Wilsson, E., Sundgren, P.-E., 1998. Behaviour test for eight-week old puppies—heritabilities of tested behaviour traits and its correspondence to later behaviour. Appl. Anim. Behav. Sci. 58, 151–162.

Wright, H.F., Mills, D.S., Pollux, P.M., 2011. Development and validation of a psychometric tool for assessing impulsivity in the domestic dog (Canis familiaris). Int. J. Comp. Psych. 24, 210–225.

Zeigler-Hill, V., Highfill, L., 2010. Applying the interpersonal circumflex to the behavioural styles of dogs and cats. Appl. Anim. Behav. Sci. 124, 104–112.

# Chapter 8

# When the Bond Goes Wrong: Problem Behaviours in the Social Context

Daniel Mills,[1] Emile van der Zee,[2] and Helen Zulch[1]

*[1]Animal Behaviour Cognition and Welfare Group, School of Life Sciences, University of Lincoln, Riseholme Park, Lincoln, UK, [2]School of Psychology, University of Lincoln, Brayford Pool, Lincoln, UK*

## 8.1 INTRODUCTION

Although the study of intraspecific bonds can trace its origins to early ethologists such as Lorenz's famous geese imprinting studies, its broader embrace as a scientific discipline is a much more recent phenomenon, with the first international conference held around half a century later in Pennsylvania in 1981 (Katcher & Beck, 1983). Despite recent advances (e.g., Dwyer et al., 2006), there is still no consensus on how the human–dog relationship should be conceptualised or even systematically examined, nor how the nature of the human–dog bond should be explained (e.g., see Anderson, 2007). This poses challenges when trying to provide a coherent framework for the ways in which the relationship can go wrong, and there is a need for more empirical work which tests the predictions of current ideas. Dogs appear to be very sensitive to the emotional state (or at least the broad emotional valence; see Racca et al., 2012) of humans (affective empathy), but this does not require that they understand or even have insight into the thoughts of others (cognitive empathy/theory of mind). Unfortunately, the distinction between these two capacities is often missed, and it may even be assumed, especially by owners, that the latter is a prerequisite for the former. As a result, the social cognitive abilities of dogs may be greatly overestimated (Kaminski & Nitzschner, 2013), and the risk of unmet expectations in human–dog interactions is greatly increased. There may be misunderstandings about a dog's ability, inappropriate interpretations of its behaviour, and/or actions on our part, which are stressful for the dog to cope with and which can ultimately be a source of tension in the relationship. As illustrated throughout this chapter, the risk of problems will depend on our expectations of the relationship, so we start by identifying three types of relevant human–dog relationships, after which

The Social Dog. http://dx.doi.org/10.1016/B978-0-12-407818-5.00008-5

we discuss theoretical origins of the human–dog relationship as well as several dimensions of it, before we focus in more detail on specific problems that may occur within our relationship with dogs. It is only by considering the nature of the human–dog bond that we begin to see where problems may arise.

## 8.2 AN INITIAL FRAMEWORK FOR CONSIDERING HUMAN–DOG SOCIAL RELATIONSHIPS

Bidirectional relationships can be differentiated into those that primarily develop from the compatible emotional needs of the subjects (attachments) and those which originate from an alliance based on common interest (affiliations) (Weiss, 1998). Examples of the former include the mother–offspring relationship and other relationships characterised by a biologically mandated emotional dependency (attachment bond); by contrast, the emotional bond (affiliative bond) that exists within an affiliation is secondary to the common interest of the subjects and might include, for example, the social alliances ('friendships') that form between juveniles which allow them to engage in social play together with minimal risk of harm. However, in the case of the dog, there are also a large number of relationships which are more unidirectional, being based primarily on the needs of the owner (such as working partnerships, but also laboratory animals and certain pet–owner relationships) and, as such, enforced by humans for this purpose. In these situations, people are in a position of ultimate control over the maintenance and dissolution of the relationship, regardless of the interests of the dog. Nonetheless, there may still be a bond (i.e., agent that binds the two together), but again this is secondary to the force which brings the subjects together. So, in the context of dog–human relationships, a third category needs to be distinguished, which is characterised by the instrumental value of the dog to its owner (service relationship). To this list, we should add negative relationships, since in some situations dogs may be considered to be pests (a threat to human health and well-being, as occurs with strays, especially where rabies is endemic). This requires no bond between the two species, but nor are the subjects disinterested in each other, and so a relationship still exists between them. The human aversion to the presence of the dog may not be reciprocated by the dog, and it might be that this is a useful model of the endpoint of some closer social relationships in which the dog is ultimately abandoned or given up to a shelter.

A mature, attractive relationship may be held together by several types of bonds. For example, where there is a working partnership, affection towards the dog and concern for its interests can still exist as a strong feature. By distinguishing between the three relationships featuring a bond, we can focus on the most prominent reason for a relationship's existence as distinguished from other reasons that may hold at the same time. This is important because a different relationship holds different expectations and so is potentially at risk from different threats.

Problems can arise in any close relationship. For example, recent work (reviewed in Chapter 6) has established that the relationship between a dog and its owner offers the opportunity for the development of secure attachment (Bowlby, 1969). This means that the owner can function as a source of both safety and security for the dog, acting as a buffer against certain stressors in the environment. Although this type of attachment can help to protect the dog against the expression of certain problem behaviours, such as inappropriate fear responses, it may, in some circumstances, also predispose the dog to experiencing difficulty with coping when separated from the attachment figure, as discussed in further detail later in this chapter.

This example of a characteristic which may be beneficial in one context being problematic in another raises the wider fundamental question: What is it that makes a relationship problematic? The very use of the term 'problematic' in this context indicates that the relationship is not optimal, and so a cause for concern about the welfare of at least one individual within it. Nonetheless, although the well-being of those involved is obviously of importance, a concern for welfare is not the only issue that may make a relationship suboptimal. Relationships are characterised by the behaviours occurring within them, and so the root of any problem must lie within the constituent behaviours, i.e., a behaviour problem. This does not mean that the problematic behaviour which underpins the disturbance in the relationship is pathological in a medical sense (i.e., a qualitatively distinct disorder of the central nervous system—malfunctional *sensu;* Mills, 2003). Rather, it may reflect expressions of normal (albeit possibly extreme) cognitions and emotional responses in a way that is problematic to the owner. These may or may not be appropriate from the dog's perspective. Ultimately, it is how behaviours are perceived by the humans involved which leads to them (and the associated relationship) being labelled as a problem; i.e., it is the failure of a dog's behaviour to meet or fulfil owner expectations in one or more aspects of the interactions occurring between them which causes the relationship to be viewed as problematic by the owner. The expectations an owner has will depend on the nature of the relationship, as alluded to previously, so we consider this in the next section.

## 8.3 DIMENSIONS WITHIN HUMAN–DOG RELATIONSHIPS

Relationships and the expectations within them change over time; but whereas a service relationship will be primarily unidirectional, i.e., defined by the ability of one individual to fulfil the behavioural needs of the other with respect to the function it performs, an interpersonal (i.e., non-service) relationship is more bidirectional, i.e., defined to a greater extent by the interaction between the agents within it. However, despite different emphases, it should be remembered that all human–dog relationships are necessarily bidirectional and dynamic in their quality to a greater or lesser extent, and even among service dogs, such as military working dogs, the quality of the interpersonal relationship between handler and dog impacts on the functional performance of the dog (Lefebvre et al., 2007).

Hinde (1976) has provided a useful framework for describing relationships, and we suggest that this can be usefully adapted to characterise both successful and problematic human–dog relationships. He suggests there are at least eight dimensions to consider when defining a relationship, which additionally help to explain possible sources of tension within it that can lead to the disruption of the bond. In the latter instance, it is worth re-emphasising that it may be how the behaviour within these dimensions is perceived by the owner, rather than the objective reality of how the situation conflicts with the owner's needs, which may be key to the definition of the relationship as problematic. The dimensions Hinde describes are summarised in Table 8-1 and discussed further here:

1. The content of interactions within the relationship. This is perhaps the most obvious feature of a relationship and refers to what is done together. Certain types of activities may co-occur to evidence the function of relationship, with the quality of interaction defining a style within this. It might be argued that where the primary function of a relationship is for the dog to tolerate petting (physical, non-sexual gestures of intimacy), then this should be described as the pet dog–owner relationship, and the degree to which this occurs and appears to be mutually enjoyed may reflect the closeness of this relationship. However, often other attributes to the relationship exist between a dog in the home and its owner, reflecting variation in breadth of the relationship between these two individuals. This arises from the second dimension to be considered.

**TABLE 8-1** Three Different Types of Human–Dog Bonded Relationships and Eight Different Dimensions within These Relations Characterising the Human–Dog Bond*

| Relationships | Emotional involvement/ emotional dependency (attachment) | Dimensions within Relationships | Content of interactions Diversity of interactions Reciprocity versus complementarity of interactions |
|---|---|---|---|
| | Common interest (affiliations) | | Quality of interactions Frequency and patterning of interactions Intimacy |
| | Working partnership | | Cognitive perspectives of interactions Multidimensional qualities |

*Relations and dimensions are not mutually exclusive and can therefore overlap. This framework makes it possible to identify the source of different kinds of (perceived) problem behaviours resulting from within the relationship.

2. The diversity of interactions contained within the relationship. In the previous example of the pet dog–owner relationship, the relationship was very narrowly defined (on the basis of a single type of interaction); however, a relationship may be much broader. 'Pet' dog owners typically engage in a range of other interactions with their dogs, such as asking the dogs to respond to specific cues, which can affect their satisfaction within the relationship (Clark & Boyer, 1993). Although 'obedience' is another element of the content of the interactions described in the previous section, its inclusion on top of 'petting' adds to the diversity within the relationship, which is a different quality. Increased diversity demands increased physical and psychological ability, and expectations need to be realistic. For example, owners who work long hours may enjoy hill-walking and so choose a dog capable of joining them in this activity. However, they also want to enjoy calm and relaxed petting interactions with their dog when they get home at night after a hard day's work. Each of these activities is a realistic expectation on its own, but together, this diversity means specific measures may be required to keep the dog occupied when the owner is working, to ensure the dog can cope with these contrasting demands and reduce the risk of problematic behaviour. Pet dog–owner relationships are often multifaceted in this way (multiplex *sensu;* Hinde) and need to be managed accordingly. Therefore, the breadth of the relationship is an important dimension in its own right because it defines the range of activities that the owner expects to be shared within it. Relationships of different breadth may also be associated with fulfilling different needs and conferring different benefits.

3. Level of reciprocity versus complementarity of interactions within the relationship. In a reciprocal interaction, individuals show similar behaviour; thus, a dog may appear to reciprocate an owner's affection towards it with behaviours perceived to be equally loving. By contrast, in a complementary interaction, each may take on a different but compatible role; for example, an owner may take on the role of caregiver and the dog as care-receiver. The reasons for such a relationship are varied, but in some instances it seems likely that owners derive psychological benefit from the sense of 'feeling needed' in such a relationship. This implies that the success of the relationship depends on the willingness and degree to which individuals exhibit the expected reciprocity or complementarity. For example, secure attachment (see Chapter 6) and dominance–subordinance relationships depend on complementary responses by the individuals concerned. In such circumstances, power and control within the relationship are not evenly distributed, but resistance by either party to the role expected of it is likely to be a source of tension.

4. The quality of the interactions within the relationship. This refers to the style of interaction, with sensitivity often reflected by the latency to respond to the behaviour of each other, so that the exchange is highly synchronised and has a 'flow' to it. As reviewed by Smuts, in Chapter 4 of this volume, synchronicity between play partners may be a useful proxy for predicting the strength of

the affiliative bond between them. However, this may not always be a positive flow because resistance may also 'flow' within an escalating series of antagonistic gestures that may ultimately result in a physical attack (by either party). In dogs, this series of expressions has been described as the ladder of aggression (Shepherd, 2002), although the relationship between the elements on each rung of the ladder may not be ordinal as implied by the metaphor. While aggression is often thought of as a single behaviour, from an interaction's perspective, aggressive behaviour is perhaps more usefully considered as a style of response that may be applied to many functionally different types of behaviour, such as resource or mate acquisition or control of a situation. It involves risking harm to achieve a goal which is blocked by another individual, but an alternative strategy might be to tolerate this situation. Humans and dogs naturally express their affection towards conspecifics in very different ways, and a failure to appreciate this can lead to problems in the quality of interaction between an owner and his/her dog. Needless to say, if an owner's expressions of attraction towards his/her dog (such as attempts to hug it) are rejected (perhaps because these expressions are perceived as scary by the dog) in ways that are perceived to be overtly aggressive (such as through snarling or snapping), a serious threat to the relationship arises. This is discussed in more detail later, but it is worth noting here that perhaps in many situations dogs merely tolerate this type of interaction and do not enjoy it, so the quality of this type of interaction may be relatively poor. This can be managed simply by teaching owners how to behave in more enjoyable ways towards their dog.

5. Frequency and patterning of interactions within the relationship. The rate at which certain behaviours or exchanges relevant to a relationship occur or the order in which they occur may be very important dimensions to the nature of the relationship. For example, as discussed later, certain attention-seeking behaviours by a dog may be highly desirable to owners, but if these become too frequent, then they may be regarded as a nuisance and the relationship may then be deemed problematic by the owners. The behaviour itself does not necessarily differ between acceptable and problematic attention seeking, and indeed the behaviour may be intermittently reinforced by the owners in certain situations; it is the frequency of the manifestation of the behaviour that is important to consider here.

6. The intimacy of a relationship. Dog–human relationships are often characterised by varying levels of disclosure by the owner, which reflect a high level of intimacy. The extent to which owners share their secrets, worries, and other intimate concerns with their dog, without fear of betrayal, has been identified as an important feature of the relationship (Serpell, 1983), and may bring important psychological benefits. However, somewhat surprisingly, closer examination of this dimension appears to have been largely overlooked as a specific predictive factor of some of the benefits that may extend from dog ownership. From the dog's perspective, it might be that activities such as communal sleeping with the owner could reflect the closeness of the personal association perceived, but this has yet to be explored empirically.

7. Cognitive perspectives of the interactions. The type and extent of cognitive processes attributed by owners to their dogs' behaviour are central to the perceptions of their relationships. For example, owners may suggest their dog is jealous of a new baby as a result of observing a number of hostile interactions with the dog in the presence of the child. These interactions may be used to justify their own actions towards the dog and reframe their own relationship as a result, even though the dog may be simply trying to respond to the altered attention given to it since the child was born or to the novel stimuli, which the baby presents, such as crying. When a problem such as this has arisen, it is essential to identify and address these cognitive perspectives of the interactions occurring not only to recommend appropriate interventions to manage the problem behaviour, but also to re-establish a healthier relationship between the owners and their dog.

8. Multidimensional qualities. The style of responding across a range of functionally different interactions between two individuals may also be used to describe the general quality of the relationship that exists between them. From the owner's perspective, this may often take the form of an expression of the dog's personality relevant to his style of interacting. For example, the owner may describe him as a very loving or unselfish dog, implying these are stable aspects of the dog's relationship with others and expressed in a diverse range of interactions. At the other extreme, a dog may be described as unpredictable or moody, and this is often problematic unless a reason for this behaviour can be identified, such as chronic pain.

Each of these dimensions contributes to the success of a dog–owner relationship in different ways, but each also has the power to destroy the relationship. It is therefore important to be systematic in the analysis of interactions and expectations of owners when evaluating a relationship that has become problematic due to the behaviour of the dog. However, it is also central to a scientific approach to understanding how we can build more successful relationships. By being more precise in how we define relationships, we can more clearly identify the most important elements to be attended to and develop more specific and efficient programmes to capitalise on the associated benefits of the social dog as a result. This applies not only to the companion animal but the working dog too. For example, Lefebvre et al. (2007) found that dogs who lived in the handler's home and who also engaged in a sport activity with their owner showed fewer signs of poor welfare (pacing, destructiveness, howling, etc.) when penned at work. In addition, Horvath et al. (2008) found a difference in cortisol levels between groups of working dogs depending on the style of play involved, which reflected differences in authoritarianism in the relationship. Those handlers whose play style included higher levels of affiliative behaviour and fewer corrections correlated with dogs with lower post-play cortisol levels. It is also worth noting that Haverbeke et al. (2008) commented on a poor handler–dog relationship contributing to poor performance of military working dogs and how an intervention which they believe enhances the relationship improves both the dog's behaviour and welfare (Haverbeke et al., 2010). Unfortunately, though, at no point is the nature of the handler–dog relationship defined in a manner that allows its assessment between groups or over time. By contrast, Serpell and Hsu (2001) found that guide dog puppies who were more highly 'attached' to their handlers were at greater risk of being rejected later for protectiveness/ aggression towards strangers. Systematic definition of the relationship between a working dog and its trainer and/or handler, along the lines indicated here, would be beneficial for not only potentially increasing the predictive power of such observations, but also for designing and assessing interventions which may improve aspects of working dogs' performance, behaviour, and welfare and reduce the problems which can arise from impairments in any of these.

## 8.4  ORIGINS OF TENSION WITHIN THE RELATIONSHIP: THEIR EXPRESSION AND CONSEQUENCES

A successful relationship depends on the needs and expectations of those involved being met in an acceptable way. In the case of the human–dog relationship, this requires a degree of mutual compromise. Problems arise when there is a conflict of interests in the interaction perceived between pet and owner, and these problems can threaten the bond between the two. It is owners who have ultimate control over the breaking of the physical aspects of the bond (such as cohabitation), since they have the power to abandon the dog or have it euthanised. Although this relinquishment may be the ultimate result of a problematic relationship, it is important to realise that, from a psychological perspective, the bond may be broken or damaged long before this action is taken. This fracturing of the bond may also have repercussions for the welfare of both dog and owner long before actual relinquishment occurs.

If we accept that a behaviour problem is defined as any behaviour posing a problem to the dog's owner, this means that the problem arises from the perception of the dog's behaviour. This may be an accurate reflection of what the dog is doing or an inference about the underlying emotion or motivation behind the dog's activities (Morris et al., 2008), which may not even be consistent with our current knowledge of the cognitive abilities of dogs; for example, an owner may erroneously perceive the dog that is urinating in the house when it is left alone is being spiteful. Problematic inferences may be made by owners in the absence of evidence or even evidence of the absence of certain psychological abilities in the dog (e.g., Horowitz, 2009). This may be due to a strong tendency to anthropomorphise the dog's behaviour (Morris et al., 2000).

There is a danger that evidence of co-evolution of two highly social species (dogs and humans) within a single ecological niche is taken to suggest the evolution of similar cognitive abilities (Schleidt & Shalter, 2003). However, as reviewed by Kaminski and Nitzschner (2013), dogs appear to be extremely sensitive to only certain human communicative gestures without necessarily possessing much broader social cognitive abilities, including theory of mind.

Nonetheless, dogs appear to have an apparent ability to produce socially compatible behaviour patterns in response to our moods, and this is often cited as an important part of the special relationship that develops with them (Serpell, 1983). However, this may arise more from affective empathy (the ability to respond in an appropriate way in relation to the emotion expressed by another) rather than cognitive empathy (the ability to appreciate what another individual is thinking), since it has been shown recently that dogs are perceptually sensitive to differences in heterospecific emotional arousal (Racca et al., 2012). This affective sensitivity is not surprising because the core emotions have an ancient neurological origin (Panksepp, 1998), and their communicative value is one of their defining features (Scherer, 1984). Emotional sensitivity may therefore offer a more parsimonious explanation for what, on the surface, appears to be highly sophisticated social cognitive abilities. Recognising this difference and the limitations of canine social cognition is important when it comes to ensuring clients have realistic expectations, especially when adjustments of their perception or belief are central to the correction of a behaviour problem.

A range of perspectives have been used to help understand the psychological bond between a dog and human (Kidd & Kidd, 1987), and some of these are more useful than others when considering how it may become problematic. Evolutionary theories, which focus on the adaptive advantages of the relationship, focus primarily on why the bond may have arisen (e.g., Herzog & Burghardt, 1987) but may also indicate an intrinsic value within it. However, Naturalistic Theory (Kellert & Wilson, 1995) emphasises the importance of contact in reducing stress, and the selection of traits in dogs, which appear to provide psychological support to, and minimise the risk of conflict with, humans. From this perspective, it is suggested that humans may be predisposed to respond in certain physiologically beneficial ways to dogs, such as the release of endorphins and oxytocin, when petting a dog (e.g., Odendaal & Lehmann, 2000; Odendaal & Meintjes, 2003), and when this response is not forthcoming, the relationship may be challenged at a fundamental level. By contrast, the Affect Theory of Social Exchange (Lawler, 2001) predicts that the emotional bond arises from the rewards provided by the social exchanges that occur within the relationship, and so it is the specific needs of the owner and the dog's role in fulfilling these that provide the basis for the relationship. Accordingly, if the dog is unable to meet these needs of the owner, then the emotional bond will be put under strain. These frameworks for conceptualising the human–dog relationship place differing emphasis on the factors which may predict breakdown, i.e., failure to relax versus failure to be rewarded from interaction, but, to date, the relative importance of one of these over the other has not been tested empirically. However, there is indirect evidence to suggest that naturalistic theory is insufficient to explain the bond that exists and why it may break down. Many studies suggest behaviour problems, in general, increase the risk of relinquishment (e.g., Arkow & Dow, 1984; Patronek et al., 1996; New et al., 1999); a closer inspection by Corridan et al. (2010) found that this was not the case for noise fears. This might be because it is difficult to reframe the distress of a dog with this problem as anything other than an expression of suffering by the dog. By contrast, other problems, which may be equally important from a welfare perspective (e.g., separation problems), can be construed (albeit incorrectly) as something more antagonistic towards the owner (spitefulness, etc.) and so facilitate the breaking of the emotional bond more readily. These results are more consistent with the predictions of the Affect Theory of Social Exchange.

When one is investigating the causes of behaviour problems that challenge the bond, there is rarely a single cause in a developmental sense (though there may be specific triggers in a regulatory sense). Rather, problems typically arise as a result of certain risk factors, such as level of habituation to novelty when young, which relate to the animal's behavioural and emotional predispositions (such as fearfulness and how this can be expressed) in certain circumstances (Mills et al., 2012). Central to understanding these risks is the differentiation of the 'obedient' from the 'well-behaved' dog, as well as the behavioural, cognitive, and affective expectations of the owner. In the next section, we consider some specific examples of problems directly related to disturbance of the human–dog relationship, which can be divided broadly into those originating from the structure of the relationship and those originating from (mis)communication within it.

## 8.5 COGNITIVE AND AFFECTIVE CONSIDERATIONS OF SOME MANIFESTATIONS OF A PROBLEMATIC RELATIONSHIP

In this section, we consider the behavioural psychobiology of a range of behaviour problems associated with a compromised human–dog relationship and the implications this has for their successful management.

### 8.5.1 Problems Related to the Structuring of the Relationship

Several studies (e.g., Serpell, 1983) indicate that one of the most important features for owners of the companion dog relationship relates to the apparent affection shown by the animal. This aspect of the interaction between dog and owner arises from both the ability of dogs and humans to form complex affectionate social relationships and the way this is shaped through the interactions that occur between them. In this section, we focus on two ways in which this can give rise to a problematic relationship. In the first, we critically appraise the role of a core biological process that has recently attracted substantial scientific interest—the attachment bond; and in the second, we focus on the inadvertent shaping of a behavioural style in the form of attention-seeking behaviours.

### 8.5.2 Attachment-Related Problems

Bowlby (1958, 1969) described attachment as an enduring psychological bond that serves to improve an infant's chances of survival by keeping it close to its mother. This bond has several operationally definable characteristics: attached individuals seek to maintain proximity and contact with the attachment figure, attached individuals become distressed when involuntarily separated and show signs of pleasure upon their return, attachment figures act as a safe haven to which the attached individual will return when frightened by the environment, and attachment figures act as a secure base from which the attached individual can move off and engage confidently in activities. Many of these features can be assessed within the Strange Situation Test (SST) developed by Ainsworth (Ainsworth et al., 1978), and this has been adapted for dogs (Topal et al., 1998, and see Chapter 6 for a review). However, the clinical value of this work is not straightforward, and some have even suggested that separation-related problems in the dog are not, in fact, associated with attachment between a dog and its owner (Parthasarathy & Crowell-Davis, 2006).

A significant problem with investigating the link between attachment and separation problems is the frequently described diagnosis of 'separation anxiety'. In general, a clinical behavioural diagnosis often requires an inference about the cognitive and affective processes underlying a given problem. As such, the behaviour of the animal is merely a sign of an underlying process and should not, in itself, be confused with a diagnosis. There may be a range of emotional and cognitive factors which may differ between individuals and which give rise to 'physical or behavioural signs of distress exhibited by the animal only in the absence of, or lack of access to, the client' which are the necessary signs of 'separation anxiety' (Overall, 1997). For example, Mills et al. (2012), on the basis of our understanding of affective neuroscience (Panksepp, 1998), argue that the signs associated with this problem may arise as a consequence of activation of one (or more) of the following primary negative affective processes:

- The desire to reinstate contact with an attachment figure following separation (PANIC *sensu;* Panksepp, 1998)
- The frustration associated with containment (RAGE *sensu;* Panksepp, 1998)
- The fear associated with aversive stimuli and associations with the current environment, such as the occurrence of thunderstorms (FEAR *sensu;* Panksepp, 1998)

In addition, some of the signs such as destructiveness can also occur in the absence of distress (and so should not be considered to imply distress on their own), but rather they may reflect the activation of positive affective systems, such as investigation of the environment (SEEKING *sensu;* Panksepp, 1998).

This means what is commonly described as separation anxiety should not be interpreted as a diagnosis, but rather as a description of presenting signs. If we wish to manage it effectively and efficiently, we must investigate the evidence relating to which specific emotional systems may underpin the expression of the problem. To do this, we can borrow from the component process theory of emotion (see Scherer, 1984), using evidence triangulated from four different sources:

1. The association between the response and the anticipated or actual arrival or removal of specific stimuli
2. The physiological arousal profile, which supports action being taken in association with the triggering event
3. The wider behavioural tendencies (rather than just the specific response) shown at the time of arousal, e.g., to escape
4. An evaluation of the communicative elements of the animal's state (such as the use of howling, which is directed towards social contact or careful ethological analysis of facial features concurrent with the problem behaviour)

Unless a diagnosis is made at the motivational-emotional level, data on other behavioural and environmental associations with separation-related problems are of very limited value (e.g., Flannigan & Dodman, 2001; McGreevy & Master, 2008). Even if this is appreciated, misleading conclusions about the role of attachment in separation-related problems may still arise from failing to appreciate that the SST measures the style of attachment and not its intensity as is sometimes implied (Parthasarathy & Crowell-Davis, 2006). The term 'separation-related problem' is preferable to separation anxiety when trying to describe the presenting manifestation of problematic behaviour when the dog is left alone, as it is perhaps less likely to imply that a diagnosis has been made. Accordingly, some, but not all, separation-related problems involve a specific problem in the attachment bond between the dog and owner, but the value of tests like the SST in revealing this is seriously limited. Finally, it should be noted that regardless of the involvement of attachment, these problems can lead to disruption of the bond, due to the impact that the associated behaviours have on the owner's quality of life.

### 8.5.3 Attention-Seeking Problems

The Affect Theory of Social Exchange (Lawler, 2001) suggests that pleasurable elements of the interaction that occurs between the dog and its owner are likely to be reinforced. However, this could paradoxically lead to problems, as the reinforcement of behaviours that may be considered desirable in one context may lead to their expression in other situations in which they are inappropriate and thus may result in undesirable behavioural expressions that can challenge the bond. Data relating to this phenomenon are, however, very limited.

The only specific study of attention-seeking behaviour in dogs known to the authors is an online survey of 130 owners (Mills et al., 2010, and unpublished). Twenty-five behaviours and 20 scenarios involving attention-seeking behaviour, such as whining or pawing at the owner, were identified from a series of interviews prior to the development of this online questionnaire, in order to minimise the use of open responses in the survey. When the results were analysed, as expected, there was no association between the mechanism used to gain attention and the sex of the dog (because the prevalence of behaviours was simply the product of arbitrary reinforcement). The most commonly indicated context in which attention-seeking behaviour occurred was when the owner was sitting down nearby and the dog wanted to fuss or to play (70%), followed by when the dog was hungry (57%), and when the dog needed the toilet (55%). The only other context in which the majority of dogs were reported to seek attention from their owner was when the owner was interacting with other dogs (52%). These preliminary results are testament to the highly playful and social nature of the relationship between dogs and their owners, but the results point to some other potentially important considerations in relation to the social cognitive abilities of dogs and the bond that develops or breaks down. Twenty-six commonly occurring attention-seeking behaviours were identified, of which the most common, but not always endearing, were stands and stares (71%), hovers around and follows (65%), wags tail (62%), nudges with head/muzzle (60%), barks (55%), puts head on lap (55%), and whines/squeaks/whimpers and paws (both 51%). Twenty percent of owners reported that the most endearing behaviour shown by the dog could also become annoying in certain circumstances, and 10% reported that their dog would growl to get attention—the latter being identified as the most annoying behaviour shown by the dog as an attention-seeking behaviour. According to these reports, the most effective behaviours used by dogs to get attention were barking and putting the head on the lap, with barking also being the most commonly rated annoying attention-seeking behaviour of dogs (36%). Because the data also show that dogs changed their strategies when the owner was engaged in different tasks/levels of attention, they suggest that dogs are able to discriminate social contexts, such as the attention focus of the owner, and adapt their behaviour accordingly (Schwab & Huber, 2006). If an individual dog adapts its strategy in a manner deemed appropriate by the owner—for example, seeking attention at times when it is likely to be met with a positive response and refraining at times when the owner is obviously engaged in another activity—it seems reasonable to suggest that animals who are more skilled at this sort of adaptation will not only be more likely to be perceived as well behaved (i.e., fitting in with their owner) but may also elicit a stronger bond with the owner (i.e., by using endearing behaviours in the right circumstance and inhibiting these at other times). We hypothesise that greater cognitive flexibility in the regulation of social behaviour (rather than the inflexible expression of reinforced behaviours) may be an important correlate of a successful bond between owners and their dogs. Further work examining these predictions in an experimental setting are currently under way in our lab. If the predictions are born out, then this would suggest that those working to rehome dogs would do well to focus on using behaviour modification techniques that build cognitive flexibility in their animals to maximise the chance of a successful partnership.

Although many owners may equate well-behavedness with obedience, the two are quite different. A well-behaved dog chooses to perform appropriate responses in different situations, does not disturb the owner, and anticipates his/her behaviour, adjusting its responses accordingly; an obedient dog does what is asked, even at times of conflict of interest between the owner and the dog. The well-behaved dog is preferable (although there are times when obedience is essential for the dog's safety) because the dog's behaviour prevents conflict from arising in the first place and so can be expected to result in a much more enriching relationship for most owners. The development of this ability in the dog does, however, require a different view of management—one that goes beyond behaviourism with its focus on obedience training to one that embraces the more cognitively complex world of dogs in order to develop life skills (Zulch & Mills, 2012). Indeed, some innovative training protocols described recently in the literature shed an interesting light on the cognitive abilities of dogs (e.g., Topal et al., 2006; Cracknell et al., 2008), and further investigation of these is warranted for both academic and practical reasons.

## 8.6 COMMUNICATION WITHIN THE RELATIONSHIP

Communication within the human–dog relationship requires an efficient exchange of information between two species, both of which have their own communicative systems, and both of which have their own prejudices in processing information extracted from the environment and from each other. We begin with a brief overview of compatible and incompatible behavioural and cognitive tendencies between dogs and us and how these tendencies can give rise to problems, but we also provide some pointers to solutions to improve the communication process. As an example of how communication problems can lead to an identification of how the human–dog bond can go wrong, we look at the way in which obedience failure is perceived.

### 8.6.1 Obedience Failure

Although, as mentioned in the previous section, a well-behaved dog is not the same as an obedient one, the failure of a dog to respond to an owner's request can be the prelude to conflict and ultimately a breakdown in the bond between the two. These requests for obedience are typically made verbally by the owner with the expectation that the dog understands the cue that the owner believes has been used. However, a command typically includes both verbal and nonverbal elements (even if these are not recognised by the owner), and thus there is a necessity for the dog to process both elements efficiently (Mills, 2005; Kaminski et al., 2012). In addition, when the owner makes a judgement regarding the dog's obedience in response to the cue, he/she, of necessity, needs to process and correctly interpret signals delivered by the dog. It is at both ends of this dyadic human–dog interaction that things can go wrong and can lead to either correct or incorrect perceptions of disobedience.

Research has shown that both children and adults are prejudiced to interpret teeth baring by the dog as laughing (Meints et al., 2010). Such an interpretation can lead children to hug a dog that is perceived to be happy and being bitten as a result. This example illustrates that it is not necessarily the case that a dog that bites is disobedient (i.e., violates an implicit or explicit rule that biting is not permitted), but that our reading of the signals emitted by the dog is less than perfect, and that it is us who need training (Meints & de Keuster, 2009), as much as the dog may need training to prevent biting humans. In fact, it has also been shown that the way in which humans and dogs read each others' emotions differs (Racca et al., 2012). Whereas, for example, the dog focuses on the left side of the human face for negative emotions and the right side of the human face for positive emotions, 4-year-old children look at the left side of the face for both negative and positive emotions, suggesting that different brain mechanisms in the dog and humans are responsible for reading emotions in people's faces. Humans, unlike dogs, have a fusiform facial area that is biased towards the right cerebral hemisphere and dedicated to facial processing. This is important because the use of different brain mechanisms for evaluating emotions by the two species is not conducive to an efficient understanding of each other's non-verbal communicative intentions. It is therefore easy to draw erroneous conclusions about the involvement of disobedience on the part of the dog when, in reality, there is a misunderstanding based on species-specific non-verbal communication strategies.

There is also considerable opportunity to inappropriately infer disobedience when a dog fails to respond as desired to verbal human-to-dog communication, for example, unrecognised sound properties linked to a command, which while insignificant to the owner, may be important to the dog (Fukuzawa et al., 2005a). Mills et al. (2005) found that in relation to just the command word, changes in the emotional content of the command affected performance, with a gloomy tone of voice resulting in the most variable response among dogs.

When one is considering response to instructions including verbal commands, there are many additional important influences on the tendency to respond as the owner may wish. Training history and the nature of the task are just two of those discussed here. When dogs were taught to bow or spin using specific verbal-visual combinations, they biased their response towards the visual cues when the two were in conflict (Skyrme & Mills, 2010). However, dogs trained to retrieve objects by name attended more to verbal cues than to pointing gestures if both cues were in conflict (Grassmann et al., 2012), and the performance of dogs trained to 'sit' and 'come' were influenced more by the nature of the verbal command than the possibility of eye contact (Fukuzawa et al., 2005b). In addition, the latter study points out that the quality of the sound used (tape recordings versus natural commands) and the distance of the person uttering the commands in relation to the dog have an influence on performance. The context sensitivity of verbal commands on performance was also observed by Braem and Mills (2010), who showed that embedding commands in between other words or changing the normal physical location in which a command is given can significantly affect performance, both during and after learning. One therefore needs to understand the dog's learning history and its sensitivity to contextual factors in relation to commands in order to judge whether or not the dog is disobedient, even when following the plainest commands such as 'sit' or 'come'.

These results suggest that the learning of the contiguity or perceived contingency between a verbal command and its consequences is quite context specific and prone to behavioural and cognitive prejudices linked to both species in the communicative relationship.

Learning how to generalise a verbal command is another example of how processing differences at both ends of the human–dog dyad can potentially influence the perception of obedience failure. When humans learn to refer to objects, they are prone to the so-called shape bias: when we hear a word referring to an object, both children and adults assume that a word is linked to the shape of an object. For example, when learning that a nonsense word such as 'dax' refers to a particular object, humans tend to generalise the meaning of that word to all objects that have the same shape, while at the same time ignoring differences in texture and size (Landau et al., 1988). Van der Zee et al. (2012) have shown that this is unlikely to be the case for the dog. Having taught a 5-year-old border collie called Gable with a history of word-learning several nonsense words that referred to objects, they discovered that Gable generalised the meaning of these words to objects that were of the same size if Gable was briefly exposed to such nonsense word–object combinations before testing, but that Gable tended to generalise the meaning of words referring to objects that were of the same texture if he was exposed to nonsense word–object combinations for a longer time before testing. This difference in generalising word meaning has potential consequences when teaching dogs to interact with objects. If we focus on teaching the dog from the perspective of our own capacity to distinguish between different shapes in a verbal context, we may set the dog an unnecessarily difficult or even impossible task, potentially leading to an assumption that the dog is being disobedient. We thus need to be sensitive to the information-processing abilities that are connected to giving verbal commands, information processing abilities that may well differ for both parts of the human–dog dyad. More research into this area of cognition and how it impacts on communication is necessary to determine whether dogs disobey, or whether we fail to take into account the species-specific difference between ourselves and the dog when teaching or instructing dogs.

## 8.6.2  Aggressive Behaviour

One of the problem behaviours that owners find most difficult to deal with is 'aggression', and thus this is a problem that commonly leads to a fractured bond, including the ultimate breakdown, relinquishment, or euthanasia (Denenberg et al., 2005). Owners of dogs exhibiting aggressive behaviour to those outside the family (both conspecifics and heterospecifics) are under societal pressure to alter the animals' behaviour, whilst aggressive behaviour towards family members and other household pets is particularly difficult to live with on a daily basis. Misunderstandings within the relationship between dog and owner can be triggers of aggressive behaviour by the dog towards its owner—for example, conflict over perceived access to resources or the dog's denial of certain interactions on the part of the owner (grooming, petting). However, in addition, the damaged relationship that stems from aggressive expressions by the dog, whether this is directed towards the owner or others, can be characterised by owner behaviour, such as attempts at physical punishment, which feeds back into the cycle in a negative manner. Inappropriate owner responses that create this cycle frequently derive from the misconceptions about the nature of aggressive behaviour. Therefore, in this section, we examine the most common motivations for the behaviour, discuss it within the context of wider social relationships, and explore how people can better understand dog communication in situations where aggressive behaviour may result so that they can respond more appropriately to avoid escalation of the situation.

Aggression has no agreed scientific definition and is not a diagnosis, nor emotion, nor even an objectively distinguishable behavioural act, but rather it is a perceived style of responding. It is a label given to behaviour which appears to cause harm or have the potential to cause harm. For this reason, the term 'aggressive behaviour' is perhaps preferable to 'aggression' to describe this perception, because it requires further qualification in terms of context and motivational-emotional basis. Harm can happen incidentally within the context of play and is an integral part of predation, and in this situation, these acts may be referred to as 'play aggression' or 'predatory aggression'; but in these contexts, the behaviour does not have a communicative function, unlike when it occurs within an ongoing social interaction. Therefore, these forms will be excluded from further discussion in order to focus on the subset of aggressive behaviours that are sometimes referred to as affective aggression, which serve to remove another individual from the immediate environment through an agonistic display.

Dogs may be motivated to express aggressive behaviour in a number of situations associated with differing emotional states; for example, a dog that is anxious or fearful (FEAR *sensu;* Panksepp, 1998) may resort to fighting rather than flight or freezing; aggressive behaviour might also be shown when a dog is frustrated in its ability to reach a desired goal (RAGE *sensu;* Panksepp, 1998); aggressive behaviour may also be shown proactively, for example, in relation to chasing away others (SEEKING *sensu;* Panksepp, 1998); in addition, aggressive behaviour may occur in response to the sensory affect of pain (Panksepp, 1998). Over time dogs can learn that an aggressive response reliably gains them a desired outcome, and so the response, like other learned actions, can become habitual, with little specific emotional content, although a display of heightened arousal may be maintained.

It is possible that more than one primary emotional system is involved in any given situation (e.g., frustration combined with fear when the route to safety is blocked), and emotional incongruence (e.g., as occurs during approach-avoidance conflict) may lead to a dog displaying aggressive behaviour in an attempt to gain control and create a predictable outcome. However, there is no specific evidence of which the authors are aware to suggest that dogs are capable of aggressive acts in response to more cognitively complex emotions such as jealousy or as a result of complex cognitive scheming such as a desire to become a dominant individual. To date, all such encounters reported to the authors in the course of their clinical work can be explained by reference to simpler processes, such as current environmental contingencies. Unfortunately, empirical data allowing a thorough scientific analysis of aggressive behaviour in dogs are lacking, and so much of what is described is based on personal, including clinical, experience and abstraction from other species. However, we can minimise the risk of bias by ensuring our reasoning stays within the bounds of our knowledge of the cognitive abilities of the social dog.

Much human-directed aggressive behaviour could be avoided if people took into account the different preferences for interaction style between the two species and improved their communication (see Mills & Zulch, 2010, for further details). For example, gestures which humans find appropriate—such as face-to-face greetings, making extended eye contact, and reaching out to shake hands or hug—may all be interpreted by dogs as threatening gestures. Additionally, humans, who believe that dogs who express aggressive behaviour are challenging them for social status will frequently punish dogs for growling when they are simply indicating their unease with the situation without immediately resorting to overt physical harm. Such actions by humans are counterproductive and not only may reinforce the dogs' unease, but also may lead to the suppression of these warning signals in future, resulting in an unpredicted bite at a later date.

Like other social species, dogs tend to avoid overtly aggressive behaviour as a first option in social situations because this risks an immediate escalation to physical attack and the potential for injury, together with the possible damage to relationships on which the dog depends. For this reason, dogs have evolved a complex repertoire of body language and facial expression to try to avoid conflict where possible. Subtle signs that a dog is trying to distance itself from another include averting its gaze, moving away, lying down and raising a hind leg, exhibiting body stillness and tension, tucking its tail, flattening its ears, and pulling the lips back. Yawning and lip licking are also frequently seen in this context, but especially, in the authors' experience, at times of emotional or motivational conflict (Figure 8-1). Failure to read these gestures for what they are, or even worse, misinterpreting gestures of appeasement as a sign of the dog feeling guilty, are likely to lead to inappropriate responses on the part of the human in the situation and hence lead to escalation of the behaviour resulting in lunging, snapping, and/or biting. Likewise, dogs also use subtle signs to indicate their desire to acquire a resource, space, etc., from another, which need to be recognised for their communicative value and responded to appropriately in order to avoid escalation to aggressive conflict. These signs include orientation, including staring towards a resource, proximity to it, and posture during an encounter (an up-and-forwards body posture encourages deference from others; see Figure 8-2).

**FIGURE 8-1**  A back-and-down body posture is often a sign of deference. Lip licking may indicate emotional conflict.

**FIGURE 8-2**  An up-and-forwards posture indicates interest and may be extended to indicate a desire to acquire a resource. Failure to recognise this may lead to conflict and aggressive behaviour.

Effective management of aggressive behaviour depends on understanding and differentiating the underlying cognitive and affective processes which result in a given behavioural expression (Mills et al., 2012). Though limited, our knowledge can be applied in a systematic manner to enable us to make appropriate inferences and predictions in a clinical setting, which can then be tested against progress. A suggested approach to enable us to do this is as follows:

1. Examine the context in which the aggressive interaction occurs from the perspective of identifying potential eliciting stimuli as well as those maintaining the behaviour—i.e., the consequences of the behaviour for the dog.
2. Evaluate this information in the context of the level of arousal expressed by the dog, the sequencing of its behaviour, and its communicative function—i.e., its body language— to determine the most likely emotional state or states involved.

**3.** If possible, elicit descriptions of alterations in the behaviour over time, including changes in owner response to a range of signals given by the dog in the trigger as well as other situations.
**4.** Generate and test hypotheses about the dog's behaviour without putting others at risk. Stronger testing is based on falsification rather than corroboration, where possible.

This systematic analysis of the aggressive behaviour should enable a clinician to more accurately assess the processes underlying the display and therefore implement a strategy for behaviour change that will not only modify the behaviour but also re-establish the relationship between dog and owner.

## 8.7 CONCLUSIONS

In this chapter, we have drawn on direct information where available, but also indirect sources, in order to propose a rational framework for describing relationships and examining how they can go wrong. At the heart, this is the need to understand the types of conflict of interest and expectation that arise when dogs and people share the same home, and this requires a deeper understanding of both the natural history but also the natural psychology of both species which underpin their biases and limits.

At the core of many relationship problems lies this conflict of interests; however, it is likely that conflicts occur more on an individual than a species basis. For this reason, further work is required to better define and describe the dimensions of the human–pet/working dog relationship so that individuals can best define their ideal relationship, recognise where and why things go wrong within it, and engage in appropriate remedial actions.

Greater scientific engagement aimed at increasing our understanding in this regard will involve fundamental research on both social cognition and affect in the dog as well as the dynamics of human–dog interaction and the needs which dogs continue to meet in a partnership that spans more than 15,000 years. This will enable us to build more effective, enduring, and compassionate relationships, in this most long-standing of partnerships.

**Future Directions**

Future exploration of problem behaviours in the dog should be considered.
- Bidirectionally: in the human–dog bond, two parties may contribute to (perceived) problem behaviour
- By further defining the human–dog relationship and its dimensions (for which we have made a start)
- By taking into account the cognitive, emotional, and behavioural prejudices of both species because these, for example, play a role in their communication
- By taking into account the types of conflict that may arise and mutual expectations

## ACKNOWLEDGEMENTS

We would like to acknowledge the support of ESF through their CompCog Research Networking Programme, which has helped facilitate important interdisciplinary links that have informed our development of some of the issues discussed in this chapter.

## REFERENCES

Ainsworth, M.D.S., Blehar, M.C., Waters, E., Wall, S., 1978. Patterns of attachment: a psychological study of the strange situation. Lawrence Erlbaum Associates, Hillsdale, NJ.

Anderson, D.C., 2007. Assessing the human-animal bond: a compendium of actual measures. Purdue University Press, West Lafayette, IN.

Arkow, P., Dow, S., 1984. The ties that do not bind: A study of human-animal bonds that fail. In: Anderson, R.K., Hart, B.L., Hart, L.A. (Eds.), The pet connection: its influence on our health and quality of life. University of Minnesota Press, Minneapolis, MN, pp. 348–354.

Bowlby, J., 1958. The nature of the child's tie to his mother. Int. J. Psycho-analysis 39 (5), 350.

Bowlby, J., 1969. Attachment and loss. Vol. I: Attachment. Penguin Books, London, UK.

Braem, M.D., Mills, D.S., 2010. Factors affecting response of dogs to obedience instruction: a field and experimental study. Appl. Anim. Behav. Sci. 125 (1), 47–55.

Clark, G.I., Boyer, W.N., 1993. The effects of dog obedience training and behavioural counselling upon the human-canine relationship. Appl. Anim. Behav. Sci. 37, 147–159.

Corridan, C.L., Mills, D.S., Pfeffer, K., 2010. Comparison of the prevalence of behavioural problems in samples of owned and relinquished companion dog populations in the UK. In: Scientific Proceedings Veterinary Programme, BSAVA Annual Congress. BSAVA Publications, Gloucester, UK.

Cracknell, N.R., Mills, D.S., Kaulfuss, P., 2008. Can stimulus enhancement explain the apparent success of the model-rival technique in the domestic dog (*Canis familiaris*)? Appl. Anim. Behav. Sci. 114, 261–272.

Denenberg, S., Landsberg, G., Horwitz, D., Seksel, K., 2005. A comparison on case referred to behaviourists in three different countries. In: Mills, D. (Ed.), Current issues and research in veterinary behavioural medicine—Papers presented at the 5th International Veterinary Behavior Meeting. Purdue University Press, West Lafayette, IN, pp. 56–62.

Dwyer, F., Bennett, P.C., Coleman, G.J., 2006. Development of the Monash dog owner relationship scale (MDORS). Anthrozoos 19 (3), 243–256.

Flannigan, G., Dodman, N.H., 2001. Risk factors and behaviors associated with separation anxiety in dogs. J. Am. Vet. Med. Assoc. 219 (4), 460–466.

Fukuzawa, M., Mills, D.S., Cooper, J.J., 2005a. The effect of human command phonetic characteristics on auditory cognition in dogs (*Canis familiaris*). J. Comp. Psychol. 119 (1), 117.

Fukuzawa, M., Mills, D.S., Cooper, J.J., 2005b. More than just a word: non-semantic command variables affect obedience in the domestic dog (*Canis familiaris*). Appl. Anim. Behav. Sci. 91 (1), 129–141.

Grassmann, S., Kaminski, J., Tomasello, M., 2012. How two word-trained dogs integrate pointing and naming. Anim. Cogn. 15 (4), 657–665.

Haverbeke, A., Laporte, B., Depiereux, E., Giffroy, J.M., Diederich, C., 2008. Training methods of military dog handlers and their effects on the team's performances. Appl. Anim. Behav. Sci. 113 (1), 110–122.

Haverbeke, A., Rzepa, C., Depiereux, E., Deroo, J., Giffroy, J.M., Diederich, C., 2010. Assessing efficiency of a Human Familiarisation and Training Programme on fearfulness and aggressiveness of military dogs. Appl. Anim. Behav. Sci. 123 (3), 143–149.

Herzog, H.A., Burghardt, G.M., 1987. Are we ready for a theory of human animal relationships? Anthrozoos 1, 145–146.

Hinde, R.A., 1976. On describing relationships. J. Child Psychol. Psychiatry 17 (1), 1–19.

Horowitz, A., 2009. Disambiguating the "guilty look": Salient prompts to a familiar dog behaviour. Behav. Proc. 81 (3), 447–452.

Horváth, Z., Dóka, A., Miklósi, Á., 2008. Affiliative and disciplinary behavior of human handlers during play with their dog affects cortisol concentrations in opposite directions. Horm. Beh. 54 (1), 107–114.

Kaminski, J., Nitzschner, M., 2013. Do dogs get the point? A review of dog–human communication ability. Learn. Motivation 44 (4), 294–302.

Kaminski, J., Schulz, L., Tomasello, M., 2012. How dogs know when communication is intended for them. Dev. Sci. 15 (2), 222–232.

Katcher, A.H., Beck, A.M., 1983. New perspectives on our lives with companion animals. University of Pennsylvania Press, Philadelphia, PA.

Kellert, S.R., Wilson, E.O., 1995. The biophilia hypothesis. Island Press, Washington, DC.

Kidd, A.H., Kidd, R.M., 1987. Seeking a theory of the human/companion animal bond. Anthrozoos 1 (3), 140–157.

Landau, B., Smith, L.B., Jones, S.S., 1988. The importance of shape in early lexical learning. Cogn. Dev. 3 (3), 299–321.

Lawler, E.J., 2001. An affect theory of social exchange. Am. J. Sociol. 107 (2), 321–352.

Lefebvre, D., Diederich, C., Delcourt, M., Giffroy, J.M., 2007. The quality of the relation between handler and military dogs influences efficiency and welfare of dogs. Appl. Anim. Behav. Sci. 104 (1), 49–60.

McGreevy, P.D., Masters, A.M., 2008. Risk factors for separation-related distress and feed-related aggression in dogs: additional findings from a survey of Australian dog owners. Appl. Anim. Behav. Sci. 109 (2), 320–328.

Meints, K., de Keuster, T., 2009. Brief report: don't kiss a sleeping dog: the first assessment of "the blue dog" bite prevention program. J. Pediatr. Psychol. 34 (10), 1084–1090.

Meints, K., Racca, A., Hickey, N., 2010. How to prevent dog bite injuries? Children misinterpret dogs' facial expressions. Proceedings of the 10th World Conference on Injury Prevention and Safety Promotion, 21–24 September, London, UK. Inj. Prev. 16 (suppl. 1), A68.

Mills, D.S., 2003. Medical paradigms for the study of problem behaviour: a critical review. Appl. Anim. Behav. Sci. 81 (3), 265–277.

Mills, D.S., 2005. What's in a word? Recent findings on the attributes of a command on the performance of pet dogs. Anthrozoos 18, 208–221.

Mills, D.S., Beral, A., Lawson, S., 2010. Attention seeking behavior in dogs—what owners love and loathe! J. Vet. Behav. 5, 60.

Mills, D.S., Dube, M.B., Zulch, H., 2012. Stress and pheromonatherapy in small animal clinical behaviour. Wiley-Blackwell, Chichester, UK.

Mills, D.S., Fukuzawa, M., Cooper, J.J., et al., 2005. The effect of emotional content of verbal commands on the response of dogs. In: Mills, D., et al. (Eds.), Current issues and research in veterinary behavioural medicine–Papers presented at the 5th International Veterinary Behavior Meeting. Purdue University Press, West Lafayette, IN, pp. 217–220.

Mills, D., Zulch, H., 2010. Appreciating the role of fear and anxiety in aggressive behavior by dogs. Vet. Focus 20, 44–49.

Morris, P., Fidler, M., Costall, A., 2000. Beyond anecdotes: an empirical study of "anthropomorphism". Soc. Anim. 8, 151–165.

Morris, P.M., Doe, C., Godsell, E., 2008. Secondary emotions in non-primate species? Behavioural reports and subjective claims by animal owners. Cogn. Emotion 22, 3–20.

New, J.C., Salman, M.D., Scarlett, J.M., Kass, P.H., Vaughan, J.A., Scherr, S., 1999. Characteristics of dogs and cats and those relinquishing them to 12 US animal shelters. J. Appl. Anim. Welfare Sci. 2, 83–95.

Odendaal, J.S.J., Lehman, S., 2000. The role of phenylethylamine during positive human dog interaction. Acta. Veterinaria Brno. 69, 183–188.

Odendaal, J.S.J., Meintjes, R.A., 2003. Neurophysiological correlates of affiliative behaviour between humans and dogs. Vet. J. 165 (3), 296–301.

Overall, K.L., 1997. Clinical behavioral medicine for small animals. Mosby-Year Book, Inc, St. Louis, MO.

Panksepp, J., 1998. Affective neuroscience: the foundations of human and animal emotions. Oxford University Press, Oxford.

Parthasarathy, V., Crowell-Davis, S.L., 2006. Relationship between attachment to owners and separation anxiety in pet dogs (Canis lupus familiaris). J. Vet. Behav. Clin. Appl. Res. 1 (3), 109–120.

Patronek, G., Glickman, L., Beck, A., McCabe, G.P., Ecker, C., 1996. Risk factors for relinquishment of dogs to an animal shelter. J. Am. Vet. Med. Assoc. 209, 575–581.

Racca, A., Guo, K., Meints, K., Mills, D.S., 2012. Reading faces: differential lateral gaze bias in processing canine and human facial expressions in dogs and 4-year-old children. PloS One 7 (4), e36076.

Scherer, K.R., 1984. On the nature and function of emotions: a component process approach. In: Scherer, K.R., Ekman, P. (Eds.), Approaches to emotion. Erlbaum, Hillsdale, NJ, pp. 293–317.

Schleidt, W.M., Shalter, M.D., 2003. Co-evolution of humans and canids; an alternative view of dog domestication: homo homini lupus? Evolution Cogn. 9, 57–72.

Schwab, C., Huber, L., 2006. Obey or not obey? Dogs (Canis familiaris) behave differently in response to attentional states of their owners. J. Comp. Psychol. 120 (3), 169.

Serpell, J.A., 1983. The personality of the dog and its influence on the pet-owner bond. In: Katcher, A.H., Beck, A.M. (Eds.), New perspectives on our lives with companion animals. University of Pennsylvania Press, Philadelphia, PA, pp. 57–63.

Serpell, J.A., Hsu, Y., 2001. Development and validation of a novel method for evaluating behavior and temperament in guide dogs. Appl. Anim. Behav. Sci. 72 (4), 347–364.

Shepherd, K., 2002. Development of behaviour, social behaviour and communication in dogs. BSAVA Manual of Canine and Feline Behavioural Medicine. BSAVA, Gloucester, 8–29.

Skyrme, R., Mills, D.S., 2010. An investigation into potential overshadowing of verbal commands during training. J. Vet. Behav. Clin. Appl. Res. 5 (1), 42.

Topál, J., Byrne, R.W., Miklósi, A., Csányi, V., 2006. Reproducing human actions and action sequences: "Do as I do!" in a dog. Anim. Cogn. 9 (4), 355–367.

Topal, J., Miklósi, A., Csanyi, V., Doka, A., 1998. Attachment behaviour in dogs (Canis familiaris): a new application of the Ainsworth (1969) Strange Situation Test. J. Comp. Psychol. 112, 219–229.

van der Zee, E., Zulch, H., Mills, D., 2012. Word generalization by a dog (Canis familiaris): is shape important? PloS One 7 (11), e49382.

Weiss, R.S., 1998. A taxonomy of relationships. J. Soc. Pers. Relationships 15 (5), 671–683.

Zulch, H., Mills, D.S., 2012. Life skills for puppies: laying the foundation for a loving, lasting relationship. Veloce Publishing Ltd, Dorchester, UK.

# Social Cognition

# Chapter 9

# Social Learning in Dogs

Péter Pongrácz

*Department of Ethology, Biological Institute, Eötvös Loránd University, Budapest, Hungary*

## 9.1 INTRODUCTION: FROM SOCIAL BEHAVIOUR TO THE SOCIAL DOG

Social species offer perhaps the most fascinating and intriguing forms of behaviour for lay enthusiasts of zoology as well as for scientists. Although group-living animals face basically the same challenges of survival as solitary species, many new and unique behavioural phenomena evolved specifically for the management and maintenance of group living or for regulating inter- and intra-group interactions. Another large cluster of behaviours that characterises social species also can be observed in solitary animals, but in a different form; the reason is that some forms of behaviour had the opportunity to change to serve a social lifestyle in a more adaptive way.

Many animals inherit a sometimes amazingly detailed 'set of skills and instructions' for survival as a legacy of their evolutionary past; however, the environment is rarely so constant even for short-living species that a fine-tuning or adjustment would not be necessary in their behaviour for more effective coping. The nervous system provides the major means and tool for being able to change one's behavioural responses to freshly emerging stimuli and situations from the environment, regarding both its inanimate and living components. When an animal reacts to stimuli differently from the way it did before, besides the possible ontogenetic processes, we face the phenomenon of learning. In general, learning is the complex capacity of detecting and processing information, storing some aspects of it in the neural system, and retrieving the stored information in a form of new or altered behavioural actions. Of course, learning has many forms and mechanisms, which may differ from each other considerably on the level of neural complexity and type; memory span requirements; or in the aspects of what, when, how, and from whom an animal can learn. One of the many ways to categorise learning is to separate the individual strategies or mechanisms from the social ones.

Distinguishing between social learning and the individual forms of information acquisition is not merely an arbitrary step—although at first glance there does not seem to be a big difference whether an individual changes its behaviour

The Social Dog. http://dx.doi.org/10.1016/B978-0-12-407818-5.00009-7

as an effect of learning something through its own experiences by trial-and-error, or from observing the activity of another animal. Social learning often involves an extra step for the observer animal: it should be attracted to the primary actor (or, as it is usually called, 'the demonstrator') or to its activity to allow for learning some of the aspects of its actions. Although there are specific cases of social learning (for example, transmission of food preference in very young animals) when merely the properly timed exposure of the 'observer' is enough for information transfer to occur, in most cases social learning depends on the existence of the specific social attraction (or even a social bond) that directs the observer's attention to the action of the demonstrator, or to the demonstrator itself.

Dogs have provided some of the most dynamically proliferating research subjects of social behaviour in the past 20 years. Although 'man's best friend' was a long-time target for scientists such as Pavlov when developing the foundations of experimental behaviour sciences, social behaviour of dogs gained renewed interest only because their special relationship with humans was considered as being not a disadvantage anymore in proper ethological experiments, but as a source of dogs' most unique species-specific characteristics. The fact that dogs' behaviour is not only easy to test in the presence of humans but that it manifests itself in its most natural form in the human environment made dogs a favourite subject for many ethologists who were interested in the higher end of cognitive mechanisms and capacities in a strongly social species.

Sometimes dogs are referred to as 'artificial' (Miklósi & Topál, 2013) or 'man-made' animals because they do not have an extant wild-living form in nature; on the contrary, the vast majority of dogs live in more or less close relationship with humans and show basically the same level of domestication. However, this does not mean that the behaviour of dogs is not interesting from an evolutionary aspect. The social behaviour and cognitive skills of dogs are rooted in their heritage from their wild ancestors, but these behaviours and skills were formed more or less to comply with their new (social) environment, that of humans. This chance for studying such an exciting co-existence between two highly social species—the challenge to test such capacities in dogs that made and makes this relationship work—provides the specific attraction for many scientists to do 'dog research' nowadays. Various forms of interspecific communication, social attraction and attachment behaviour (e.g., Topál et al., 1998), or eavesdropping on third-party interactions (Marshall-Pescini et al., 2011) are just randomly chosen examples of the extent of the complexity of the dog-human relationship.

In this chapter we examine some aspects of social learning in dogs, showing different approaches scientists have made to discover and understand the mechanisms, function, ontogeny, and evolution of this behaviour in this species. In the initial part of the chapter, we outline the function of social learning, taking examples from many different species to highlight the overall importance of this type of behaviour for social animals. We then move on to give an overview of the underlying mechanisms falling under the umbrella term 'social learning' and show how researchers have made attempts to tease such mechanisms apart through careful experimental studies. In the main part of the chapter, we move on to outline the importance of studying social learning in dogs and what has been discovered in terms of its underlying cognitive mechanisms and function.

## 9.2 SOCIAL LEARNING AS AN ADAPTIVE MECHANISM

In general, social learning is considered as an adaptive mechanism when the costs of individual learning are too high (for example, in a case of potentially poisonous foods), or there is a high risk for useful inventions being lost without social transmission to other individuals (Galef & Giraldeau, 2001). It is very important that one should be aware of the advantages (and sometimes disadvantages) of social learning from the ecological point of view, before delving deeper into the research of the mechanism of such a phenomenon. According to Laland (2004), the possible adaptive functions of social learning can be categorised on the basis of a simple question: whose behaviour does the observer copy and when? Following this line, learning from a demonstrator is advantageous, if

- *The demonstrator's solution is more effective than the observer's own.* For example, detouring around an obstacle is a more effective behaviour than trying to dig under it (Pongrácz et al., 2001).
- *The asocial (trial-and-error) learning is too costly.* For example, in mountain gorillas (*Gorilla beringeri beringeri*) it would be too laborious to learn individually how to process the otherwise well-protected nettle before consumption (Byrne, 2003).
- *It is safer to choose the demonstrator's choice instead of an unsecure alternative.* For example, rats (*Rattus norvegicus*) choose the food they can smell on the breath of their companion instead of an unknown one (Galef & Wigmore, 1983).

However, observers can be picky about whom to copy. Galef and Laland (2005) described cases in which the 'quality' of the demonstrator was revealed from the choices of the observers:

- *Copying the other observers can bias the popularity of a demonstrator.* For example, female quails (*Coturnix japonica*) tend to prefer those males that were chosen by other hens before them (White & Galef, 2000).
- *Successful demonstrators are more popular.* For example, evening bats (*Nycticeius humeralis*) follow those conspecifics that proved to be successful hunters (Wilkinson, 1992).
- *The behaviour of older demonstrators can be preferred by the observers.* For example, female guppies (*Poecilia reticulata*) choose those males that were previously chosen by the older females (Kirkpatrick & Dugatkin, 1994).

Still remaining with the adaptive value of social learning, we can sort the plethora of cases into functional categories. In other words, we can sort them according to 'what' the observers learned from the demonstrator:

- *Mate choice and sexual behaviour can also be subjects of socially transmitted information.* For example, in cowbirds (*Molothrus ater*) young males learn the most effective courtship song by observing the response of the adult females to their tentative song fragments (Freeberg, 1996).
- *Predator avoidance is another important component of survival that can be enhanced in social species.* Besides the well-known dilution and confusion effect of larger groups of prey animals in the case of a predatory attack, learning of potentially dangerous predators can be adaptive by following the behavioural response of the experienced group members. For example, it was found in blackbirds (*Turdus merula*) that young birds learn to identify dangerous predators by joining the mobbing choir of the older blackbirds (Vieth et al., 1980). In rhesus monkeys, the young animals learn from seeing the alarm reaction of adult specimens that they have to be afraid of snakes (Cook & Mineka, 1989).
- *The route that leads to some kind of target area also can be learned from conspecifics.* Besides the well-known exemplars of how honey bees direct their hivemates to a rich nectar source with their informative dance (e.g., Riley et al., 2005) and how geese learn the migratory route from the elder birds during their first autumn travel (Reed, 1999), guppy fish serve another good example for social route learning. These group-living fish tend to follow the socially acquired route to feeding areas even if this route is energetically more costly than a simpler, more straightforward access path (Laland & Williams, 1998).
- *Social learning can help in choosing the right (digestible, or non-poisonous) type of food, usually in herbivorous or omnivorous species.* Wild rabbit pups, for example, learn the composition of the maternal diet very early, from *in uteri* effects and through their mother's milk (Altbäcker et al., 1995).
- *The most suitable food-obtaining technique also can be transmitted by social learning, as it was found in wild-living chimpanzees (*Pan troglodytes*) in different areas of Africa.* Because particular termite species behave (fight) differently against their intruders, such as chimpanzees, these apes use such 'termite fishing' tools and methods that best fit the needs of catching and eating these insects the most effectively (Whiten et al., 2001). These tool-using techniques are transmitted between generations of chimpanzees through social learning.
- *Acquiring complex behavioural sequences (such as making or using tools) and transmitting cultural norms and traditions, especially in our own species, humans (*Homo sapiens*), also would not be possible without social learning.* Basically, the whole 'cultural evolution' is based on the between-generation social knowledge transfer (Castro & Toro, 2004). Humans, even at a very young age, possess the capacity of paying attention to and learning from the specifically configured 'pedagogical' (or teaching) cues and behaviour

of knowledgeable demonstrators (Meltzoff, 1996; Gergely & Csibra, 2003). Across the literature on social learning, imitation has been and sometimes is still thought to be a primarily human-specific mechanism of information transfer (Shea, 2009). As we will see later, imitative learning has all the key features of how one can acquire new and, many times, complex behavioural sequences, which characterise the majority of human activities. As humans are capable of imitative learning from a very young age (Wood et al., 2012), researchers have hypothesised that imitation does not necessarily require the understanding of the goals and reasons of the demonstrator, but to become successful imitators, young infants should be interested in (and capable of) following others' behaviour as exactly as possible and for the copying as a reason in itself (Jones, 2009).

Obviously, this list is far from being complete; however, it can give a hint to the complexity and variability of the functions of social learning as it emerged in many different points of evolution. Social learning is definitely not the privilege of those species that are considered to be the 'most intelligent' or capable of high-level cognitive processes. When we discuss the different manifestations of social learning in dogs, we should never forget to try to find the way the particular behaviour fits to the ecology of dogs; in other words, adaptive function is worthy of discussion even if our primary goal is testing, for example, the mechanism of the behaviour.

## 9.2.1 Mechanisms of Social Learning

Whiten and Ham (1992) defined social learning with a simple but straightforward and still valid sentence. In their words, when social learning happens, 'B learns some aspect of the behavioural similarity from A' (p. 248). Of course, simple definitions have a disadvantage of missing many of the finer details, and they also leave ample room for interpretations. If we remember the colourful array of examples of functionally different cases of social learning from the previous section, it is reasonable to ask, then, exactly *which aspect or detail* of the demonstrator's behaviour was learned by the observer. The answer can be also interesting, whether the similarity between the demonstrator's and observer's actions is close or somewhat looser (the so-called fidelity question). Finally, one can ask exactly *how* the observer learned from the demonstrator, and this question takes us to the cognitive aspect of social learning that deals with, for example, an intriguing problem: whether the observer understood the goal (if there was any) of the demonstrator. Those empirical or theoretical attempts that try answering the questions of 'what' and 'how' the observers learned during a social transmission process, we can classify as studies dealing with the mechanism of social learning.

More than a handful of mechanisms were described in the decades' worth of research on social learning. Sometimes the distinction between them seems to be quite arbitrary, or the definitions show the strong influence of the researchers who were interested mostly in particular aspects of cognitive processes behind one or two selectively chosen mechanisms. It is not surprising that there is an ample number of summarizing papers dealing with defining and explaining the (sometimes different) set of mechanisms of social learning (e.g., Galef, 1988; Heyes, 1993; Byrne & Russon, 1998; Zentall, 2006). It is not our task in this chapter to choose one or another author's system of classification; instead, we will try to give a short collection of mechanisms, differentiating them by the factors that define particular details of the learned action or the supposed cognitive process involved in the observer (Table 9-1). Then we will show that for a successful distinguishing characteristic between two candidate mechanisms, the researchers tried to estimate the extent of matching between the demonstrator's and observer's actions, the ability of incorporating their own as well as learned components into the behavioural sequence, and insight into the 'goal' of the demonstrator.

**TABLE 9-1** Mechanisms (Subcategories) of Social Learning (Column on the Left) and Their Typical Characteristics (Top Row)[a]

| | Same Object | Fidelity of 'Copying' | Understanding of Goal | Reward Is Needed | New Action | Known Action |
|---|---|---|---|---|---|---|
| **Social Influence (No Social Learning)** | | | | | | |
| Contagion | Yes | n/a | No | No | No | Yes |
| Social facilitation | Yes | n/a | No | Yes | No | Yes |
| **Simpler Social Learning Mechanisms** | | | | | | |
| Observational conditioning | Yes | High | No | Yes | Yes | No |
| Response facilitation | n/a | High | No | No | No | Yes |
| Enhancement | Yes[b] No[c] | Medium | Not necessary | Yes | Yes | n/a |
| **Complex Cognitive Understanding Is Required** | | | | | | |
| Emulation | No | Low | Yes | Yes | Yes | No |
| Imitation | Not necessary | High | Yes | Yes | Yes[d] | No |

[a]For further explanation, read sections 9.2.1.1–9.2.1.5.
[b]Local enhancement
[c]Stimulus enhancement
[d]Imitation requires either an action that was not in the repertoire of the observer or a new sequence of individual actions in which the sequence is obviously one that has never been tried.
(Source: Pongrácz, 2009, with modifications)

### 9.2.1.1 Distinguishing between Social Learning and Social Influence

In some cases, when two or more animals are together, their behaviour may show considerable similarity, and one can even discover a 'demonstrator' among them that seemingly initiated the action and an 'observer' that seemingly took over the exemplar from the 'demonstrator'. However, one cannot be sure whether learning, in the sense this term is commonly used, happened in all cases. Some argue that behavioural congruence can develop without learning among individuals that are present simultaneously at the right time and place. Zajonc (1965) called the process 'social facilitation', when the mere presence of a demonstrator increased the motivational level of the 'observer', but the 'demonstrated' action (i.e., what the demonstrator was actually doing) was not important from the aspect of resulting behaviour. Obviously, there are ways to distinguish between social facilitation (or, as it is sometimes called, either 'social influence' or 'behavioural contagion') and 'true' social learning. An easy rule of thumb method is to look at the long-term consequences of the transmitted action in the observer individual. If the particular behaviour occurs later on its own (i.e., in a similar context but without the similar behaviour seen from a demonstrator), then one can assume that the observer most likely learned at least some aspects of the action. However, if the particular behaviour emerges again only when there is a demonstrator behaving similarly, more likely the situation is a case of social influence (or contagion).

### 9.2.1.2 Learning through Various Forms of Social Enhancement

When social learning most likely does not require deeper understanding of the demonstrator's goal, or the connection between the action and its consequence, the underlying mechanism is usually described as some form of enhancement. The term 'enhancement' refers to the effect of demonstration, in which the proportion of the observer's behaviour directed toward the location or object of the demonstrator's activity increases (Spence, 1937). This then results in a similar action to that of the demonstrator. The term 'local enhancement' (Roberts, 1941) refers to those instances in which an animal is attracted to a place or object by the current presence of a conspecific at that place or by residues of the demonstrator's activity at that location, e.g., odour cues. Stimulus enhancement has a more general effect, as in those cases when the observer is attracted to all objects that have a similar physical appearance to the one which was manipulated by the demonstrator.

Local and stimulus enhancements act on particular objects of the environment, which were 'marked' by the action of the demonstrator. In the case of response facilitation (Byrne, 1999, 2002; Byrne & Russon, 1998), a previously known, familiar motor pattern (behaviour) of the observer is primed in a new context. A few authors have emphasised the lack of higher cognitive background behind this mechanism and labelled cases of response facilitation as 'mimicry' (Tomasello & Call, 1997).

### 9.2.1.3 Observational Conditioning

The mechanism known as 'observational conditioning' is possibly more common than we previously thought and is quite likely responsible for many cases of so-called true imitation (Heyes, 1993). It is important to note that the observer should see the demonstrator's action yielding the reward, because this can reinforce the demonstrator's action as a stimulus. The special feature of observational conditioning is that observing is enough to learn the action, and the observer does not need to do the action itself beforehand.

### 9.2.1.4 Emulation

During emulation, the observer's action will not necessarily be identical, or even similar, to the one of the demonstrator if we consider only the behaviours involved. Instead of cloning the motor pattern, observers recognise the problem to be solved and may alternatively develop their own solution (Tomasello & Call, 1997). As emulating observers are able to realise the nature of the problem, in experimental conditions they don't need a demonstrator if the 'action' can be shown to them with use of an automatic device (this is the so-called ghost control for goal emulation).

### 9.2.1.5 Imitation

Once the term 'imitation' was used merely as a synonym for almost any kind of copying-like behaviour in animals. Only in the 1950s did the realization start to emerge that behavioural similarity between the observer and demonstrator can be achieved through many different mechanisms, which can be distinguished along a few basic characteristics. For example, the demonstrator can draw the observer's attention (1) to a particular location through its activity, (2) to a particular problem or solution of a problem, (3) or directly to its behaviour (or behavioural sequence). Imitation became one of the many social learning mechanisms, and in the beginning researchers mostly concentrated on the form of behaviour (in other words, to the motor component of the behaviour) for separating imitation from the other types of social learning. According to Thorpe (1963), who came up with perhaps the first explicit description, true imitation has been defined as 'the copying of a novel or otherwise improbable act or utterance, or some act for which there is clearly no instinctive tendency' (p. 135). It is clearly noticeable that the main emphasis lies in the 'novelty' of the transmitted behaviour, what is in sharp contrast with, for example, contagion (see the previous section) that is the triggering of an action that was definitely in the repertoire of the observer beforehand. Three decades after Thorpe, Heyes (1993) still concentrated on the motor part of the behavioural action, when giving this definition of imitation: 'Imitation means learning something about the form of behaviour through observing others, while other kinds of social learning are learning about the environment through observing others' (p. 1003). This definition is useful because it provides insight into those 'tactics' by which researchers usually try to prove whether a case of social learning was (or was not) imitation. This is a type of backward, or exclusion-based, protocol, characterised somewhat sarcastically though wisely by Zentall (2004): 'Imitation is a form of social learning that remains when one has ruled out or controlled for all of the alternative mechanisms (mimicking, response facilitation, stimulus and local enhancement) that might contribute to the higher probability of a copied response' (p. 18).

## 9.3 THE DOG AS A MODEL FOR STUDYING CON- AND HETEROSPECIFIC SOCIAL LEARNING

As one can see in the previous section, a considerable part of social learning research is about the transmission of food preferences, or ways and means of foraging techniques. Although we will show an example later, when dogs learned how to look for food by learning from the behaviour of others, social transmission of food-obtaining techniques was investigated usually in species, where foraging requires sophisticated skills of handling and manipulating objects often of a complex nature. Traditionally, subjects of these kinds of experiments are found among particular taxa of birds [Corvidae, like ravens, *Corvus corax* (Bugnyar & Kotrschal, 2002); parrots, like keas, *Nestor notabilis* (Gajdon et al., 2004); budgerigars, *Melopsittacus undulatus* (Heyes & Saggerson, 2002); or starlings, *Sturnus vulgaris* (Campbell et al., 1999)]; rodents, especially the Norway rat (e.g., Galef & Whiskin, 2001); and different species of the primates [e.g., marmosets, *Callithrix jacchus* (Caldwell & Whiten, 2004); chimpanzees (Custance et al., 2001; Horner & Whiten, 2005)].

It is easy to understand that dogs differ from the above-listed species in many respects. Dogs may feed on a varied diet; however, they consume bulky rather than meticulously 'wrapped' food items—in other words, their food is usually easy to consume. Dogs did not evolve with skills for holding or manipulating small, complex objects; their paws and mouth are not especially dextrous. Therefore, dogs were not an appealing subject for those scientists who were interested in the fine details of transmitting food-obtaining techniques or the mechanisms of social learning because there are well-proven species available to be tested among similarly reliable experimental designs, such as problem boxes and lockable food containers (or 'artificial fruits').

However, dogs provide another intriguing option for social learning research, and it is based mostly on their evolutionary origins through domestication. It has been argued that dogs, as an 'artificial species', were evolved during their long domestication history for the role of the most successful heterospecific companion of humans (Miklósi & Topál, 2013). It is also widely accepted that the typical behavioural features of dogs can be investigated within the human social sphere as the dogs' natural environment. As Topál and colleagues (2009b) summarised, dogs succeeded in becoming man's best friend through acquiring or developing a set of skills called 'social competence'. Among these skills, the interspecific attachment behaviour (Topál et al., 1998; see also Chapter 6), attention to human (ostensive) communication (Topál et al., 2009a; see also Chapter 12), and attention to humans in general (Miklósi et al., 2003) could be mentioned as the most basic elements. Regarding de Waal's 'Bonding- and identification-based' model for observational learning (de Waal, 2001), one could expect that individuals of a species (dogs) that form strong emotional bonds (attachment) with individuals of the other species (humans), and stay focused on human activity and communication, will show evidence of social learning at the interspecific level. From this aspect, dogs gained relative popularity in recent social learning research as the model of dog-human social transmission of different kinds of behavioural patterns, with an evolutionary explanation based on the common evolutionary past of the two species.

In the following sections of this chapter, we take a closer look at a few 'focus points' of research work done on the broad topic of social learning in dogs. At first, we concentrate on papers that targeted particular mechanisms of social learning. We start with contagious yawning—a phenomenon that is considered very 'low level' social learning because it (seemingly) happens automatically after the subject is exposed to a yawning demonstrator. However, as we will see, being able to yawn contagiously has deeper implications for the capacity of empathy and emotional matching. From the simplest, we move on to perhaps the most complicated issue of social learning mechanisms: imitation. Luckily, researchers developed various empirical techniques (as well as various definitions) for grabbing this elusive phenomenon, and we can review the results of several experiments performed on dogs' imitative capacity. We see how the researchers tackled 'motor-imitation' in dogs, trying to tease apart concurrent mechanisms such as emulation and stimulus enhancement from 'true imitation'; another section shows the results of those studies that employed the 'Do as I do' technique, teaching their subjects the basic rule to follow the demonstrator's actions; and finally we review the studies done on rational imitation in dogs, where the researchers claimed that dogs, just as human babies, can recognise the reason behind the actions of a demonstrator. After reviewing the results on dogs' imitation, we summarise a long series of experiments using the detour paradigm, which provides insight into the complexity and possible interplay among several social learning mechanisms during problem solving. In later sections, we describe such research reports that can shed some light on the functional aspect of social learning in dogs. In other words, these results talk about the adaptive value of particular cases of social information acquisition for dogs. We will see that dogs learn to look for food when they notice another dog was probably successful in such a task. We review the results of some of the papers that tested for the existence of social eavesdropping in dogs—a very useful capacity for extracting relevant information about the social relationships in a multimember group. The last section is about the effect of a dog's rank in hierarchy on its performance in various social learning tasks, which proves that besides the details and circumstances of the demonstrated action, the social attributes of the observer also can be decisive factors during social learning.

## 9.3.1 Social Influence, Contagion, Social Learning: Is There Any Difference?

Behavioural synchronisation was observed among dogs on several occasions through the decades of ethological research. Some of these behaviours are connected to feeding. For example, dogs in pairs ran faster to get a food reward than those running alone (Vogel et al., 1950; Scott & McGray, 1967). Group feeding enhanced the volume of consumed food in dog puppies (Ross & Ross, 1949). The term 'contagious behaviour' emerged as a mechanism for explaining the spread of barking among dogs living in hearing distance from each other (Whiten & Ham, 1992).

Recently, another 'classic' contagious form of behaviour, yawning, gained interest in the research of socially transmitted behaviours in dogs. Contagious yawning is considered as a form of behavioural synchronisation among humans (e.g., Provine, 1986). Theories stem from the fact that human groups have a particularly high level of social complexity, and some authors state that the presence of contagious yawning is connected to the capacity of feeling and expressing empathy for group members (e.g., Norscia & Palagi, 2011). This kind of affective empathy can involve automatic or unconscious motor-mimicry mechanisms copying others' behaviour and a consequent afferent feedback (resulting in similar inner state to the one of the model). As indirect proof of this theory, it was found that contagious yawning is diminished in those individuals who suffer empathy-related disorders (e.g., Helt et al., 2010). In humans, the neural mechanism controlling contagious yawning is also thought to be related to those capacities that are responsible for the 'conscious empathy', in other words, understanding others' feelings without necessarily mimicking them (e.g., Platek et al., 2003). This notion gained support from those studies, which discovered that the capacity for contagious yawning emerges in children with similar timing to other important features of a cognitive mind (such as attributing false beliefs to others, or the theory of mind; e.g., Anderson & Meno, 2003; Singer, 2006).

It is still unclear whether contagious yawning can be connected to conscious (or cognitive) empathy in non-human animals. Cognitive empathy is considered to emerge phylogenetically and ontogenetically with other 'indicators of mind', and among others, it requires a capacity for self–other differentiation and perspective taking and that the subject represents the state of another's feelings without the necessary involvement of emotional matching (Preston & de Waal, 2002). Contagious yawning was found in several animal species, and it would be difficult to prove the parallel existence of the previously mentioned components of 'indicators of mind' in each and every one of them. However, the connection between contagious yawning and affective empathy is still possible in such species as the chimpanzee (Campbell et al., 2009), gelada baboons (*Theropithecus gelada;* Palagi et al., 2009), or even the budgerigars (Miller et al., 2011), where familiarity is a very important element of social relationships among individuals. Indeed, just as in humans (Norscia & Palagi, 2011), the intensity or occurrence of contagious yawning was found to be related to the strength of social bonds between group members in chimpanzees (Campbell & de Waal, 2011) and in baboons (Palagi et al., 2009).

Dogs became popular candidates in the research of contagious yawning when Joly-Mascheroni and colleagues (2008) showed that about three-quarters of their test population of adult dogs yawned after witnessing an unfamiliar human's yawnings. Knowing the strong emotional bonds between dogs and humans in their mixed-species groups (Topál et al., 1998; Miklósi & Topál, 2013), researchers were quick to draw the conclusion that the contagion of yawning from humans to dogs may be another sign of interspecific (emotional) synchronisation. As further evidence of affective empathy lying behind the spread of yawning from humans to dogs, Silva and colleagues (2012) showed that dogs were more likely to yawn contagiously if they heard their owner's yawns than the sound of unfamiliar humans' yawning. However, some serious criticisms were also raised concerning contagious yawning in dogs. Other researchers could not repeat the original results of Joly-Mascheroni et al. (2008), eventually challenging the whole theory of the phenomenon either in the interspecific or the intraspecific (i.e., dog-dog) domain (Harr et al., 2009; O'Hara & Reeve, 2010). As a most likely explanation for exactly why dogs did yawn in the study of Joly-Mascheroni et al. (2008), the theory of 'tension yawning' was suggested. Yawning can be the sign of social stress in dogs (Beerda et al., 1998); therefore, O'Hara and Reeve (2010) concluded that if dogs are placed in an experimental situation where they feel stressed while observing a human model (or a video recording of a human model) who is yawning, they may produce contagious-looking yawns just because of the stress. Similar, non-empathy-related concerns were expressed regarding the auditory playback study of Silva et al. (2012). As dogs are able to recognise their owners' voice (Adachi, 2007), they were probably mildly confused or stressed from the fact that their owner obviously was not there where the yawns were played back at the same time. This mild stress may have resulted in more frequent yawning behaviour in the dogs when they heard their owners, but not when they heard the strangers yawning (Madsen & Persson, 2013). Some of the newest evidence of contagious yawning in dogs came from the latter authors, who tested the effects of ontogeny and the familiarity with the model on dogs' yawning behaviour. Madsen and Persson (2013) found that only dogs older than 7 months showed contagious yawning; this behaviour was clearly identifiable because dogs did not yawn if they observed motorically similar non-yawning actions (gaping) of their human models. However, familiarity to the model did not affect the occurrence of contagious yawning in this well-designed experiment, where strong emphasis was put on the non-stressing context of the testing environment.

Some researchers did not simply assume that dogs may have different stress (or arousal) levels during particular conditions of the previously reviewed experiments about contagious yawning. Philips Buttner and Strasser (2013) tested shelter dogs with an unfamiliar human demonstrator's yawns and additionally measured the saliva cortisol levels of their subjects before and after the trials to compare the stress levels of those dogs that yawned and those that did not after witnessing the demonstrations. Although the researchers found only a relatively minor proportion (20%) of their sample as contagiously yawning, these dogs showed higher stress hormone levels than the non-yawning dogs (Philips Buttner & Strasser, 2013). Although we may consider the choice of shelter dogs in this experiment as not the best solution, because these animals cannot be regarded as proper representatives of average companion dogs living in a well-balanced dog-human relationship with their owners, the results can be regarded as possible indicators of the effect of stress to contagious-looking yawning in dogs. In another recent study (Romero et al., 2013), the researchers managed to collect convincing evidence of the effect of familiarity on contagious yawning in dogs, with a parallel exclusion of stress-related bias of the results. They tested adult companion dogs with live demonstrations of yawns (including sound) and control gaping (without sound) by the dogs' owners as well as unfamiliar assistants. Romero and colleagues (2013) found that dogs were more likely to yawn when they observed their owners' yawns. The authors also performed heart rate measurements on the dogs during the test trials, and as dogs' heart rate did not differ between the trials with the owner and the unfamiliar person, Romero et al. concluded that the difference in the yawning rate was not caused by the different stress levels when facing the owner or the assistant.

In conclusion, contagious yawning is more likely to truly exist in dogs at least in the interspecific modality. Dogs yawn when they see a human's yawning, and probably the sound of a yawn is also enough to elicit contagion. Familiarity with the human model is seemingly not always necessary for the contagious yawning in dogs, and it emphasises the importance of the special bond between dogs and humans, which results in relatively flawless cooperation even between non-familiar individuals of the two species (for experimental evidence of this last statement, see any empirical study in which dogs were successfully tested in the presence of an unfamiliar experimenter). Obviously, contagious yawning needs not only the presence of a model individual, but also the demonstration of the yawning behaviour; therefore, contagious yawning can be distinguished from simpler forms of social influence. Some authors (such as Madsen & Persson, 2013) even call contagious yawning 'low-level imitation', which most probably stems from the high fidelity of copying of a behaviour, considered often as a necessity for imitation (see Table 9-1). However, as contagious yawning lacks most of the other 'ingredients' of imitation (thought to be necessary by most of the researchers in this field)—such as performing a novel behaviour or understanding of the goal of the action—the learning mechanism involved in contagious yawning most probably could be called, rather appropriately, 'mimicry' (Tomasello & Call, 1997) or 'chameleon effect' (Chartrand & Bargh, 1999). Harder still is to draw a functional conclusion of the studies on interspecific contagious yawning between dogs and humans. In other words, whether the capacity to yawn contagiously in dogs shows their competence in (unconscious) empathy for others (including humans) or this behaviour can be explained in much more parsimonious ways is still unknown. Yoon and Tennie (2010) gave a brief review of the theoretical questions and implications of the possible background mechanisms behind contagious yawning. According to this review, although the capacity of empathy would be an adaptive way for dogs to be able to predict and respond to human inner states and behaviour, the simpler (cognitively less demanding) unconscious mimicry would provide an equally interesting possibility for understanding contagious yawning between dogs and humans. Following the reasoning of Yoon and Tennie, as unconscious mimicry (at least in humans) is modulated by affiliative and bonding intentions from the mimicker's side, the capacity in dogs to yawn after seeing a human's yawning may be the sign of similar affiliative tendencies on the interspecific level. The implications of this possibility lead much further than the actual quality of the social bonds in the given dog-human dyad; initial affiliative tendencies between humans and the dogs' ancestors might have played an important role in the early domestication process of the dogs as well.

## 9.3.2 Imitation, the 'Holy Grail' of Social Learning—But Will We Find It in Dogs?

Imitation did not become the most frequently investigated (and debated) mechanism of social learning just because it is relatively hard to distinguish in practice from the others. As scientists refined their methods and theories regarding how to differentiate among forms of socially transmitted behaviours, it became more and more obvious that 'true' cases of imitation exist mostly (or, as it was thought for a longer period, *only*) in humans. It has been assumed that imitation is a necessary attribute for cultural transmission of the numerous skills that are inseparable from human culture (e.g., see Tomasello, 1990; Meltzoff, 1996). As imitation was considered more and more as one of the hallmarks of human-specific cognitive capacities, new attention arose for investigating its possible existence among non-human animal species. The principal interest in imitation therefore is at least two fold. First, from the point of view of cognitive sciences, if a species is capable of imitative learning, that behaviour could be the indicator of the existence of other high-end cognitive capacities as well that also were considered as typically 'human' (i.e., understanding others' intentions, having a self–other distinguishing awareness). Second, from comparative evolutionary considerations, finding other-than-human species capable of imitating through observation can give an insight into the emergence of this capacity through evolution, ultimately taking us closer to the understanding of the origins of human cognitive capabilities.

As the complexity of any animal social system is far behind that of humans, imitation was considered a possibly unique feature of human social learning. As a 'necessary' addition, authors started to emphasise the cognitive aspect of imitation, requiring that the observer needs to understand the goal of the demonstrator for a consequent successful imitative act (Tomasello & Call, 1997). Obviously, adding the feature 'understanding of the goal of the demonstrator' to the requirements of imitation made empirical testing especially difficult in non-human animals. Over the years, several ingenious methods were devised to gain insight indirectly into the hypothesised mental processes of the animal subjects. Generalization (e.g., in chimpanzees; Björklund et al., 2002) and incomplete demonstration (Elsner, 2007) both use the paradigm that if the observer understands the goal of the demonstrator, the observer will be able to reconstruct the whole behavioural sequence either with the employment of different objects or with the help of 'snapshots' of the demonstrated sequence or even from witnessing an unsuccessful demonstration. Of course, this kind of testing for imitation raises the question of how one could distinguish the result from the case of emulation, where authors emphasise also the convergence of the goals of the observer and demonstrator (Tomasello & Call, 1997).

### 9.3.2.1 Grabbing the Movement: Attempts for Testing 'Motor-Imitation' in Dogs

One way to test whether the observer imitated the action of the demonstrator (given that the action is a relatively new one for the subjects) is to prove that the observer directly chose the exact same solution (the motor component of the action) that was used by the demonstrator, given that possible alternative solutions were also available. Since the early 1990s, the so-called bidirectional control procedure (Heyes & Dawson, 1990) has been the standard method for challenging the observer animals with a task that offers a chance to prove the occurrence of imitation. Principally, the same approach is also used in the two-action tests (see, for example, Campbell et al., 1999). In both cases, the main idea is that we offer two basically identical solutions (performed by the demonstrator) to the observer. Obviously, the demonstrator shows only one of the two solutions to a given observer. If the observer later performs the same action as the demonstrator did, we conclude that this happened through imitation. In the bidirectional method, the same object can be moved up and down, or left and right (e.g., a protruding handle that releases the reward if it is moved left or right), whereas in the two-action protocol, the same solution to a problem can be achieved by using one of the two distinct behavioural acts (e.g., a locking mechanism which can be opened by mouth or hand). These methods provide an appropriate control for excluding stimulus enhancement (because stimulus enhancement is equally present whether the observer performs one or the other action on the same object), and with an additional 'ghost control', we can also rule out goal emulation. The ghost control (Fawcett et al., 2002) is based on moving the experimental device in the presence of, but without the actual involvement of, the demonstrator in the action. According to this theory, if the observer can still follow this kind of 'demonstration', it can learn how the environment works without the actual participation of a social agent (the demonstrator); therefore, it emulates rather than imitates the action. Of course, we can add a note here that ghost controls (assuming a positive result) *do not rule out* imitation if the observer otherwise also follows the demonstrator's action, but they show the capacity of the observer to use emulative methods, too.

Bidirectional/two-action protocols were applied on dogs more or less successfully in the past. At the beginning, a problem arose as to whether to use conspecific or heterospecific (human) individuals as demonstrators. As the goal in these studies was directed more towards testing whether dogs do or do not imitate a specific action and the researchers were less interested whether this capacity was affected by domestication, the obvious choice (just as in other species) would be a trained dog in the demonstrator's role. However, because companion dogs generally easily form good social relationships with unknown human beings, several researchers opted for using human demonstrators in their experiments, saving time and effort from the longer process of training suitable dog demonstrators. Obviously, designing an experimental device equally suitable for humans and dogs to manipulate is rather challenging, considering the anatomical differences between the two species. Results of these experiments can be also difficult to interpret because, for example, if the dog performs an action by grabbing a lever with its mouth, although the action was demonstrated by a human using his/her hand, it is puzzling whether the dog switched from its paw to its mouth because of emulation or because of a mental transfer of executive means based on understanding the similarity of the two solutions. Perhaps the best approach would be to use both a human and a dog demonstrator in these kinds of experiments, as doing so would allow researchers to draw interesting parallels (or find out interesting differences) regarding how dogs copy actions demonstrated by one or the other species.

The first study trying to use the bidirectional control protocol in dogs was performed by Kubinyi and colleagues (2003b). They used an opaque box to hide a tennis ball (a preferred toy for most dogs). The ball could be released from the box by pushing a protruding lever to either side. Among several control conditions, dogs could observe while the human demonstrator pushed the lever always in the same direction through a few consecutive trials. Although dogs used the lever more often when they witnessed the demonstration than when they did not, and they pushed the lever after the demonstration even if the ball did not roll out from the box during the human's action, in this experiment the researchers did not find evidence that dogs would copy such details of the action like the *direction* of the demonstration. The conclusion of this experiment was therefore that dogs can definitely follow human demonstration on the level of stimulus enhancement, but in this case imitation was not found. Compared to the positive results of experiments with bidirectional control on other non-primate species (such as with pigeons, *Columba livia;* Klein & Zentall, 2003), we also have to note that although the pigeons had to perform an action that directly freed their way to the reward, in Kubinyi and colleagues' experiment, the dogs were faced with a task in which pushing the lever led to the emergence of the ball through a rather obscure (hidden, indirect) way. These kinds of small details can also affect which mechanism a subject will rely on when trying to solve a problem.

A few years later, it was found that dogs may follow more closely the demonstrator's behaviour in carefully designed bidirectional protocols. Miller and colleagues (2009) used a sliding door device, where the reward was hidden behind a small door that was easy to slide to the left or to the right. With proper controls for odour cues and also for goal emulation, they employed both human and trained dog demonstrators. Interestingly, dogs showed different strategies in the case of the two demonstrator types. In both cases, the subjects copied the direction of the demonstrator's action; therefore, their behaviour exceeded the level of stimulus enhancement. However, while they performed similarly in the case when the human demonstrator executed the task or it was done with remote control in her presence, the subjects did not copy the direction of door removal when it was done with remote control in the presence of the dog demonstrator. This difference leads to the conclusion that dogs may more readily imitate the actions seen from a conspecific than from a human demonstrator, and in the latter case, they may instead use emulative solutions. Again, we should not forget that the anatomical differences may affect these kinds of results; while the dog demonstrators (and the observer dogs) used their muzzle to slide the door, the human demonstrator made the same action with her hand. For the subjects, the human's action may need an initial 'translation' to their own capacity to perform a similar movement with a different body part, which maybe enhanced the tendency for emulating in this context.

Recently, Pongrácz and colleagues (2012) used a two-action test for dogs with human demonstration. In this experiment, dogs could obtain a tennis ball if they tilted a tube set horizontally around its turning axis (see Figure 9-1). The tube could be operated either by pushing ('action 1') it downwards at any of its ends or pulling ('action 2') it downwards by a rope hanging from both ends of the tube. During a control experiment, it was established that dogs prefer the pushing-the-tube method over the rope pulling. The tests consisted of three trials, and the demonstrator performed either rope pulling or tube pushing before each trial. Demonstrations happened on the same side in the case of a particular dog. The results showed that dogs preferred the method shown by the demonstrator, as they operated the tube with the demonstrated method significantly more frequently compared to the preference level established during the control trials. Interestingly, dogs did not copy the side where the demonstrator performed the action. In this experiment, ghost control was not included; however, as both actions involved a downwards movement of one end of the tube, because dogs did follow the demonstrated type of action, emulation was not likely the mechanism behind their behaviour. Local enhancement can be ruled out by the

**FIGURE 9-1**  Experimental setup for testing dogs in the two-action test by Pongrácz et al. (2012). In the foreground, the tilting tube device is visible. A tennis ball was hidden to the middle of the horizontally attached tube. The tube could be tilted to either side around an axis in the middle. Height of the tube was adjusted accordingly to the size of the dogs. Two methods (actions) were demonstrated respectively to the experimental group: (1) tilting the tube by pushing it downwards on one side, and (2) pulling the tube downwards by grabbing one of the ropes hanging from its ends. *(Photo: P. Pongrácz)*

fact that the dogs were not faithful to the demonstrated side. The authors carefully avoided claiming that the dogs would imitate the demonstrated action, however. They stated that what they found might be a special case of 'generalised stimulus enhancement', as dogs preferred to touch the part of the equipment (rope or tube) that was operated by the demonstrator, but they did it with considerable flexibility, not strictly copying the side+object combination of the demonstration (Pongrácz et al., 2012).

### 9.3.2.2 Learning How to Learn: Imitation through the Matching-to-Sample Paradigm

A common feature of the previously discussed experiments was that the researchers always designed one specific device in each case that was expected to provide suitable means to detect whether the observer's action was a close enough copy of the demonstrated behaviour. During these kinds of experiments, researchers concentrated on the motor pattern of the behaviour, and broader cognitive or ecological consequences of the found social learning mechanisms were rarely considered. Although a close match between the observer's and the demonstrator's action is usually required for considering imitation as the mechanism for the social transmission of a behaviour, we also should not forget that animals will 'use' imitation with a great flexibility for reaching an adaptive advantage through this way of learning. Obviously, flexibility in the case of imitation will manifest itself not on the motor pattern of the copied behaviour, but in the cases and contexts where imitation is being employed.

Exactly this type of contextual flexibility during imitation became the target of those researchers who tested dogs with the matching-to-sample paradigm, which is also called the 'Do as I do' task. This method was successfully employed not only on great apes (e.g., Custance et al., 1995; Call, 2001) but also on dolphins (Herman, 2002); therefore, it seemed likely that the anatomical differences between dogs and humans would not necessarily prevent the successful imitation of the demonstrated actions. In general, the procedure consists of two phases. First, the subject is trained to perform a small set of familiar actions, which are matched to those of a demonstrator, after a simple command (Do it!). In the second phase, the animal is tested with novel actions that were never demonstrated to it before, and the only connection between the novel task and the previously trained ones is the 'Do it!' command. This way, we can tell that if the observer shows a significant level of copying even during the novel tasks, it learned a *rule*—that the task is the copying itself, independent of the goal or context of the actual behaviour. In a way, this phenomenon is an interesting combination of the motor-imitation theory (where the near-exact copying of a novel action is required) and imitation through understanding the goal of the demonstrator. Topál and colleagues (2006) were the first to prove that dogs could be capable of using a human behaviour action as a cue for displaying a functionally similar behaviour. A 4-year-old trained assistant dog learned to perform nine different actions after the presentation of the human demonstrator, in weekly 20-minute training sessions for 10 weeks. It should be noted that because of anatomical differences, human and dog actions were only partially equivalent in motor terms but were functionally similar. For example, the matching action to the human demonstrator jumping into the air with two feet was the dog standing on its two hind legs. Other trained actions included turning around the body axis, barking, jumping over a horizontal rod, putting objects into a container, carrying an object to the owner, and pushing a rod from a chair to the floor. In the second phase, the dog was tested with complex novel action sequences. Three identical plastic bottles were put on six predetermined places on the floor. The owner picked up one bottle from one place and transferred it to one of the five other places. After the 'Do it!' command, the dog was able to duplicate the entire sequence of moving a bottle from one particular place to another. One should remember that the number of possible combinations in this task was quite high; therefore, the fact that the subject still followed the demonstrated method excludes other explanations such as emulation (Topál et al., 2006).

With the 'Do as I do' method, it was proven that dogs can learn the rule that enables them to copy a human demonstrator's action preceding the 'Do it!' command. Although the results of Topál et al. (2006) provided robust evidence that dogs copy the demonstrated actions with a great flexibility (as much as the anatomical differences between the two species made it possible), theoretically, one still could hypothesise that other-than-imitation mechanisms could result in similar social transmission of behaviours. At least in the case of single actions (unlike the tasks in the second experiment of Topál et al., 2006), theo-retically, contagion or response facilitation could also trigger similar actions to the demonstrated ones. For distinguishing these (lower-order) social learning mechanisms from imitation, it would be suitable if one could prove that dogs can perform the demonstrated behaviour with considerable delay. This phenom-enon is defined as 'deferred imitation' (Klein & Meltzoff, 1999), and it provides evidence of the ability to store the seen action in the memory and then retrieve and perform it later with high fidelity, hence suggesting it has actually been learned rather than being performed immediately after the demonstration. Using the 'Do as I do' method has another advantage when testing deferred imitation: because the actions are usually not tied to some kind of device that the subject has to operate, the chance that the subject will use emulation instead of imita-tion is rather limited.

Testing another single canine subject, Huber and colleagues (2009) used the method employed by Topál et al. (2006) and confirmed the results found by the earlier study. However, they did not find convincing evidence for deferred imitation; on the contrary, the performance of their subject quickly deteriorated with the length of delay they inserted between the demonstration and the 'Do it!' command. The dog could not reliably copy the demonstrated behaviour over as short as 5 seconds in this experiment, leaving a puzzling question behind: that is, whether the matching-to-sample method in dogs is efficient only for a very limited time after the demonstration.

Clear evidence that dogs truly are able to imitate novel actions of a human demonstrator with considerable delay (up to 10 minutes) came a few years later, when Fugazza and Miklósi (2013) tested eight subjects with a purposefully modified version of the 'Do as I do' paradigm. These authors suspected that when Huber et al. (2009) repeated Topál and colleagues' (2006) method with a longer delay inserted before the 'Do it!' command, the dogs failed to copy the originally demonstrated behaviour because they were taught the rule that they should copy the action that was demonstrated *immediately before* the command. Fugazza and Miklósi (2013) therefore included a training phase in their protocol in which the dogs were *taught to wait* for the 'Do it!' command after the demon-stration. During the test phase, not only were considerably long delays included between the demonstration and the command, but different kinds of control measures were also taken. For example, the dogs had to wait behind a screen so they could not stare at the object they would have to interact with and therefore could not keep their mind active on the demonstration by looking at the object during the interval. The owner gave other commands during the delay to distract the dog; the 'Do it!' command was given by an unfamiliar assistant who did not know the action that the dog was supposed to perform (as a 'Clever Hans' control). All in all, dogs performed highly above the chance level in all these conditions (Fugazza & Miklósi, 2013). The authors concluded that a particular form of declarative (or non-procedural) memory should exist in dogs for imita-tive actions, because in their experiments the subjects did not have experience with the actions, nor did they have visual access to the object during the delay. These two conditions ensured that when the 'Do it!' command was given, the dogs could rely only on the retrieval of the memorised action from before, and their behaviour was not triggered by the sight of the object or their own earlier motor actions.

Seeing the success of the 'Do as I do' technique in eliciting imitative learning from dogs raises some intriguing questions. First, why do dogs readily imitate postural and target-related actions much easier with the matching-to-sample method than after demonstrations in the two-action or bidirectional control tasks? The answer may include the longer, elaborate training used for the 'Do as I do' technique, which is missing from the methodologies reviewed in the preceding section. But this still would not explain the success of dogs in the second phase of the 'Do as I do' experiments, when the previously untaught actions were performed. Most likely, when dogs are faced with a two-action test, for example, they simply do not concentrate on the exact details of the demonstration because the goal (result) of the demonstrator's action is more interesting for them. Contrary to this, the 'Do as I do' technique is about teaching the dog the rule of concentrating on the particular behaviour of the demonstrator, which often is not connected to any specific 'goal' (such as an appearing ball or consumable reward). A second interesting issue is whether the talent of dogs that was revealed in the 'Do as I do' experiments is being utilised for other activities in the 'everyday' life of a companion dog? It would be worthwhile to investigate further whether the matching-to-sample mechanism also appears in more natural contexts.

### 9.3.2.3 Reasonable Minds: Imitate Only When It Is 'Necessary'

Modern ethological research of dog behaviour often refers to the effect of human selection (through domestication, for example) on the behaviour and cognition of dogs (Topál et al., 2009b), but it is also common that the cognitive capacity of dogs is investigated from a comparative approach, with corresponding features of the human (and especially with the infant) mind. The reason behind this is that it is often assumed that for being successful in the human social environment, dogs would benefit from such cognitive abilities, which (at least on a rudimentary level) resemble their human partners' abilities (Miklósi & Topál, 2013). This approach of comparing the human and canine minds can be referred as the 'theory of evolutionary analogy' (Miklósi et al., 2004), which is different from the homology-based resemblance between the cognition of apes, for example.

Not surprisingly, in the past decade, several experiments were conducted on dogs, in which their behaviour and cognitive functioning were compared (directly or indirectly) to the performance of infants of different ages (e.g., object permanence: Watson et al., 2001; attachment: Topál et al., 1998; perseverative errors upon ostensive cues: Topál et al., 2009a). From the aspect of canine imitation, the most interesting study was performed by Gergely and colleagues (2002) on 14-month-old children. What they found they called 'rational imitation', because the young infants did not imitate the demonstrated action if the circumstances indicated that the demonstrator performed the action only because a more reasonable solution was impossible for her. However, the children imitated the action if the demonstrator performed the action without an apparent constraint; in other words, the situation that could be interpreted as the 'rule' was to do the action as the demonstrator did. Gergely et al. (2002) put a big lamp on a table, such that the demonstrator sat facing the infant on the other side of the table, with the infant sitting in the lap of his/her mother or father. The lamp could be turned on by pushing it downwards (see Figure 9-2). The demonstrator used her forehead to do this (an obviously awkward action, compared to simply pushing it by hand). There were two conditions: the demonstrator either held a blanket with both of her hands around her shoulders (as if she would when cold), or she had her hands on the table beside the lamp while she performed the head-pushing action (the blanket was around her shoulders in this case, too). According to the description of 'rational imitation', the subjects copied the head movement only in those cases when the demonstrator had her hands on the table. This experiment proved that even very young infants can detect the goal of a behaviour, and according to Gergely and Csibra (2003), to be able to do this, they do not need to possess the capacity of attributing mental

**FIGURE 9-2**   Two ways of demonstrating how to light the lamp by pushing it with the forehead (Gergely et al., 2002). The subjects were children who were sitting in their mother's lap on the opposite side of the table, facing the demonstrator. The demonstrator had a blanket around her shoulders. In the 'Hands occupied' condition (a), the demonstrator held the blanket across her chest with both of her hands. In the 'Hands free' condition (b), the demonstrator placed her hands on the table, beside the lamp. *(Photo: Institute of Cognitive Neuroscience and Psychology, Research Centre for Natural Sciences, Hungarian Academy of Sciences, Budapest, Hungary)*

states to others (the demonstrator, for example), but they need to evaluate the observable facts, such as the situational constraints, the action, and the consequent change in the environment (i.e., the light turns on).

This experiment gave Range and colleagues (2007) the idea to transplant the theory of rational imitation to their experiment with dogs. Using a trained dog demonstrator, they wanted to know whether or not the subjects would incorporate some of the contextual details of the demonstrator's behaviour to their choice of imitating. Just like in the experiment by Gergely et al. (2002), in which the infants could see that in one of the conditions the demonstrator was not able to operate the device with her hands (because she was holding the blanket around her shoulders), Range et al. (2007) formulated two conditions such that the demonstrator dog performed the same action but in one of the conditions it did so 'freely', whereas in the other its behaviour was reasonable only because of an obvious 'constraint'. The demonstrator always pulled a lever downwards with its paw, resulting in a food pellet falling out from a dispenser above. In one of the groups, the demonstrator additionally held a ball in its mouth while performing the lever pulling. Most importantly, the subject dogs also were tested in a control group without any demonstration, and their preference for using their paws or mouth in operating the device was established as a reference level (as discussed earlier in the experiment of Pongrácz et al., 2012). The results of the demonstration groups were very interesting: compared to the reference level, where dogs preferred to use their mouth for operating the device, subjects increased their paw actions *when they saw the demonstrator without the ball* in its mouth. When the demonstrator acted with the ball in its mouth, the subjects instead chose the originally preferred mouth action. Range et al. (2007) called this phenomenon 'selective imitation' and drew parallels between the dogs' behaviour and the infants' imitative choices in the experiment by Gergely et al. (2002). Based on the results, Range and colleagues argued that dogs, just like young infants, understood that the demonstrator was not able to operate the device with its mouth (the more reasonable solution) when the ball prevented it to do so. Therefore, the dogs kept performing the preferred mouth action. However, when the demonstrator used its paw without constraints (no ball in mouth), the dogs imitated this action, just like the infants imitated the action of the human demonstrator when her hands were otherwise not prevented from performing the most obvious solution in Gergely and colleagues' experiment (2002).

Although the experiment by Range and colleagues (2007) was based on a straightforward and elegant method, at least one important detail was left imbalanced between the two demonstration groups, which opened the way for raising alternative explanations for what was found. While in the study by Gergely et al. (2002), the blanket (which was the 'object of constraint') was on the shoulder of the demonstrator in both conditions, the crucially important ball was present for only that group of dogs in the paper by Range et al., where the demonstrator actually held it in its mouth. Unfortunately, the researchers did not include the ball in the other demonstration in some kind of an 'inert' way (such as attached to the handle or the rope of the food dispenser). Therefore, a seemingly unlikely but still possible alternative mechanism could be responsible for the mouth actions of the subjects in the case of the ball-in-mouth group: namely, that the mouth action of the subjects was not initiated by their selective imitative capacity, but more simply, as a kind of mouth-action triggering mechanism following the sight of the demonstrator's mouth holding the ball. Kaminski and colleagues (2011) performed an experiment in which they added the missing element to the protocol by Range et al., using a ball fixed to the pull-down lever of their device in one of several conditions (in another condition, the demonstrator dog held the ball in its mouth). Importantly, in this second study, the subjects did not show selective imitation (or, as Kaminski et al. named the phenomenon, 'rational imitation'). When the subjects did not see any demonstration, and also when there was a demonstration but the ball was not present at all, the observers operated the device randomly with their mouth or with their paw. However, when the ball was present (in the mouth of the demonstrator or fixed to the device), the observers instead used their mouth. This means, according to Kaminski et al. (2011), that the sight of the ball itself may initiate a mouth action form the observers. Based on the second study's results, one could conclude that unlike infants, dogs do not imitate the demonstrated action selectively regarding the constraints by which the demonstrator was influenced. However, as Huber et al. (2012) noted, even the Kaminski et al. (2011) paper lacked some important details in regard to being able to come to a fully satisfying conclusion on this question. Most importantly, the dogs in the demonstration by Kaminski et al. did not show the initial mouth bias in the control (without demonstration) context, which, according to Huber and colleagues, was a key factor in the earlier Range et al. paper for finding the significant effect of (selective) imitative response of the subjects. In other words, if the subjects did not prefer using their mouth when operating the device without demonstration, it was much harder to detect that after a demonstration of paw usage, the subjects will use the demonstrated action significantly more often. At this stage, we can only hope that one day somebody will perform the perfect combination of the two studies (Range et al., 2007; Kaminski et al., 2011) to provide a more full picture about 'wise imitation' in dogs.

### 9.3.3 Detouring around the Fence: Obscure Mechanisms and Clear Advantages of Social Learning

As we mentioned earlier, finding a really suitable task for dogs in social learning tests is not easy due to, for example, the anatomical constraints of dogs in manipulating objects. Even with devices that are fairly simple to operate, researchers may have to pretrain their subjects to be able to start the 'real' experiment (as Range et al., 2007, or Kaminski et al., 2011, did, for example). Perhaps the longest and most complete series of experiments on social learning in dogs was based on a simple problem-solving paradigm called the detour experiment (Pongrácz et al., 2001, 2003a,b, 2004, 2005, 2008). Making a detour does not require any specific skill from the subject other than *at first moving away* from the (inaccessible) target and then, after the animal has reached the farthest point of the obstacle, *moving back* on its other side to the target. This navigation around the obstacle is called 'detour', and it was found a very suitable task for challenging the individual and social learning capacities of dogs.

### 9.3.3.1 Dogs Have Difficulties with Detouring: On Their Own

Dogs were tested in detour (or detour-like) tasks only sporadically in the past. Free-living wolves were observed to use detour manoeuvres during pack-hunting (Mech, 1970), when some individuals left the trail following fleeing prey and, with a detour manoeuvre, cut ahead of it. However, the detour tasks used in behavioural research do not much resemble the highly dynamic chase of prey. The only study regarding how dogs perform detours around an obstacle we know about from the early years of ethology was performed by Buytendijk and Fischel (1932). They found that dogs can improve their performance in this kind of task through trial-and-error learning. Wolves are thought to be good in detour tasks (Scott & Fuller, 1965; Frank & Frank, 1985), and recently shelter-kept dingoes were also found to be efficient in negotiating detour problems on their own (Smith & Litchfield, 2010).

Pongrácz and colleagues designed a detour task (see Figure 9-3) in which dogs had to walk around a V-shaped transparent wire mesh fence (3 meters long with 1 meter tall 'wings', set up in an 80-degree angle) in order to obtain a reward (the reward being a toy or food, depending on the dog's preference; if the

**FIGURE 9-3**   The V-shaped fence used by Pongrácz and colleagues for testing dogs in the detour paradigm. The fence was made of steel frame covered by fine wire mesh. One side was 3 meters long and 1 meter tall. In most cases, the target reward was placed to the inside corner of the fence; therefore, dogs had to perform an inward detour to obtain it.

subject preferred both types of reward, the toy was used). The experiments were conducted in open areas near dog schools, and the participants were volunteers participating in the dog training courses. Each dog was tested only once in a series of 1-minute-long trials. If the dog obtained the target within 1 minute, the trial was terminated sooner. The owner, who remained at the starting point, was allowed to encourage the dog. The learning performance was characterised by measuring the duration for reaching the target (latency).

When one is testing for social learning, it is very important to know how the subjects perform in the task *without demonstration,* in other words, whether they can learn it on their own via trial and error. Pongrácz et al. (2001) tested the dogs in two versions of the detour test without demonstration: the dogs had to detour six times consecutively either from the outside to the inside of the fence, or the opposite way, from inside to outside. In both versions, the seventh trial was the opposite of the previous six, and this last trial served to see whether dogs can generalise their information about detouring around the fence. The results showed that dogs had serious difficulties on their own with the inward detours (only 16% of the subjects were able to detour the fence under 30 seconds in the first trial); however, they could master the task very easily in the outward trials. When they had to get behind the fence from outside, their detour latencies did not improve significantly during the first five trials. Experience with the outward detours did not help either: when the dogs had six successful outward trials, they could not generalise this information to a quick seventh detour in the opposite direction. All in all, this experiment showed that the inward detour around a V-shaped fence is a hard enough task for the dogs to expect detectable effect from social learning after demonstrating detours for the subjects.

### 9.3.3.2 Detouring Becomes Easy When a Demonstrator Is Available

In different experiments, dogs were tested with either a human or an unfamiliar dog demonstrator. Demonstrating the detour always included at first an inward walk from the starting point (where the subject was held by its owner) around one side of the fence to the inner corner and then an outward walk back to the starting point. While the human demonstrator carried the target conspicuously in his/her hand while walking inward, and he/she left it at the corner before returning to the starting point, the dog demonstrator carried the target out from the corner back to the starting point. As the results showed, this difference in the topography of the movement of the target did not affect the efficacy of social learning during these tasks (Pongrácz et al., 2001, 2003a). Importantly, the human demonstrator kept the subject's attention on his/her action by repeatedly calling the dog's name during the walk in and outward along the fence. The owner released the dog after the human or the dog demonstrator finished the action. During the 1-minute-long trials, the owner was allowed to encourage the dog to solve the task, but he/she had to stay at the starting point and was not allowed to give directional help (such as pointing with hand) for the detour.

Each test consisted of three trials, with the first serving as a baseline (no demonstration), followed by two trials with a demonstration before each.

The results showed that (especially compared to their very slow improvement when only trial-and-error learning was available) dogs utilised the demonstration with considerable ease. The detour latencies dropped significantly between the first and the later two trials. This happened both in the case of the human (Pongrácz et al., 2001) and the dog demonstrator (Pongrácz et al., 2003a). Familiarity of the demonstrator did not have an effect, as dogs learned with the same efficiency from their owner and the unfamiliar human experimenter (Pongrácz et al., 2001).

### 9.3.3.3 Social and Individual Experience: Which Is the Stronger?

Galef and Whiskin (2001) tested rats for the strength of individual experience with novel food flavours versus socially transmitted food preference. They found that social learning had the longest effect (i.e., the rats preferred the flavour of the demonstrated food over another novel flavour), if (1) the subjects did not have previous experience with the particular flavours; (2) their access to the alternative flavour remained limited after the demonstration; and (3) the alternative (non-demonstrated) flavour did not belong to a nutritionally better food. The interaction between the socially acquired and the individually collected knowledge has been also tested with the detour paradigm in dogs. The results of Pongrácz and colleagues (2001, 2003b) are discussed here in parallel with the findings of Galef and Whiskin (2001), as this comparative approach provides important information about the general features of social learning in very different species and contexts.

As we have seen, previous experience with detouring did not help the dogs when they needed to transfer their knowledge about outward detours to the inward direction. However, when Pongrácz et al. (2001, 2003b) tested *whether dogs follow the demonstrator's route* to the target behind the fence, it was found that in most cases dogs did not prefer the side of the fence where the demonstrator approached the target. Instead, dogs showed a significant preference to the direction in trials 2 and 3 that they followed in the first trial (without demonstration). This result revealed a very interesting interplay between socially acquired and individually collected information: although dogs learned the effective (i.e., fast) detour after witnessing the demonstration, this did not affect their choice of side along the fence, which was instead based on their individual experience from earlier. This fact was further supported by the next experiment, in which the previous experience with detours was systematically tested. Dogs were either provided with a demonstration from the first trial, or the results of those dogs that could not detour the fence in trial 1 (and therefore did not have experience with successful detours) were separately analysed. Additionally, Pongrácz et al. (2003b) provided unilateral/unambiguous detour demonstrations in particular groups (the demonstrator used the same side of the fence for walking behind and coming back from the corner). The results showed that dogs preferred to choose the demonstrator's route (1) if they witnessed unambiguous demonstrations, and (2) if they did not have their own experience with detouring the fence (either because the demonstration started from trial 1 or the dog could not detour the fence in trial 1). Therefore, we can conclude that trial-and-error experience has a different effect on particular aspects of dogs' behaviour in the detour task. Finer details of the detour (such as the exact route) are affected more strongly by individual experience; however, social learning seems to be the more influential factor for embracing the 'whole concept' of detouring (Pongrácz et al., 2003b).

Finally, the detour paradigm provided an opportunity to compare the longevity of socially acquired problem-solving techniques against a new, easier solution to negotiate the fence. This experiment, just as when the rats were given the opportunity to a calorically superior but not socially reinforced food, tested the subjects' behavioural flexibility in a case of duelling solutions of the same task. Pongrácz and colleagues (2003a) formed two groups, depending on the number of initial trials with demonstrations of detours (one versus three). After this phase, a door was opened on the front section of each wing of the fence, which gave quick and easy access to the target placed inside the corner. The doors remained open for three consecutive trials in both groups. No more detour demonstrations were employed; however, in the second and third 'open door' trials, the experimenter conspicuously reached through one of the doors when placing the target in the inner corner. In the same vein, the doors offered a much more straightforward solution for the dogs; most of the subjects kept on detouring if they received three detour trials with demonstration previously. Based on independent control trials, it was obvious that if the dogs were faced with the open doors in the first trial, they almost always used this easier solution. This experiment showed that a socially acquired, successful solution has a considerably long-lasting effect in dogs, even if they would be able to try another, even more efficient action. What can be the reason (or adaptive function) for this seemingly odd behaviour? Probably, for the dogs, paying attention to and learning from the humans are, in general, advantageous. Obedience, trainability, and the capacity for learning from the human-provided behavioural templates could be very important features during the (spontaneous) selection for a suitable companion. In the previous example, the human demonstrator at first repeatedly emphasised the detour around the fence as the effective solution to the problem, which may cause the dogs to be temporarily ignorant of other, alternative solutions. The next experiment showed even more convincingly that the human demonstration has specific features that influence the behaviour of dogs in social learning contexts.

### 9.3.3.4 What Makes a Human Demonstrator Typically 'Human'?

To this point, we have talked about 'human demonstration' or simply 'demonstration of the detour', seemingly ignoring the fact that the action performed by the demonstrator was more complex than walking around the wings of the V-shaped fence (Pongrácz et al., 2001). If we collect the factors possibly influencing the dogs' behaviour during this process, the most obvious are (1) the demonstrator left a trail of odour with his/her steps; (2) the demonstrator moved the target around the fence (i.e., the target itself can lead the dog to the solution of the problem); and (3) the demonstrator used so-called ostensive (attention-getting and maintaining; see, for example, Topál et al., 2009a) signals during the demonstration. The effects of these factors were systematically tested, such that experimental groups were formed in which the demonstration was 'stripped' one by one of these possible clues for the dog (Pongrácz et al., 2004).

Alongside a group using the conventional, complete demonstration sequence, four more groups were formed. In the 'scent only' group, the dogs did not see the demonstration, but the demonstrator left several trails along one wing of the fence to the target (unlike in any of the previous experiments, where the demonstrator walked around the whole fence ten times before any tests were started, just to wash all scent trails together). In the 'walk only' group, the dogs saw the demonstrator performing the detour, but he/she did not use ostensive cues and made the detour without the target. In the 'walk with target' group, the demonstrator carried the target behind the fence, but he/she still did not talk to the dog in the meantime. And finally, in the 'walk and talk' group, the demonstrator performed the action using ostensive signals but not carrying the object. The results showed that the 'human demonstration' was not effective unless the demonstrator used ostensive signals. The subjects' performance did not exceed the no-demonstration control level if they had to rely on the scent trail or the non-talking demonstrations. However, when the demonstrator walked along the fence without the target but calling the dog's attention, this demonstration was similarly effective to the 'complete' version (Pongrácz et al., 2004). These results are especially interesting if we consider that ostensive cues are obviously not necessary for the dogs in acquiring information from a dog demonstrator (Pongrácz et al., 2003a, 2008). The results of this experiment are in accordance with the findings of Erdőhegyi et al. (2007) and Topál et al. (2009a), who found that if human ostensive cues are available, dogs tend to choose the solution which was reinforced by these cues, even if this choice was obviously not the correct one. Similar results also were found by Marshall-Pescini and colleagues (2012), who showed that dogs tend to choose the smaller amount of food against the large amount if a human demonstrator expresses her liking for it by using ostensive vocalizations and simultaneous touching of the small amount of food. Importantly, in this study, stimulus enhancement alone (touching the smaller amount of food) was also enough to significantly influence the dogs' behaviour (Marshall-Pescini et al., 2012). However, in the detour test, the lack of ostensive cues in the case of the human demonstrator also made the detouring behaviour of the demonstrator ineffective (Pongrácz et al., 2004).

### 9.3.3.5 Social Learning Mechanism behind the Fence

As we have already mentioned, scientists are usually interested in the exact mechanism that governs acquisition of new knowledge during social learning. The results of the detour experiments clearly showed that dogs do learn from the action demonstrated by a human (or dog); the mere presence of humans, even if they were encouraging the dog during the control (no demo) trials was not enough to improve the subjects' performance (e.g., Pongrácz et al., 2001). Therefore, we can exclude 'simple' social facilitation, but is it possible to get closer identification of some specific social learning mechanism?

Perhaps one of the most intriguing questions is *what* exactly the dogs did learn from the demonstrated detour. Because walking around an obstacle is not a new motor action for the dogs, imitation can be ruled out almost surely, even if one considers that in particular circumstances (no *a priori* experience with detouring, and unambiguous demonstration), dogs copied the demonstrator's route with higher fidelity (Pongrácz et al., 2003b). As the differences in the human and the dog demonstrator in relation to the target (the human takes the target behind the fence; the dog retrieves it from behind the fence) did not seem to influence the observers' performance, the suggestion is that goal emulation and observational conditioning are also less likely to explain the results. Two further hypotheses remain as possible explanations. According to one of them, the demonstrator's action enhanced the far end of the wings of the fence as a key element of the environment in performing a successful detour. Because dogs usually did not follow the demonstrator's route, this behaviour rules out local enhancement; therefore, this hypothesis suggests that the main mechanism in dogs' social learning of a detour is *stimulus enhancement.* According to the other hypothesis, the key feature of the demonstrator's action would be making a detour that goes around the edge of the fence. In this case, dogs learned along the terms of *response facilitation* (Byrne, 1999) by assuming that behaviour (navigating around physical obstacles via a detour) is a specific behaviour pattern that could be primed by observation (in contrast to jumping over the fence). Mersmann and colleagues (2011), who also tested dogs with the detour paradigm, concluded that it is unlikely that dogs would learn through response facilitation, and they argued for the stimulus enhancement as the single mechanism responsible for social learning during the detour task. Besides human demonstration, Mersmann et al. (2011) also used an inanimate object (a little cart with wheels pulled on string) to show the dogs how to get around the fence. Unlike the human demonstrator, the cart did not approach the target behind the fence but disappeared behind a screen well behind the target after the 'successful detour'. In this case, the dogs' performance was higher than in the control (no demo) group in both the human and the nonanimated demonstration groups. According to the authors' conclusion, response facilitation (and goal emulation) was unlikely because the cart did not 'show' the exact route to the target; however, the dogs were still successful in the nonanimated demonstration group (Mersmann et al., 2011). The problem with this approach is two fold.

On one hand, there is no reason to assume that dogs necessarily use the same mechanism for learning (i.e., stimulus enhancement) in each condition in these experiments. A good example for this is to remember that (1) dogs are faithful to the direction they discovered around the fence; however, (2) they learn to detour quickly after seeing the demonstrator; and if they do not have their own experience, (3) they also learn the direction from the demonstrator (Pongrácz et al., 2003b). On the other hand, when the dogs learn how to navigate around the fence, it is reasonable to assume that the crucial moment or element occurs when the demonstrator passes by the end of the fence. After this, the problem is basically solved because the subject can get straight to the target without any further obstacles in its way. Therefore, response facilitation still can be one of the candidate mechanisms in the case of learning a detour. One can hypothesise that the crucially important 'response' that the demonstrator's behaviour facilitates in the observer is 'passing by the far end of the obstacle' (in other words, detouring around the far end). In this case, the demonstrator basically does not need to approach or touch the target for a successful transmission of a solution to this problem. If one considers that in most cases the human demonstrator carried the target when detouring inward and then left it in the corner of the V-shaped fence (e.g., Pongrácz et al., 2001, 2003a,b, 2004), if the most important detail of the demonstration was the demonstrator's interaction with the target, it most probably would not have helped the dogs too much in learning to detour. Namely, the last interaction between the target and the demonstrator happens when the demonstrator places the target in the corner—a location that the dog has *to move away* from at first during the detour. This means that the far end of the fence is the most relevant detail to learn to pass by, and this can happen through both stimulus enhancement and response facilitation.

In conclusion, it is fairly safe to say that dogs can learn during performing and by witnessing detours with means of possibly several interacting individual and social learning mechanisms, and it is always the function of the circumstances and context (and perhaps the individual subject itself) which mechanism will be the main factor during the process.

## 9.3.4 Placing Social Learning in an Ecological Perspective: Do Dogs Benefit from It?

It is often tempting to conduct experiments that try to define, identify, and separate particular mechanisms of social learning (e.g., Kubinyi et al., 2003b; Mersmann et al., 2011); similarly challenging is the task of running comparative research among different species when ethologists aim to find the common roots of particular capacities involved in social learning (e.g., Homer & Whiten, 2005). However, if we remember Tinbergen's famous four questions regarding behavioural research (mechanism, ontogeny, function, evolution), we should not forget the one about the *function* of the behaviour. Several studies on social learning have discussed the possible adaptive (and sometimes disadvantageous) consequences of socially transmitted behaviours (e.g., Coussi-Korbel & Fragaszy, 1995; Galef & Laland, 2005), and some examples were included at the beginning of this chapter. It is important to stay alert and keep trying to find those ecological and evolutionary constraints and environmental and social contexts that may require particular forms of social learning to develop and occur in dogs.

### 9.3.4.1 Finding Food by Extracting Socially Transmitted Cues from a Demonstrator

Social learning is proved to be a relevant factor in transmitting food preferences (e.g., in rabbits: Altbäcker et al., 1995; rats: Galef & Wigmore, 1983; or humans: Jansen & Tenney, 2001). The adaptiveness of this phenomenon is easy to understand in such species, where the presence of potentially poisonous or indigestible food makes it risky for the individual to learn on its own. Poisoning can be a relevant threat even for carnivores, especially because humans attempted (and still attempt more or less regularly) to eradicate them this way. Regarding some of the wild-living canids, Nel (1999) reviewed particular cases in which socially transmitted bait-avoiding tactics might contribute in the local success of particular members of the Canidae family. For example, juvenile black-backed jackals (*Canis mesomelas*) seem to learn how to avoid a type of baited poisonous trap, the so-called cyanide gun, in Africa (Brand & Nel, 1997). Other forms of social learning help animals to find palatable food by means of extracting information from the cues of their successfully foraging companions, or more simply, by following them. It was shown in several species that being able to follow resourceful foragers is an adaptive tactic (e.g., in the house sparrow, *Passer domesticus*: Barta & Giraldeau, 1998; in pigs, *Sus scorfa*: Held et al., 2000; or in chimpanzees: Hirata & Matsuzawa, 2001). It also was shown that dogs are able to follow another dog and find a hidden piece of food in this way (Cooper et al., 2003).

Heberlein and Turner (2009) investigated whether dogs are able to find the location where another dog (the demonstrator) has already found (and eaten) a piece of food, but without being allowed to follow their companion during its initial foray to the food. The authors assigned an observer dog to each demonstrator (the pairs were usually living in the same household). While the observer was held on leash, the demonstrator was allowed to visit one of four possible locations where it had previously found food in the pre-training phase. In half of the cases, the demonstrator received food again in the usual place, while in the other half of the tests, the previously baited location was empty. After the demonstrator visited the hiding place (and if it found food there, ate it), the observer was allowed to interact with the returning demonstrator. In the test phase, the observer could visit any of the four locations without the presence of the demonstrator. According to the results, the observers did not visit the hiding place that the demonstrator previously had visited more often whether or not the demonstrator found food there. However, in the group in which the demonstrator actually ate behind the screen, the observers initiated snout contact more frequently with the returning demonstrator, and in general they were keener to search around the room than in the group in which the demonstrator did not find food (Heberlein & Turner, 2009). These results show that dogs are aware of their partner's food searching (and finding) activity and may try to locate the food source if they notice the success of another dog. In the future, an interesting task will be to investigate whether olfactory, visual, or acoustic cues play the most important role in this behaviour.

### 9.3.4.2 Task Eavesdropping: Third-Party Imitation

In ethology, eavesdropping is usually referred to as a process in which an individual gains information about others' hierarchical or reproductive status by observing their interaction as an outsider, or 'third party' (e.g., Bonnie & Earley, 2007). It was shown that dogs are also capable of interspecific eavesdropping. In a study by Kundey et al. (2011), dogs could observe two humans who were competing over food. Later, the dogs chose the 'less competitive' human to take food from when they had the opportunity to do so. In another study, Marshall-Pescini et al. (2011) found that dogs were able to recognise which human was generous with her partner in a begging-for-food interaction, later choosing the sharing experimenter over the selfish one. Social eavesdropping therefore has a clear adaptive value for the observer because it can extract important information from others' interactions without being involved in a possibly costly trial-and-error session with unfriendly partners. Social learning is usually discussed in contexts in which the observer's behaviour is influenced directly by the demonstrator's action or some kind of alteration in the environment caused by the demonstrator. In general, these cases can be regarded as 'second-party' copying, a term which refers to the one-to-one nature of these events. However, it is also possible that observers may learn from seeing another observer's act of copying from a demonstrator, a phenomenon that can be regarded as a kind of 'eavesdropping of a task'. This process is thought to be especially important in human culture, and not surprisingly these kinds of 'third-party transmissions' were also found in great apes (Tomasello et al., 1989; Tomasello, 1999). Task eavesdropping would have an adaptive value because it theoretically speeds up the spreading of advantageous behavioural inventions across a population of highly social species. However, aside from humans, third-party copying of behaviours was only very sporadically encountered in animals.

Probably the best known example of employing eavesdropping in behavioural acquisition is the model-rival training developed by Irene Pepperberg for teaching grey parrots (*Psittacus erithacus*) to perform different kinds of tasks, usually involving verbal utterances and categorization of objects by their shape, colour, etc. (e.g., Pepperberg & Brezinsky, 1991). The core of the rival training is a triad of one demonstrator and two pupils, where one pupil (observer) is getting active training in a task, including rewards after the good 'answers', while the other observer is merely a close witness to this process. It was found that grey parrots can learn very successfully in this way, basically being passive observers of a companion that is getting trained. Not surprisingly, the success of rival training on parrots initiated research on dogs also; McKinley and Young (2003) found that dogs can be trained with this method just as effectively as with the traditional operant conditioning method. Importantly, the task they used for the dogs was similar to the one the grey parrots were trained for: learning a name and retrieving a new object from a choice of other targets (toys). A behaviour that is directed to some kind of (living or inanimate) object in the environment is called 'transitive'. Most cases of socially acquired behaviours fall into this category. Some authors (such as Tennie et al., 2009) consider the previously discussed experiment by Range et al. (2007) to be a third-party imitation of a transitive act (i.e., pulling a handle).

Intransitive actions (those that do not involve a target object from the environment) are seldom encountered among the investigated cases of non-human social learning. Obviously, particular forms of vocal imitation in birds (e.g., Pepperberg & Brezinsky, 1991) can be regarded as intransitive, as well as the acquisition of some of the non-object-related hand signals during the language projects with great apes (Gardner & Gardner, 1969). With dogs, we know at least two studies in which the observer animals copied intransitive actions of the demonstrator. In an experiment by Kubinyi and colleagues (2003a), dog owners were asked to include a new and 'reasonless' addition to their usual daily dog-walking route. This part of the walk happened right before the owner and his/her dog arrived home. Depending on the circumstances of the owner's home entrance, instead of entering the door or gate of the home, the owner started a short extra walk to the next corner on the street or to the next level in the house via a staircase. It was found that after several repetitions (in some cases more than 50 occasions) dogs started to incorporate the new section into their walking routine, a behaviour that became obvious as the dogs took the leading position instead of hesitating at the door when the owner started the extra route (Kubinyi et al., 2003a). The authors called this phenomenon 'social anticipation', assuming that this capacity is adaptive for the dogs when they have to accommodate new social habits and rules in the companionship of humans. The other experiment is the previously discussed 'Do as I do' study by Topál et al. (2006), in which the experimenters included several intransitive actions to the training and testing regime of their subject. These researchers found that (with regard to its anatomical constraints) the dog efficiently imitated such behaviours of the demonstrator, such as jumping into the air or turning around his vertical axis.

However, besides the rival training experiment of McKinley and Young (2003), no convincing third-party social learning experiments were known in dogs until Tennie and colleagues (2009) tried to prove that dogs may be able to imitate intransitive actions seen performed by a demonstrator-observer dyad. In our opinion, the experiment by Range et al. (2007) cannot be regarded as a clear case of task eavesdropping because although the experimenter was the one who gave the command to the demonstrator to operate the device, the context of this act had very little resemblance to a training, or 'teaching', setup. In the study by Range et al., the experimenter instead played only an 'enabling', or permitting, role, marking the moment when the demonstrator dog had to perform the action. On the other hand, Tennie et al. (2009) constructed a real 'eavesdropping' context, in which the demonstrator dog performed repeatedly intransitive actions on commands given by the experimenter; meanwhile, the observer dog watched this event from behind a Plexiglas screen. Several groups were formed on the basis of whether or not the action in question was known for the observer, whether ostensive cues were given by the experimenter to the observer, and whether or not the observer received rewards during the test. The results were negative in the sense that observers did not learn to perform the witnessed action on the command that the experimenter used during the demonstration. It is early to conclude that dogs are simply unable to imitate intransitive actions in an eavesdropping situation. There are a few details of the experimental design in the study by Tennie et al. that may have caused the dogs to fail in the task. The intransitive manner of the task makes it difficult for a third-party observer to (1) concentrate on the interaction of an unfamiliar human-dog dyad and (2) extract the exact goal of the demonstrated action. If the participants were familiar, the results probably would be different; plus, we know from the study by McKinley and Young (2003) that transitive actions can be transmitted in a rival-training context. The other problematic point with the experiment by Tennie et al. (2009) is that the demonstrator dog sometimes made errors during the experiment, and in these cases the test was continued by adding new demonstration trials to it so as to 'compensate' for the faulty ones. It is not hard to imagine that the observer dogs may have had difficulty extracting which action they were supposed to perform because, from the other side of the Plexiglas window, the faulty and correct demonstrations could be hard to distinguish.

In conclusion, we can note that investigating third-party social learning (eavesdropping) in dogs is one of the most interesting and intriguing possibilities, and the importance of this approach lies in the fact that these kinds of situations are very typical in the natural social environment of average companion dogs. The crucial detail for future experiments is choosing the proper context and/or task for the observer dogs. It is very important that the 'scenario' performed by the two demonstrators should contain such elements that make it irresistibly interesting for the observer. This can be either the target object the demonstrators are interacting with or the demonstrators themselves or probably the competitiveness of the context. Most probably, dogs will not perform well in an experiment in which the action of the two demonstrators or the demonstrators themselves are not or are just moderately relevant for the observer dog.

### 9.3.4.3 Social Learning and Social Structure: Rank-Dependent Behavioural Transmission

Unlike in the laboratory, where the experimental conditions are set up for mostly dyadic (or sometimes triadic; see preceding section) interactions for examining the phenomenon of social learning, in nature, behavioural transmission is usually not restricted strictly to one demonstrator and observer. Besides such obvious cases of 'group-learning' as when a clutch of young chicks learn together from their mother what to eat and what not (e.g., Nicol & Pope, 1996), the process, direction, and effectiveness of information transfer are often affected by the social dynamics within the group. One of the interesting questions from this aspect is whether social rank would have an effect on how social learning takes place among individuals. According to Coussi-Korbel and Fragaszy (1995) and Laland (2004), the strength of social bonds between the individuals, as well as the clarity of social rank structure, affects the occurrence of social learning (the stronger the bonds and the clearer the dominance relationship between two individuals, the more likely social learning will happen). It is easy to understand that in most social groups there is a difference among individuals regarding how effectively they can serve as demonstrators—in other words, how adaptive it is for others to learn from particular group members. Based on several different species, often the dominant individuals are found to be the more successful foragers and problem solvers on their own (e.g., in house mice, *Mus musculus*: Barnard & Luo, 2002; chimpanzees: Chalmeau & Gallo, 1993; brown lemurs, *Lemur fulvus*: Anderson et al., 1992; domestic chickens: Nicol & Pope, 1999). Based on this, it is reasonable to assume that when social learning takes place, most probably the dominant ones will be the demonstrators; in other words, it is a more adaptive strategy to copy a higher-ranked individual than a subordinate one.

Although the existence, nature, and consequences of social hierarchy among dogs has gained interest on both the pages of scientific and popular literature (Bradshaw et al., 2009; see also Chapters 3 and 4), surprisingly few empirical studies have dealt with the possible effects of dominance rank on dogs' behaviour aside from the problems with aggressive behaviour (Borchelt, 1983). Social learning is only one of the candidate behavioural phenomena that would be worthy to examine as a function of hierarchy among dogs.

Dogs are very special from the respect that unlike other group-living, highly social animals, they can be introduced to and tested with unfamiliar humans and dogs with considerable ease. While it would be almost impossible that unfamiliar chimpanzees, macaques, or even budgerigars could be used together as observer-demonstrator dyads, most (properly socialised) dogs meet and interact with unfamiliar individuals without any problem on a daily basis. However, when the effect of social rank is the question, we need somehow to characterise the rank of the individual dogs, and it can be done more reliably if the given subjects live together with at least one more dog that has the same owner. In two independent studies, Pongrácz and colleagues (2008, 2012) employed a short questionnaire for surveying the social rank of their subjects, in relationship to their behaviour with other dogs of the same owner. The four questions purposefully avoided the terms 'dominant' and 'subordinate' because the authors did not want to bias the owners based on their subjective feelings about these categories. Therefore, the questions targeted such simple interactions among the dogs as licking each other's mouth, competing over food, barking at strangers at the gate, and winning occasional fights among the dogs. From these, licking the other dog's mouth was considered a typically subordinate behaviour, while taking away the other dog's food, winning a fight, and barking first or more at strangers (Scott & Fuller, 1965) were considered behaviours typically shown by dominant dogs. Based on the answers, the majority of the subjects could be categorised as dominant or subordinate individuals.

In both experiments, dogs were tested with unfamiliar demonstrators and also without demonstration as a control. Somewhat surprisingly, neither of the studies found any difference between the performance of dominant and subordinate dogs when they had to solve a problem on their own. Pongrácz et al. (2008) used the detour paradigm to find out whether the subjects' social rank has an effect on the learning success from both a human and a dog demonstrator. Dogs were tested in three consecutive trials, and before each trial, the demonstrator showed how to detour around the V-shaped fence. When the demonstrator was a human, both dominant and subordinate dogs learned the task successfully, and the dominant dogs showed somewhat faster acquisition than the subordinate ones. This result was strengthened further by the findings of another study, in which a different, two-action test was employed (Pongrácz et al., 2012). Here, dominant dogs copied the demonstrated action more frequently than the subordinate subjects; in other words, they were again the more effective learners from the human demonstrator. However, when the demonstrator was an unfamiliar dog in a detour task, a strikingly different result was found. While subordinate dogs learned very efficiently from the demonstrator (their detour latencies were shorter even in the first trial than in the control group), throughout the three trials the dominant dogs were absolutely unable to learn from the unfamiliar dog demonstrator how to detour the fence effectively. Putting together the results from the two papers with human and dog demonstrators, we can somehow explain the behaviour of dominant and subordinate dogs using a model in which the strength of influence of a given demonstrator depends on the distance to the 'next above in rank' in the social group of the subject. According to this theory, dogs will learn mostly from the nearest higher-ranked individual. If humans are regarded as the 'highest authority' in a mixed dog-human group, dominant dogs most keenly and effectively exploit the behaviour of a human demonstrator. The subordinate dogs, however, pay more attention to the behaviour of other dogs (even if they are unfamiliar), because in their own group there is a dominant dog between them and the owner.

The results of the two experiments show that dominance rank has an effect on dogs' behaviour in contexts that involve active competition (or aggression), but probably even more importantly, the position in the hierarchy of their own group affects dogs' behaviour in such interactions where the partners do not belong to the familiar group. These findings underline the unique social capacities that dogs were selected for. Unlike their wild relatives, the wolves, or, for example, the great apes, which show mostly agonistic behaviours upon meeting with non-group-member conspecifics, dogs are ready for amicable interactions with unfamiliar humans and dogs. As dogs partake in the everyday social interactions of humans with outsiders of the family group, being able to form and maintain a mostly conflict-free relationship with previously unknown individuals could be an important adaptive feature during domestication. However, as we found, dogs tended to carry over from home their rank-based attitude to social learning contexts even if the demonstrator was an unfamiliar individual. This puzzling result can indicate that perhaps the boundaries between the 'familiar group' and 'strangers' became blurry in the dogs as a result of the human efforts to create a more and more community-friendly companion animal. Therefore, such behaviours, which normally would be reasonable only within the hierarchical system of the familiar group members, are now somewhat expanded to non-group-member individuals as well.

## 9.4 CONCLUSIONS

Ethologists have performed a considerable number of empirical studies on social learning issues in dogs, mostly during the past two decades. As we can see, these efforts have yielded a rich collection of knowledge from very different directions and aspects. Dogs have proved themselves as fruitful subjects for experimenting on social learning because they provide opportunity for studying, for example, such specific mechanisms as imitation and stimulus enhancement as well as provide information about how social hierarchy may affect information transfer in mixed-species groups. The investigation of dogs, compared to other non-human species, offers many advantages: the easy access to a larger number of subjects; the ease of working with them because the dogs' natural environment includes humans and man-made artefacts; and probably most importantly, the special bond between dogs and humans that enables the human experimenter to act as demonstrator at the same time—all are clear benefits of choosing dogs as target species. Of course, testing dogs comes with a price, too. Dogs are very sensitive to human-given cues, such as ostensive communication or touching parts of the experimental device, which may cause unwanted bias in the results. Another problem is that dogs, by their nature, do not have dexterous limbs like the hands of primates or the beaks of parrots. Because many of the popular testing methods in social learning studies involve sophisticated problem boxes and fine details of movements when operating them, dogs will not, or only after excessive training will, be able to perform on an acceptable level.

Reviewing the literature, we may conclude that particular aspects of social learning in dogs received more ample attention than others from the researchers. For example, there is a plethora of more or less successful attempts to find convincing proof of the ability for motor imitation in dogs. This is, of course, not surprising, as the research of imitation is very popular in other non-human species as well, due to its suspected connection to higher cognitive capacities and also because imitation is a very typical way (young) humans acquire their skills from their conspecifics. Partly because of the aforementioned difficulties of dogs with operating complex devices, designing a proper motor imitation task is not easy in their case. Another problem can be that dogs may not even be paying too much attention to the fine details of an experimental device or the demonstrator's action; instead, they are rather goal oriented and, at least in the case of a human demonstrator, heavily influenced by the ostensive or ostension-like cues of the demonstrator. These reasons, together, may result in a more enhancement- or emulation-biased repertoire of social learning capacities in the dog. However, as the 'Do as I do' experiments showed, imitation is not unknown for dogs, if the approach to the task is based on the dogs' willingness 'to be trained' and/or follow human-set rules. Probably this is why the matching-to-sample paradigm worked so well on dogs in eliciting imitation, and perhaps the relative lack of object-bound tasks in the case of the 'Do as I do' tests also favoured the imitative performance of the subjects.

Compared to the relative abundance of experimental work on separating candidate mechanisms of social learning, we can notice the scarcity of such studies that deal with the functional or ecological aspects of social information transfer in dogs. The fact that most dogs live within the boundaries of human families does not exclude the possibility for discovering and examining such phenomena when social learning 'naturally occurs' among dogs or among dogs and humans. At the same time, there also is a fortunate access to the few existing feral dog populations, which offers a challenging but also very promising area of further research into social learning in a naturalistic environment. The discovery of how social rank can affect dogs' social learning in different contexts is just one of the many possible ways that ecological factors can influence social behaviour in this species. All in all, the lesson that should be learned from the previous 20 years' worth of researching social learning in dogs is that we should keep on being open-minded and ready to observe new phenomena about the behaviour of this fascinating species, feeding the researchers' never-satisfied striving to discover new aspects on social behaviour.

**Future Directions**

- It is still not known whether dogs learn differently from familiar and unfamiliar dogs. Testing dogs in tasks when another dog demonstrates would be good if suitable multi-dog groups would provide opportunity for developing 'familiar dog demonstrators'.
- The ontogeny of social learning mechanisms is not known in dogs. Testing different age groups on similar tasks (like in human children) could provide answers to whether puppies or juveniles learn differently than adult dogs.
- One could attempt to find a functional or ecological explanation for the success of the 'Do as I do' technique in dogs. Is there any aspect of this way of imitation in the natural behavioural contexts of dogs?
- We do not know whether teaching behaviour occurs among dogs. It is known that *Felid* and *Mustelid* species provide living prey for their offspring, thus facilitating development of their predatory behaviour. Could we find any indications for intentional maternal information transfer towards the puppies in dogs?

## ACKNOWLEDGEMENTS

Many of the previously mentioned studies were supported by the Hungarian Scientific Research Fund (OTKA K82020). The author of this chapter was supported by the János Bolyai Research Scholarship from the Hungarian Academy of Sciences. The author is thankful to Claudia Fugazza for her prompt and valuable comments on the 'Do as I do' experiments and to József Topál for providing photographs of the experiment about rational imitation in children.

## REFERENCES

Adachi, I., Kuwahara, H., Fujita, K., 2007. Dogs recall their owner's face upon hearing the owner's voice. Anim. Cogn. 10, 17–21.

Altbäcker, V., Hudson, R., Bilkó, Á., 1995. Rabbit-mothers' diet influences pups' later food choice. Ethology 99, 107–116.

Anderson, J.R., Meno, P., 2003. Psychological influences on yawning in children. Curr. Psychol. Lett. 11, 1–7.

Anderson, J.R., Fornasieri, I., Ludes, E., Roeder, J.J., 1992. Social processes and innovative behaviour in changing groups of Lemur fulvus. Behav. Processes. 27, 101–112.

Barnard, C.J., Luo, N., 2002. Acquisition of dominance status affects maze learning in mice. Behav. Processes. 60, 53–59.

Barta, Z., Giraldeau, L.-A., 1998. The effect of dominance hierarchy on the use of alternative foraging tactics: a phenotype-limited producing-scrounging game. Behav. Ecol. Sociobiol. 42, 217–223.

Beerda, B., Schilder, M.B.H., van Hooff, J.A.R.A.M., de Vries, H.W., Mol, J.A., 1998. Behavioural, saliva cortisol and heart rate responses to different types of stimuli in dogs. Appl. Anim. Behav. Sci. 58, 365–381.

Björklund, D.F., Yunger, J.L., Bering, J.M., Regan, P., 2002. The generalization of deferred imitation in enculturated chimpanzees (Pan troglodytes). Anim. Cogn. 5, 49–58.

Bonnie, K.E., Earley, R.L., 2007. Expanding the scope for social information use. Anim. Behav. 74, 171–181.

Borchelt, P.L., 1983. Aggressive behavior of dogs kept as companion animals: classification and influence of sex, reproductive status and breed. Appl. Anim. Ethol. 10, 45–61.

Bradshaw, J.W.S., Blackwell, E.J., Casey, R.A., 2009. Dominance in domestic dogs—useful construct or bad habit? J. Veterinary Beh. Clin. Appl. Res. 4, 135–144.

Brand, D.J., Nel, J.A.J., 1997. Avoidance of cyanide guns by black-backed jackal. Appl. Anim. Behav. Sci. 55, 177–182.

Bugnyar, T., Kotrschal, K., 2002. Observational learning and the raiding of food caches in ravens, Corvus corax: is it 'tactical' deception? Anim. Behav. 64, 185–195.

Buytendijk, F.J.J., Fischel, W., 1932. Die Bedeutung der Feldkräfte und der Intentionalität für das Verhalten des Hundes. Archiv. Néerlande de Physiol. 17, 459–494.

Byrne, R.W., 1999. Imitation without intentionality: using string parsing to copy the organization of behavior. Anim. Cogn. 2, 63–72.

Byrne, R.W., 2002. Imitation of novel complex actions: what does the evidence from animals mean? Adv. Study Behav. 31, 77–105.

Byrne, R.W., 2003. Imitation as behaviour parsing. Phil. Trans. Roy. Soc. B: Biol. Sci. 358, 529–536.

Byrne, R.W., Russon, A.E., 1998. Learning by imitation: a hierarchical approach. Behav. Brain. Sci. 21, 667–721.

Caldwell, C., Whiten, A., 2004. Testing for social learning and imitation in common marmosets, Callithrix jacchus, using an artificial fruit. Anim. Cogn. 7, 77–85.

Call, J., 2001. Body imitation in an enculturated orangutan (*Pongo pygmaeus*). Cybernet. Syst. 32, 97–119.

Campbell, F.M., Heyes, C.M., Goldsmith, A.R., 1999. Stimulus learning and response learning by observation in the European starling, in a two-object/two-action test. Anim. Behav. 58, 151–158.

Campbell, M.W., Carter, J.D., Proctor, D., Eisenberg, M.L., de Waal, F.B., 2009. Computer animations stimulate contagious yawning in chimpanzees. Proc. R. Soc. Lond. B. 276, 4255–4259.

Campbell, M.W., de Waal, F.B.M., 2011. Ingroup-outgroup bias in contagious yawning by chimpanzees supports link to empathy. PLoS One 6, e18283.

Castro, L., Toro, M.A., 2004. The evolution of culture: from primate social learning to human culture. PNAS 101, 10235–10240.

Chalmeau, R., Gallo, A., 1993. Social constraints determine what is learned in the chimpanzee. Behav. Processes 28, 173–179.

Chartrand, T.L., Bargh, J.A., 1999. The chameleon effect: the perception-behavior link and social interaction. J. Pers. Soc. Psychol. 76, 893–910.

Cook, M., Mineka, S., 1989. Observational conditioning of fear to fear-relevant versus fear-irrelevant stimuli in rhesus monkeys. J. Abnorm. Psychol. 98, 448–459.

Cooper, J.J., Ashton, C., Bishop, S., West, R., Mills, D.S., Young, R.J., 2003. Clever hounds: social cognition in the domestic dog (*Canis familiaris*). Appl. Anim. Behav. Sci. 81, 229–244.

Coussi-Korbell, S., Fragaszy, D., 1995. On the relationship between social dynamics and social learning. Anim. Behav. 50, 1441–1453.

Custance, D., Whiten, A., Sambrook, T., Galdikas, B., 2001. Testing for social learning in the "artificial fruit" processing of wildborn orangutans (*Pongo pygmaeus*), Tanjung Puting, Indonesia. Anim. Cogn. 4, 305–313.

Custance, D.M., Whiten, A., Bard, K.A., 1995. Can young chimpanzees (*Pan troglodytes*) imitate arbitrary actions? Hayes and Hayes (1952) revisited. Behaviour 132, 837–857.

Elsner, B., 2007. Infants' imitation of goal-directed actions: the role of movements and action effects. Acta. Psychol. 124, 44–59.

Erdőhegyi, Á., Topál, J., Virányi, Z., Miklósi, Á., 2007. Dog-logic: inferential reasoning in a two-way choice task and its restricted use. Anim. Behav. 74, 725–737.

Fawcett, T.W., Skinner, A.M.J., Goldsmith, A.R., 2002. A test for imitative learning in starlings using a two-action method with an enhanced ghost control. Anim. Behav. 64, 547–556.

Frank, H., Frank, M.G., 1985. Comparative manipulation-test performance in ten-week-old wolves (*Canis lupus*) and Alaskan malamutes (*Canis familiaris*): a Piagetian interpretation. J. Comp. Psychol. 3, 266–274.

Freeberg, T.M., 1996. Assortative mating in captive cowbirds is predicted by social experience. Anim. Behav. 52, 1129–1142.

Fugazza, C., Miklósi, Á., 2013. Deferred imitation and declarative memory in domestic dogs. Anim. Cogn. http://dx.doi.org/10.1007/s 10071–013–0656–5.

Gajdon, G.K., Fijn, N., Huber, L., 2004. Testing social learning in a wild mountain parrot, the kea (*Nestor notabilis*). Learn. Behav. 32, 62–71.

Galef Jr., B.G., 1988. Imitation in animals: history, definition and interpretation of data from the psychobiological laboratory. In: Zentall, T.R., Galef, G.B. (Eds.), Social learning. Lawrance A. Erlbaum, Hillsdale, NJ, pp. 3–28.

Galef Jr., B.G., Giraldeau, L.-A., 2001. Social influences on foraging in vertebrates: causal mechanisms and adaptive functions. Anim. Behav. 61, 3–15.

Galef Jr., B.G., Laland, K.N., 2005. Social learning in animals: empirical studies and theoretical models. BioScience 55, 489.

Galef Jr., B.G., Whiskin, E.E., 2001. Interaction of social and asocial learning in food preferences of Norway rats. Anim. Behav. 62, 41–46.

Galef Jr., B.G., Wigmore, S.W., 1983. Transfer of information concerning distant foods: a laboratory investigation of the 'information-centre' hypothesis. Anim. Behav. 31, 748–758.

Gardner, R.A., Gardner, B.T., 1969. Teaching sign language to a chimpanzee. Science 165, 664–672.

Gergely, G., Bekkering, H., Király, I., 2002. Rational imitation in preverbal infants. Nature 415, 755.

Gergely, G., Csibra, G., 2003. Teleological reasoning in infancy: the naive theory of rational action. Trends. Cogn. Sci. 7, 287–292.

Harr, A.L., Gilbert, V.R., Phillips, K.A., 2009. Do dogs (Canis familiaris) show contagious yawning? Anim. Cogn. 12, 833–837.

Heberlein, M., Turner, D.C., 2009. Dogs, Canis familiaris, find hidden food by observing and interacting with a conspecific. Anim. Behav. 78, 385–391.

Held, S., Mendl, M., Devereux, C., Byrne, R.W., 2000. Social tactics of pigs in a competitive foraging task: the 'informed forager' paradigm. Anim. Behav. 59, 569–576.

Helt, M.S., Eigsti, I.M., Snyder, P.J., Fein, D.A., 2010. Contagious yawning in autistic and typical development. Child. Dev. 81, 1620–1631.

Herman, L.M., 2002. Vocal, social, and self-imitation by bottlenosed dolphins. In: Nehaniv, C., Dautenhahn, K. (Eds.), Imitation in animals and artifacts. MIT Press, Cambridge, MA, pp. 63–108.

Heyes, C.M., 1993. Imitation, culture and cognition. Anim. Behav. 46, 999–1010.

Heyes, C.M., Dawson, G.R., 1990. A demonstration of observational learning in rats using a bidirectional control. Q. J. Exp. Psychol. 42. 59–71.

Heyes, C.M., Saggerson, A., 2002. Testing for imitative and non-imitative social learning in the budgerigar using a two-object/two-action test. Anim. Behav. 64, 851–859.

Hirata, S., Matsuzawa, T., 2001. Tactics to obtain a hidden food item in chimpanzee pairs (Pan troglodytes). Anim. Cogn. 4, 285–295.

Horner, V., Whiten, A., 2005. Causal knowledge and imitation/emulation switching in chimpanzees (Pan troglodytes) and children (Homo sapiens). Anim. Cogn. 8, 164–181.

Huber, L., Range, F., Virányi, Z., 2012. Dogs imitate selectively, not necessarily rationally: reply to Kaminski et al. (2011). Anim. Behav. 83, e1–e3.

Huber, L., Range, F., Voelkl, B., Szucsich, A., Virányi, Z., Miklósi, Á., 2009. The evolution of imitation: what do the capacities of non-human animals tell us about the mechanisms of imitation? Phil. Trans. R. Soc. B. 364, 2299–2309.

Jansen, A., Tenney, N., 2001. Seeing mum drinking a 'light' product: Is social learning a stronger determinant of taste preference acquisition than caloric conditioning? Eur. J. Clin. Nutr. 55, 418–422.

Joly-Mascheroni, R.M., Senju, A., Shepherd, A.J., 2008. Dogs catch human yawns. Biol. Lett. 4, 446–448.

Jones, S.S., 2009. The development of imitation in infancy. Phil. Trans. R. Soc. B. 364, 2325–2335.

Kaminski, J., Nitzschner, M., Wobber, V., Tennie, C., Bräuer, J., Call, J., Tomasello, M., 2011. Do dogs distinguish rational from irrational acts? Anim. Behav. 81, 195–203.

Kirkpatrick, M., Dugatkin, L.A., 1994. Sexual selection and the evolutionary effects of copying mate choice. Behav. Ecol. Sociobiol. 34, 443–449.

Klein, E.D., Zentall, T.R., 2003. Imitation and affordance learning by pigeons (Columba livia). J. Comp. Psychol. 117, 414–419.

Klein, P.J., Meltzoff, A.N., 1999. Long-term memory, forgetting and deferred imitation in 12-month-old infants. Devel. Sci. 2, 102–113.

Kubinyi, E., Miklósi, Á., Topál, J., Csányi, V., 2003a. Social mimetic behaviour and social anticipation in dogs: preliminary results. Anim. Cogn. 6, 57–63.

Kubinyi, E., Topál, J., Miklósi, Á., Csányi, V., 2003b. The effect of human demonstrator on the acquisition of a manipulative task. J. Comp. Psychol. 117, 156–165.

Kundey, S.M.A., De Los Reyes, A., Royer, E., Molina, S., Monnier, B., German, R., Coshun, A., 2011. Reputation-like inference in domestic dogs (*Canis familiaris*). Anim. Cogn. 14, 291–302.

Laland, K.N., Williams, K., 1998. Social transmission of maladaptive information in the guppy. Behav. Ecol. 9, 493–499.

Laland, K.N., 2004. Social learning strategies. Learn. Behav. 32, 4–14.

Madsen, E.A., Persson, T., 2013. Contagious yawning in domestic dog puppies (*Canis lupus familiaris*): the effect of ontogeny and emotional closeness on low-level imitation in dogs. Anim. Cogn. 16, 233–240.

Marshall-Pescini, S., Passalacqua, C., Ferrario, A., Valsecchi, P., Prato-Previde, E., 2011. Social eavesdropping in the domestic dog. Anim. Behav. 81, 1177–1183.

Marshall-Pescini, S., Passalacqua, C., Miletto Petrazzini, M.E., Valsecchi, P., Prato-Previde, E., 2012. Do dogs (*Canis lupus familiaris*) make counterproductive choices because they are sensitive to human ostensive cues? PLoS ONE 7, e35437.

McKinley, S., Young, R.J., 2003. The efficacy of the model–rival method when compared with operant conditioning for training domestic dogs to perform a retrieval–selection task. Appl. Anim. Behav. Sci. 81, 357–365.

Mech, L.D., 1970. The wolf: The ecology and behaviour of an endangered species. Natural History Press, New York.

Meltzoff, A.N., 1996. The human infant as imitative generalist: A 20-year progress report in infant imitation with implications for comparative psychology. In: Galef, B.G., Heyes, C.M. (Eds.), Social learning in animals: the roots of culture. Academic Press, New York, pp. 347–370.

Mersmann, D., Tomasello, M., Call, J., Kaminski, J., Taborsky, M., 2011. Simple mechanisms can explain social learning in domestic dogs (*Canis familiaris*). Ethology 117, 675–690.

Miklósi, Á., Kubinyi, E., Topál, J., Gácsi, M., Virányi, Z., Csányi, V., 2003. A simple reason for a big difference: wolves do not look back at humans but dogs do. Curr. Biol. 13, 763–766.

Miklósi, Á., Topál, J., 2013. What does it take to become 'best friends'? Evolutionary changes in canine social competence. Trends. Cogn. Sci. in press.

Miklósi, Á., Topál, J., Csányi, V., 2004. Comparative social cognition: what can dogs teach us? Anim. Behav. 67, 995–1004.

Miller, H.C., Rayburn-Reeves, R., Zentall, T.R., 2009. Imitation and emulation by dogs using a bidirectional control procedure. Behav. Processes. 80, 109–114.

Miller, M.L., Gallup, A.C., Vogel, A.R., Vicario, S.M., Clark, A.B., 2011. Evidence for contagious behaviors in budgerigars (*Melopsittacus undulatus*): an observational study of yawning and stretching. Behav. Processes. 89, 264–270.

Nel, J.A.J., 1999. Social learning in canids: an ecological perspective. In: Box, H.O., Gibson, K.R. (Eds.), Mammalian social learning. Cambridge University Press, Cambridge, pp. 259–277.

Nicol, C.J., Pope, S.J., 1996. The maternal feeding display of domestic hens is sensitive to perceived chick error. Anim. Behav. 52, 767–774.

Nicol, C.J., Pope, S.J., 1999. The effects of demonstrator social status and prior foraging success on social learning in laying hens. Anim. Behav. 57, 163–171.

Norscia, I., Palagi, E., 2011. Yawn contagion and empathy in *Homo sapiens*. PLoS ONE 6, e28472.

O'Hara, S.J., Reeve, A.V., 2010. A test of the yawning contagion and emotional connectedness hypothesis in dogs, *Canis familiaris*. Anim. Behav. 81, 335–340.

Palagi, E., Leone, A., Mancini, G., Ferrari, P.F., 2009. Contagious yawning in gelada baboons as a possible expression of empathy. Proc. Natl. Acad. Sci. 106, 19262–19267.

Pepperberg, I.M., Brezinsky, M.V., 1991. Acquisition of a relative class concept by an African gray parrot (*Psittacus erithacus*): discriminations based on relative size. J. Comp. Psychol. 105, 286–294.

Philips Buttner, A., Strasser, R., 2013. Contagious yawning, social cognition and arousal: an investigation of the processes underlying shelter dogs' response to human yawns. Anim. Cogn. http://dx.doi.org/10.1007/s10071-013-0641-z.

Platek, S.M., Critton, S.R., Myers, T.E., Gallup, G.G., 2003. Contagious yawning: the role of self-awareness and mental state attribution. Cogn. Brain Res. 17, 223–237.

Pongrácz, P., 2009. Social learning in dogs. In: Helton, W.S. (Ed.), Canine ergonomics—the science of working dogs. Taylor & Francis, New York, pp. 43–61.

Pongrácz, P., Bánhegyi, P., Miklósi, Á., 2012. When rank counts—dominant dogs learn better from a human demonstrator in a two-action test. Behaviour 149, 111–132.

Pongrácz, P., Miklósi, Á., Kubinyi, E., Gurobi, K., Topál, J., Csányi, V., 2001. Social learning in dogs: the effect of a human demonstrator on the performance of dogs in a detour task. Anim. Behav. 62, 1109–1117.

Pongrácz, P., Miklósi, Á., Kubinyi, E., Topál, J., Csányi, V., 2003a. Interaction between individual experience and social learning in dogs. Anim. Behav. 65, 595–603.

Pongrácz, P., Miklósi, Á., Timár-Geng, K., Csányi, V., 2003b. Preference for copying unambiguous demonstrations in dogs. J. Comp. Psychol. 117, 337–343.

Pongrácz, P., Miklósi, Á., Timár-Geng, K., Csányi, V., 2004. Verbal attention getting as a key factor in social learning between dog (Canis familiaris) and human. J. Comp. Psychol. 118, 375–383.

Pongrácz, P., Miklósi, Á., Vida, V., Csányi, V., 2005. The pet dogs' ability for learning from a human demonstrator in a detour task is independent from the breed and age. Appl. Anim. Behav. Sci. 90, 309–323.

Pongrácz, P., Vida, V., Bánhegyi, P., Miklósi, Á., 2008. How does dominance rank status affect individual and social learning performance in the dog (Canis familiaris)? Anim. Cogn. 7, 90–97.

Preston, S.D., de Waal, F.D.M., 2002. Empathy: its ultimate and proximate bases. BBS 25, 1–72.

Provine, R.R., 1986. Yawning as a stereotyped action pattern and releasing stimulus. Ethology 72, 109–122.

Range, F., Virányi, Z., Huber, L., 2007. Selective imitation in domestic dogs. Curr. Biol. 17, 868–872.

Reed, J.M., 1999. The role of behavior in recent avian extinctions and endangerments. Conserv. Biol. 13, 232–241.

Riley, J.R., Greggers, U., Smith, A.D., Reynolds, D.R., Menzel, R., 2005. The flight paths of honeybees recruited by the waggle dance. Nature 435, 205–207.

Roberts, D., 1941. Imitation and suggestion in animals. Bull. Anim. Behav. 1, 11–19.

Romero, T., Konno, A., Hasegawa, T., 2013. Familiarity bias and physiological responses in contagious yawning by dogs support link to empathy. PLoS ONE 8, e71365.

Ross, S., Ross, J.G., 1949. Social facilitation of feeding behavior in dogs: I. Group and solitary feeding. J. Genet. Psychol. 74, 97–108.

Scott, J.P., Fuller, J.L., 1965. Genetics and the social behavior of the dog. University of Chicago Press, Chicago, IL.

Scott, J.P., McGray, C., 1967. Allelomimetic behaviour in dogs: negative effects of competition on social facilitation. J. Comp. Psychol. 2, 316–319.

Shea, N., 2009. Imitation as an inheritance system. Phil. Trans. R. Soc. B. 364, 2429–2443.

Silva, K., Bessa, J., de Sousa, L., 2012. Auditory contagious yawning in domestic dogs (Canis familiaris): first evidence for social modulation. Anim. Cogn. 15, 721–724.

Singer, T., 2006. The neuronal basis and ontogeny of empathy and mind reading: review of literature and implications for future research. Neurosci. Biobehav. Rev. 30, 855–863.

Smith, B.P., Litchfield, C.A., 2010. How well do dingoes, Canis dingo, perform on the detour task? Anim. Behav. 80, 155–162.

Spence, K.W., 1937. The differential response in animals to stimuli varying within a single dimension. Psychol. Rev. 44, 430–444.

Tennie, C., Glabsch, E., Tempelmann, S., Bräuer, J., Kaminski, J., Call, J., 2009. Dogs, *Canis familiaris*, fail to copy intransitive actions in third-party contextual imitation tasks. Anim. Behav. 77, 1491–1499.

Thorpe, W.H., 1963. Learning and instinct in animals, second ed. Harvard University Press, Cambridge, MA.

Tomasello, M., 1990. Cultural transmission in the tool use and communicatory signalling of chimpanzees? In: Parker, S.T., Gibson, K.R. (Eds.), "Language" and intelligence in monkeys and apes. Cambridge University Press, Cambridge, pp. 247–273.

Tomasello, M., 1999. The cultural origins of human cognition. Harvard University Press, Cambridge, MA.

Tomasello, M., Call, J., 1997. Primate cognition. Oxford University Press, New York.

Tomasello, M., Gust, D., Frost, G.T., 1989. A longitudinal investigation of gestural communication in young chimpanzees. Primates 30, 35–50.

Topál, J., Byrne, R.W., Miklósi, Á., Csányi, V., 2006. Reproducing human actions and action sequences: "Do as I Do!" in a dog. Anim. Cogn. 9, 355–367.

Topál, J., Gergely, Gy, Erdőhegyi, Á., Csibra, G., Miklósi, Á., 2009a. Differential sensitivity to human communication in dogs, wolves, and human infants. Science 325, 1269–1272.

Topál, J., Miklósi, Á., Csányi, V., 1998. Attachment behaviour in dogs: a new application of Ainsworth's (1969) Strange Situation Test. J. Comp. Psychol. 112, 219–229.

Topál, J., Miklósi, Á., Gácsi, M., Dóka, A., Pongrácz, P., Kubinyi, E., Virányi, Z., Csányi, V., 2009b. The dog as a model for understanding human social behavior. Adv. Stud. Anim. Behav. 39, 71–116.

Vieth, W., Curio, E., Ernst, U., 1980. The adaptive significance of avian mobbing. III. Cultural transmission of enemy recognition in blackbirds: cross-species tutoring and properties of learning. Anim. Behav. 28, 1217–1229.

Vogel, H.H., Scott, J.P., Marston, M., 1950. Social facilitation and allelomimetic behavior in dogs. Behaviour 2, 120–133.

De Waal, F.B.M., 2001. The ape and the sushi master: cultural reflections of a primatologist. Basic Books, New York.

Watson, J.S., Gergely, Gy, Csányi, V., Topál, J., Gácsi, M., Sárközi, Z., 2001. Distinguishing logic from association in the solution of an invisible displacement task by children (*Homo sapiens*) and dogs (*Canis familiaris*): using negation of disjunction. J. Comp. Psychol. 115, 219–226.

White, D.J., Galef Jr., B.G., 2000. 'Culture' in quail: social influences on mate choices of female *Coturnix japonica*. Anim. Behav. 59, 975–979.

Whiten, A., Goodall, J., McGrew, W.C., Nishida, T., Reynolds, V., Sugiyama, Y., Tutin, C.E.G., Wrangham, R.W., Boesch, C., 2001. Charting cultural variation in chimpanzees. Behaviour 138, 1481–1516.

Whiten, A., Ham, R., 1992. On the nature and evolution of imitation in the animal kingdom: reappraisal of a century of research. In: Slater, P.J.B., Rosenblatt, J.S., Beer, C., Milinski, M. (Eds.), Advances in the study of behavior. Academic Press, New York, pp. 239–283.

Wilkinson, G.S., 1992. Information transfer at evening bat colonies. Anim. Behav. 44, 501–518.

Wood, L.A., Kendall, R.L., Flynn, E.G., 2012. Context-dependent model-based biases in cultural transmission: children's imitation is affected by model age over model knowledge state. Evol. Hum. Behav. 33, 387–394.

Yoon, J.M.D., Tennie, C., 2010. Contagious yawning: a reflection of empathy, mimicry, or contagion? Anim. Behav. 79, e1–e3.

Zajonc, R.B., 1965. Social facilitation. Science 149, 269–274.

Zentall, T.R., 2004. Action imitation in birds. Learn. Behav. 32, 15–23.

Zentall, T.R., 2006. Imitation: definitions, evidence, and mechanisms. Anim. Cogn. 9, 335–353.

# Chapter 10

# What Dogs Understand about Humans

Juliane Bräuer

*Max Planck Institute for Evolutionary Anthropology, Leipzig, Germany*

Many dog owners claim that their dogs can 'understand' them. In this chapter I try to answer the question of what dogs really understand about others. Because dogs do prefer humans over their conspecifics as social partners (Gacsi et al., 2005; Topál et al., 2005; Miklósi, 2007; Horowitz, 2011; see also Chapter 6 in this book, Prato-Previde & Valsecchi), I concentrate on the question of what they understand about their human companions. Do dogs know what humans can perceive? And if so, *how* do they assess what humans can perceive? Do they understand humans' goals and intentions? Are they able to attribute mental states to others, such as beliefs and knowledge?

## 10.1 MONITORING

One precondition to understanding human behaviour, perception, and mental states is 'monitoring', i.e., looking at the human and being attentive to what he or she is doing (see Emery, 2000). Indeed, dogs constantly monitor humans, similar to subordinates in a social group who always pay attention to dominants (Chance, 1967; Emery, 2000). They are highly attentive to what humans are doing.

This attentiveness towards humans is already present in dog puppies. One difference in the behaviour of dog puppies in comparison to hand-reared wolf puppies is dogs' increased tendency to gaze at the human's face (Gacsi et al., 2005; see also Miklósi et al., 2003). Thus, even when dog puppies and wolf puppies have the same experiences (because they are raised in an identical way), dogs gaze more at humans. This suggests that the exceptional attentiveness towards humans has an innate component that was probably selected for during domestication.

Dogs not only constantly monitor humans but also are able to receive important information by watching them. They have outstanding skills in using human communicative cues, especially the pointing gesture (see Chapter 11, Topál).

The Social Dog. http://dx.doi.org/10.1016/B978-0-12-407818-5.00010-3

Dogs also use social referencing. Similar to children, they seek information from the owner about an object to form their own understanding and guide their actions. In particular, when presented with an ambiguous object, they react in accordance with the information delivered by the owner. If owners express their worries, dogs inhibit their movements towards the object. But if owners show a positive attitude, dogs move closer to the object and interact with it sooner (Merola et al., 2011; Merola et al., 2012).

There is also some evidence that dogs can use information when it is not directed at them. Dogs might be able to make reputation-like inference by observing third-party interactions (Kundey et al., 2011; Marshall-Pescini et al., 2011; but see Nitzschner et al., 2012). And there is very clear evidence that they evaluate humans on the basis of direct experiences, i.e., searching for contact with a 'nice' experimenter—compared to an ignoring experimenter (Nitzschner et al., 2012).

Thus, dogs are highly attentive towards humans—and seem to be able to extrapolate information about specific humans (and their environment). But do they also monitor humans' head and eye orientation and look in the same direction? Following the gaze of others is an adaptive skill that enables individuals to obtain useful information about the location of food and also about dangers and social interactions. It is also considered to be an important step towards an understanding of mental states such as attention and intention (Baron-Cohen & Cross, 1995; Tomasello et al., 2005; Range & Virányi, 2011).

Agnetta et al. (2000) first investigated whether dogs follow human gaze by adopting a procedure that was first used with human infants (Butterworth & Jarrett, 1991). The subject sat in front of the experimenter. The experimenter first looked at the subject and gained its attention. Then she suddenly turned her head straight up or to the left/right of the subject and looked into free space for about 5 seconds. In control conditions, the experimenter looked straight at the subject for 5 seconds.

Dogs did not follow the human gaze in that setup. That is surprising, as they have been shown to be able to use human gaze in an object choice task to locate hidden food (Miklósi et al., 1998; Agnetta et al., 2000; Soproni et al., 2001; but see also Hare et al., 2002, and Bräuer et al., 2006). In the object choice task, a piece of food is hidden in one of two containers, but the dog is unable to see in which one. The experimenter then gives a cue to the correct location by gazing at it. The question here is which container the dog will then approach.

In most studies using the object choice task, dogs followed the human gaze and found the correct cup, but they did not follow human gaze into free space in the study of Agnetta et al. (2000). The authors speculated whether dogs follow human gaze only in a foraging situation—when food is involved. Strikingly, wolves, dogs' closest relatives, do follow human gaze into free space in a similar situation (Range & Virányi, 2011).

Teglas et al. (2012) recently addressed these contrasting results on dogs' gaze-following skills by using an eye tracker. Dogs were presented with a series of movies in which a human turned her head and gazed towards one of two identical containers—either in an ostensive or a non-ostensive way. In the ostensive condition, the human looked straight at the dog and addressed the subject in a high-pitched voice before she gave the cue, whereas in the non-ostensive condition, she did not look at the dog and addressed it using a low-pitched voice. Dogs' eye-gaze patterns were recorded with an eye tracker.

Dogs looked longer towards the gaze-congruent area (i.e., the area around the cup the human was gazing at) in the ostensive condition compared to the non-ostensive condition. However, there was no significant difference in first look towards the gaze-congruent area. The authors conclude that dogs' following of human gaze is context-dependent—i.e., it occurs only when the human head turning is proceeded by ostensive cues (see also Chapter 11, Topál).

Thus, there are potentially two different explanations as to why dogs sometimes use human gaze in an object choice task (Miklósi et al., 1998; Agnetta et al., 2000; Soproni et al., 2001) and sometimes do not (Hare et al., 2002; Bräuer et al., 2006) and why they do not follow human gaze into free space (Agnetta et al., 2000).

First, it may be due to the different ways in which dogs were made attentive before the human turned her head, in the different studies. This is difficult to evaluate based on the reported methods. Usually, the human experimenter simply establishes eye contact with the dog prior to delivering the cue; however some authors also state that they call the dog by name if it is not attentive (i.e., Soproni et al., 2001). But without a detailed comparison between methods, it is impossible to draw a conclusion about how 'ostensively' humans behaved to gain the dogs' attention (i.e., whether or not they called with a high-pitched voice) and if this reliably affected results in the expected direction.

A second potential explanation for the contrasting results is the different measures that Teglas et al. (2012) used. Whereas in an object choice task, it is measured which container subjects approach, Teglas et al. (2012) measured the first look of subjects and the duration of the look towards the target area. They found that dogs looked longer towards the correct cup, but they did not find an effect for the first look.

Thus, one question is how dogs make their choice to select the container in an object choice task. Do they go to the cup they first looked at, or do they approach the container where they looked longer? Here, the duration of the gaze-cue in the object choice task might become crucial for what container dogs select. For example, in the study of Bräuer et al. (2006), dogs used continuous looking more effectively than a looking cue that lasted 4 seconds. Further studies have to investigate systematically whether and how ostensive attention-getters or duration of the gaze cues account for the mixed results in dogs' gaze-following abilities.

## 10.2 PERSPECTIVE TAKING

But for the main question addressed in this chapter—i.e., what do dogs understand about humans?—it is less important under what circumstances dogs follow human gaze but more crucial what they *understand* about the human gaze. In other words, do they understand what humans can see? A first step to address this question is to consider whether they are sensitive to human attention. In three different experimental approaches, it was found that dogs behave differently depending on the human's attentional stance.

In the first approach that was borrowed from studies with primates (Povinelli & Eddy, 1996; Kaminski et al., 2004), dogs were tested in a situation in which they could beg for food. Subjects could choose between two humans who were eating based on either the visibility of the humans' eyes or the direction of their face. Dogs begged more from the attentive than from the inattentive human (Gacsi et al., 2004; Virányi et al., 2004; Udell et al., 2011; see also Udell & Wynne, 2011, and Virányi & Range, 2011). Similarly, dogs take into account the attentional state of an experimenter when they beg for help because they cannot solve a problem (Marshall-Pescini et al., 2013).

In the second experimental approach, owners gave a command to their dogs to lie down. Either the owner was facing the dog, a human partner, or neither or he was visually separated from the dog. Subjects were more ready to follow the command if the owner attended them during the instruction delivery (Virányi et al., 2004).

In the third approach, the dogs also had to follow a command. In this case, the experimenter forbade dogs to eat a piece of food on the floor. The human either looked at the dog or was distracted, having eyes closed or back turned. Dogs obeyed the command better when the human was attentive compared to all other conditions (Call et al., 2003; Schwab & Huber, 2006).

Thus, in all three approaches, dogs were sensitive to the human attention and behaved accordingly. One explanation for this sensitivity is that they simply reacted to certain stimuli such as the open eyes when assessing whether the human was attentive. In other words, dogs might have learned that they have to obey only when they can see the open eyes of the human and that begging is successful only when the eyes are visible—without truly understanding what humans can see.

To start to address this problem, in a study by Bräuer et al. (2004), the authors placed forbidden food behind three different kinds of barriers. One barrier was small, one was big, and the third one was big but had a Plexiglas window in the middle through which the food could be seen. The authors found that dogs were sensitive to whether or not the presented barrier was effective or ineffective at obstructing a human's vision. Dogs ate more forbidden food when the barrier was large compared to when it was small or when it had a window (in the latter two cases, the human could see either the approach or the taking of the forbidden food). However, these results also can be explained by dogs relying on stimuli, instead of an understanding of the human's visual access to the food. For example, dogs could have tried to avoid seeing the human during the approach (small barrier) and during taking of the food (big barrier with the window)—and that is why they ate the forbidden food preferentially when the barrier was big and opaque.

Therefore, Kaminski et al. (2009a) used a different approach in which dogs could not simply rely on different stimuli in order to solve the problem. They investigated whether dogs could take the visual perspective of a human when that differed from their own perspective. In that study, dogs were encouraged to cooperate with the human experimenter. Each of two toys was placed on the dog's side of two small barriers so that the dog could see both of them. One barrier was opaque, and the other one was transparent.

There were three conditions. In the experimental condition, a human sat on the opposite side of the barriers, such that she could see only the toy behind the transparent barrier. The experimenter then told the dog to 'Bring it here!' but without designating either toy in any way. In the Back Turned control, the experimenter also sat on the opposite side but with her back turned so that she could see neither toy; and in the Same Side control, she sat on the same side as the dog such that she could see both toys. Dogs preferred to approach the toy behind the transparent barrier in the experimental as compared to the back turned and the same side conditions.

Thus, if the human could see only one of the toys, dogs brought back precisely the one that the human could see. They did so although they themselves could see both toys. The authors concluded that dogs are really sensitive to humans' visual access, even if it differs from their own perspective (Kaminski et al., 2009a).

This conclusion is supported by another study by Kaminski et al. (2013). In this case, researchers used another approach in which dogs could not react to certain cues because contextual information and social cues were in conflict. As in other studies (see preceding descriptions), the experimenter forbade food to the tested dogs. However, in this version of the task, how the room was illuminated varied. Either the human, the food, or both were dark, or everything was illuminated. Dogs stole significantly more food when it was dark compared to when it was light. That is not surprising, as in the dark, social cues are absent. Thus, dogs could simply have stolen the food because they did not see the human and her eyes.

However, the dogs' choice to steal forbidden food and how they behaved depended on what was illuminated. Illumination around the food, but not the human, affected the dogs' behaviour. They hesitated longer to steal the food when it was illuminated compared to the condition in which only the human was illuminated. Thus, dogs did not simply take the sight of the human as a signal to avoid the food.

One could argue that dogs simply hesitate to approach forbidden food that is illuminated if the rest of the room is dark. But the authors excluded that possibility because they showed in another experiment that dogs do not hesitate to approach illuminated forbidden food when they are in private. It is therefore possible that dogs really understand that when the food is illuminated, the human can see them approaching and stealing the food.

Taken together, all these studies offer very strong evidence that dogs know what humans can see in the sense that they understand when the humans' line of sight is currently blocked or when they are not in a position to see things. Dogs have shown this understanding in different situations—when they should obey, when they beg, and when they fetch toys. That implies that their perspective-taking skills are very flexible. This is similar to chimpanzees that are also able to assess what others can see in various situations (Hare et al., 2000; Kaminski et al., 2004; Liebal et al., 2004; Bräuer et al., 2007).

However, in contrast to our closest living relatives, dogs rely on what they themselves can perceive when they assess what the human can see (Melis et al., 2006; Bräuer et al., 2013). In a study by Bräuer et al. (2013), dogs were confronted with the following task (see Figure 10-1). Forbidden food was placed in a tunnel so that they could retrieve it by using their paw. At the beginning of each trial, dogs were placed opposite the experimenter so that they could see both the experimenter and the tunnel. However, when they tried to retrieve the food from one side of the tunnel, they were unable to see the experimenter, but the experimenter could potentially see the dog's paw. Dogs could choose between an opaque and a transparent side of the tunnel, but they did not show a preference for the opaque one. Thus, they did not hide their approach when they could not see a human present. This indicates that here they use an egocentric strategy to assess what humans perceive. In other words, in that setup, dogs seem to conclude: 'If I do not see her, then she does not see me'. This is a very successful strategy—in most but not in all cases. To date, only chimpanzees have shown they can solve this problem. They do understand 'Although I cannot see the human, she can see my hand grabbing the forbidden food' (Melis et al., 2006).

Dogs are sensitive to what humans can see, but what about other modalities? Do dogs know what humans can hear? Kundey et al. (2010) first raised this question using a design that was also borrowed from studies with primates (Melis et al., 2006; Santos et al., 2006). In a pretest, dogs experienced that the human experimenter would forbid them to eat food. In the subsequent test, subjects had the opportunity to take food from one of two containers. The human experimenter now either looked straight ahead (hence she could see the containers) or placed her head between her knees facing the ground, not looking at the containers. The containers only differed in one way: one was silent both when food was being inserted and when it was removed, whereas the other was noisy. In both cups, small bells were attached across the opening with a translucent cord, but in the silent cup, the ringers of the bells had been removed. Dogs preferred the silent container when the human experimenter was not looking. This suggests that they did

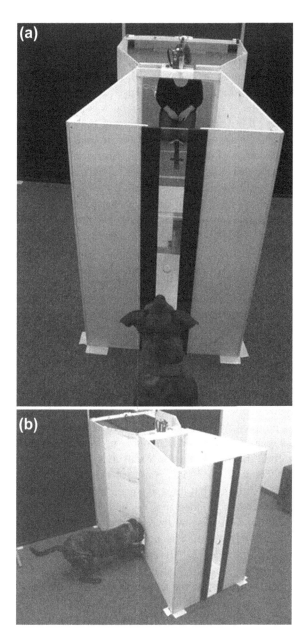

FIGURE 10-1   Experimental setup of the study of Bräuer et al. (2013): (a) apparatus from the view of the dog; (b) dog's approach to the tunnel. *(With kind permission from Springer Science and Business Media.)*

not simply have a preference for silence when they approached a potentially forbidden food, but they preferred silence if it was relevant to obtaining food unobserved by a human gatekeeper.

Raising a similar question, Bräuer et al. (2013) found that dogs preferred a silent approach to forbidden food. They used the paradigm described in the preceding paragraphs in which forbidden food was placed in a tunnel so that dogs could retrieve it by using their paw. This time both sides of the tunnel were transparent, and there was a carpet inside and around each side of the tunnel. One carpet was made of crinkly plastic foil and therefore produced a noise when the dog approached that side, while the other carpet was silent.

As in the study by Kundey et al. (2010), the experimenter did not look at the forbidden food and oriented her face downwards. Bräuer et al. (2013) found that dogs preferred the silent tunnel approach to the forbidden food. They did so although they could not see the human while they took the forbidden food; thus, they obviously had no problem remembering that the human was there.

But dogs preferred the silent approach only when this was really necessary, i.e., when the human stayed in the room and when the food was really forbidden. In the two control conditions in which the experimenter either left the room or verbally motivated the dog to take the food, dogs had no preference for the silent approach.

In conclusion, dogs are sensitive to what humans can hear. It is not simply a predisposition to be silent in critical situations which may be derived from their being carnivores (note that, for example, dogs' closest living relatives, wolves, try not to be detected when stalking prey; Peterson & Ciucci, 2003), because dogs do not always choose to approach silently; rather, they do so only when humans are present and *could* detect their forbidden approach. Overall, these studies suggest that dogs do not simply use strategies such as approaching forbidden food silently, but that their sensitivity to what humans can hear is flexible (Kundey et al., 2010; Bräuer et al., 2013).

It is, however, striking that dogs avoid the noisy tunnel but not the visible tunnel to approach forbidden food. In both cases, they could not see the human. This can be explained by the intrinsic differences between the way visual and auditory information is propagated. Whereas two subjects can have a completely different view (for example, when they stand opposite each other), both subjects will hear the same sound when they are in one room. The crucial point here is that while the dog is approaching the noisy tunnel, she/he can hear the noise herself/himself. In that moment, she can hear what the human can (and should not!) hear. In other words, also in this situation she/he can use the egocentric strategy 'When I hear the noise, then the other hears it too'(Bräuer et al., 2013).

Thus, dogs are very sensitive to humans' perception, but they most likely rely on what they themselves can perceive when they assess what the human can see and hear.

## 10.3  SEEING LEADS TO KNOWING

Dogs have developed skills to assess what humans can see, but do they also understand what humans register from their environment? In other words, do dogs understand what humans *have seen*—and more importantly—that seeing leads to knowing? To investigate this issue, two paradigms were used with dogs that were originally developed in studies with apes: the 'Ignorant helper' paradigm and the 'Guesser-Knower' paradigm.

Virányi et al. (2006) used the 'Ignorant helper' paradigm invented by Gomez (1996; see also Whiten, 2000) to investigate whether dogs are able to recognise a human's state of knowledge and ignorance depending on what she/he has seen. Two objects were hidden in a room: the dog's favorite toy and a stick, necessary to retrieve the out-of-reach toy. The helper was absent or present while the two objects were hidden. After the hiding process, the helper either knew the places of (i) both the toy and the stick, (ii) only the toy, (iii) only the stick, or (iv) neither of them. Dogs observed the whole hiding process in all conditions, but they could not reach the objects.

As other studies have shown, dogs are able to indicate the location of hidden objects without training (Miklósi et al., 2000; Kaminski et al., 2011a). The question here was what hiding place the dog would show to the helper. Dogs rarely indicated where the stick was hidden, suggesting that they did not understand that the helper needed the stick in order to retrieve the toy (Virányi et al., 2006). But dogs showed the place of the toy more often if the helper had been absent during the toy-hiding process compared to when she was present. Does that mean that dogs informed the helper about the toy because they understood that the helper had not seen where it was hidden? The helper was the owner, and her knowledge was established by her being present during the hiding of the toy the entire time or her being absent and re-entering after the objects had been placed. In all conditions, the stick was placed first and the toy was placed second. That means that in the conditions in which the helper was ignorant about the place of the toy, she re-entered immediately before the dogs started to indicate the objects (Virányi et al., 2006). Thus, dogs' increased showing behaviour in these conditions may reflect their different levels of arousal because owners reappeared rather than being evidence for understanding of past visual access (Kaminski et al., 2009a).

Topál et al. (2006) used a similar design with Philip, a highly trained assistance dog. The objects were a ball and a key. The ball was placed in one of three boxes. The helper—which was again the owner—needed the key in order to open the boxes. The question was whether Philip would indicate the place of the key only in cases when the helper was absent while the key was hidden. Indeed, he was able to adjust his communicative behaviour adequately to the different conditions; i.e., he informed the helper in 6/8 trials when he did not know where the key was hidden.

The authors state that the exact mechanism underlying this performance is not clear. One explanation is rapid discrimination learning, as Philip has sophisticated abilities for reading subtle cues of human behaviour and extensive experience with different communicative situations, in particular with his owner. This seems to be more likely than the explanation that Philip was able to recognise the relationship between the information observed by the helper and his knowledge (Topál et al., 2006).

However, Cooper et al. (2003) also reported some evidence that dogs know what humans have seen. They used the famous Guesser–Knower paradigm invented by Povinelli et al. (1990). Here, subjects can see that one of three places is baited with food, but they do not see which location. There are two informants. One informant (the knower) is present at the moment of the baiting and has full visual access to it. The second informant (the guesser) is absent during the baiting process. The question is whether subjects follow the information of the guesser or of the knower. Do they understand that only the knower is aware of the correct location of the food, as she/he has seen where it was hidden?

After the baiting, both the knower and the guesser point to different locations; the guesser always points to a wrong place. Cooper et al. (2003) reported that in the first trial 14/15 dogs (93%) chose the location pointed to by the knower (Cooper et al., 2003; Roberts & Macpherson, 2011) and ignored the pointing cue of the guesser. However, as Udell and Wynne (2011) point out, the experiment referred to in Cooper et al. (2003) has never been properly published. As the exact methods are not presented, it is impossible to assess whether alternative explanations such as experience with the experimenters or simpler behavioural cues were ruled out in that experiment (Udell & Wynne, 2011). It is striking that dogs in the study by Cooper et al. (2003) ignored the pointing gesture of a human (the guesser), since according to recent studies, they do so only in a few situations—for example, when they have seen that the target location is empty (Scheider et al., 2012) or after many trials (Petter et al., 2009; see following text).

Importantly, in the study reported by Cooper et al. (2003), the effect that dogs preferred to follow the pointing gesture of the knower was present only in the first trial. In their overall performance, dogs did not choose the knower. This drop in accuracy could be attributed to dogs' confused memories of the roles of the knower and guesser in the current trial with memories of the roles they had played on previous trials (Roberts & Macpherson, 2011). Thus, if there was an effect, it was not very strong.

Moreover, Kaminski et al. (2009a), using a novel paradigm, found no evidence that dogs understand what a human experimenter has seen in the past. In their setup, two toys were placed on the dog's side of two small barriers (see previous description), but in this experiment both barriers were opaque. The dog could witness the placing of both toys. The human experimenter sat on the opposite side of the barriers, but she witnessed only where one toy was placed because she was out of the room for the placing of the second toy.

After that, the experimenter asked the dog to fetch the toy. From the experimenter's viewpoint, neither toy was visible because the barriers were opaque—but she knew of the existence of one toy since she had seen it being placed behind one of the barriers. The question was whether dogs could also appreciate the researcher's experience with only one of the two toys. But dogs did not prefer to fetch the 'experimenter' toy. In other words, subjects did not differentiate between the two toys on the basis of whether the human had previously seen them and therefore knew about them. These results show that dogs in this study were unable to take into account what a human had seen in the immediate past.

Taken together, there is only weak evidence that dogs understand that what humans have seen leads to them knowing about it. This evidence comes from studies using the 'Ignorant helper' and the 'Guesser–Knower' paradigm. Note that these paradigms have also produced mixed results in great apes (Povinelli, 1994), and alternative explanations such as learning could not be excluded (Gomez, 1996; Topál et al., 2006). In contrast, there is clear evidence in competitive situations that chimpanzees are able to determine what a competitor has seen (Hare et al., 2001) and that seeing leads to knowing (Kaminski et al., 2008). The study by Kaminski et al. (2009a), described previously, used a cooperative setup with dogs in which dogs were asked to perform an action from their everyday lives without any prior training. The negative results in the study are particularly strong evidence that dogs are unable to understand what humans have seen in the past because the exact same paradigm was used to show that dogs do understand the human perspective in the present (see previous description).

## 10.4 INTENTIONS

Another question that arises if we want to know what dogs know about others is whether they understand humans' goals and intentions. It is highly adaptive not only to react to what humans are doing but also to anticipate what they will do. The question here is whether dogs understand human behaviour simply as bodily motion and use behavioural rules to predict future behaviour or whether they understand it as intentional goal-directed actions. In the latter case, dogs should discern directly what a human is trying (but failing) to do, what state of the environment she/he is trying to bring about, and what her/his goal is (see Tomasello et al., 2005; Call & Tomasello, 2008).

However, it is not easy to investigate whether dogs understand human intentions, as they constantly monitor humans, are able to read subtle cues of human behaviour, and have very good learning abilities. Thus, it is extremely difficult to prove that they have not simply learned that one action is usually followed by another event. One possible scenario, to make this point clearer, is the owner who grabs the leash, followed by the dog immediately running to the door. The dog seems to expect a walk. Does the dog understand that the owner has the intention to go for a walk, or has it just learned that if the leash is touched, a walk is likely?

Despite the difficulties in teasing these alternative explanations apart, there are some interesting findings relating to dogs' understanding of human intentions. These recent studies do not examine the ability to understand goals and intentions directly, but rather they investigate imitation, communication, deception, and the occurrence of helping behaviour. But in order to solve these problems, the ability to understand humans' goals and intentions would be helpful or even necessary for the dogs.

Range et al. (2007) found that dogs copy others' actions more often when those actions are the efficient solution to a problem rather than when they are not. In their study, they used a problem (i.e., operating a rod) that dogs could solve by using either their mouth or their paw. In a baseline condition, it was shown that dogs prefer to use their mouth for that action.

Dogs then saw a demonstrator dog that always used the less preferred action (the paw) to solve the problem. In one condition, this was the rational thing to do because the demonstrator dog carried a ball in her mouth, making it impossible for her to use her mouth for the action. But in the other condition, there was no obvious reason for the demonstrator dog's preference for paw usage. Range et al. (2007) found that the observer dogs preferred to use the mouth when they had observed the demonstrator using the paw whilst their own mouth was occupied by the ball, but they preferred to use the paw when they had observed the demonstrator using the paw with her mouth free. This indicates not only that dogs attend to others' goal-directed actions during demonstrations, but also that they copy others' choice of means to perform that action.

Kaminski et al. (2011b) replicated that study and added a further control condition. They did not find that dogs imitate rationally. They suggest that dogs in the study by Range et al. (2007) did not selectively attend to the irrational nature of the action but were simply distracted by the ball (Kaminski et al., 2011b).

Also in a second experiment, Kaminski et al. (2011b) found no evidence that dogs understand others' goals and attend to the means that others use to fulfill their goal. Here, an experimenter gave an unusual cue about the location of a hidden food in an object choice design. She moved her leg towards the target location. In one context, the human's hands were occupied because she was holding a heavy book, so using the leg was a rational means to communicate. In the other context, the human's hands were not occupied, making it irrational for the experimenter to use her leg. Thus, in the latter context, extension of the leg could be interpreted as a random action, not meant to communicate anything.

Dogs did not distinguish between these two conditions; they used the leg movement as a cue to find the hidden food irrespective of whether it was the rational or an irrational means to communicate. They were unable to infer that the human's goal in such situations would be using her hand to communicate—if it were free. Thus, dogs did not take into account the situational constraints faced by a human experimenter in that communicative situation.

In contrast, Kaminski et al. (2012) found that dogs distinguished random from intended movements in a communicative setting (see also Soproni et al., 2001). The main question here was whether dogs can understand the *communicative* intentions of the human, i.e., that they infer that the human wants to communicate something to them. This ability to recognise the communicative intent turns otherwise meaningless behaviours such as the human pointing gesture into meaningful communicative acts.

Kaminski et al. (2012) also used an object choice task. The human experimenter either communicated about the location of the hidden food with a communicative intent or she produced similar but non-communicative movements in the same direction (see Figure 10-2). For example, she either pointed at the correct cup and alternated her gaze between subject and cup; or she stretched out her arm and index finger cross-lateral from her body such that it mirrored the pointing cue while pretending to check the time on her watch and alternated her gaze between a clock on the wall and the correct cup. Dogs followed the intentional pointing cue more often than the non-intentional one.

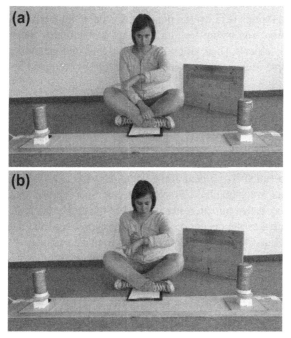

**FIGURE 10-2**   Conditions (a) 'Intentional Point' and (b) 'Non-Intentional Point' in the study of Kaminski et al. (2012). See color plate section.

Why did dogs preferentially follow the intentional cue in that study but not in the study by Kaminski et al. (2011b)? Note that here (but not in the study by Kaminski et al., 2011b), the communicative intent was established through eye contact. Indeed, one question addressed in the study by Kaminski et al. (2012) was what specific cues dogs use to determine when human communication is intended for them. They found that eye contact is the most important cue. It is possible that, for dogs, human eyes simply function as a kind of automatic trigger that raises the level of arousal and therefore the attention to the human—which then leads to greater accuracy in gesture following in situations with eye contact.

But it is also possible that dogs make a more discerning use of eye contact to read human communicative intentions. Indeed, dogs are very skillful in determining whether communication is intended for them—and they use exactly the same cues human infants use to identify communicative intent (Csibra, 2003; Behne et al., 2005).

Also, in another study, it was shown that dogs understood that a human was trying to communicate something to them (Kaminski et al., 2009b). Here, specially trained dogs were confronted with a new task. These dogs were trained to identify objects by their names and fetch them on command. In the new task, owners held up a replica of the object and said: 'Bring it here!' Dogs successfully used iconic replicas to fetch the desired object. These findings also suggest that dogs comprehended the communicative intention of the human. Holding up a replica is in itself a meaningless behaviour, but dogs understood that the human was trying to communicate to them that they should fetch the matching object. Thus, dogs have a flexible understanding of humans' communicative intentions (Kaminski et al., 2009b; see also Pettersson et al., 2011).

A different approach to investigate whether dogs understand humans' intentions was taken by Petter et al. (2009), who used a paradigm developed by Woodruff and Premack (1979) with chimpanzees. The question was whether dogs can detect human deception. They used an object choice task. Dogs could choose between two containers, one of which contained a food reward. Two different experimenters pointed at the containers. On half of the trials, a cooperative human tester pointed to the baited container, and on the other half of the trials, a deceptive human tester pointed to the empty container. Dogs were clearly sensitive to the deception; they either approached the alternate bucket from the one cued by the deceiver or refused to make a choice.

However, the dogs behaved in the same way when the cues were inanimate. When black and white boxes stood close to the containers representing the 'cooperator' and 'deceiver', dogs also learned to approach the 'cooperator' box more often than the 'deceiver' box. Dogs were equally good at discriminating between the cooperator and deceiver and at obtaining the reward whether the cues were delivered by humans or were represented by inanimate objects. The authors concluded that dogs do not understand the intentions of humans in that situation and simply learn the association between cues and their outcomes (Petter et al., 2009).

As dogs are selected to cooperate with humans, they might display their intention-reading abilities only in cooperative rather than deceptive situations. Dogs have been used for rescue, search, service, and guide purposes (Serpell, 1995; Svartberg & Forkman, 2002). But the question is whether dogs that help humans actually understand humans' goals and intentions or whether they only perform innate or trained behaviour.

Providing others with usful information is considered to be one kind of helping behaviour (Warneken & Tomasello, 2009). Kaminski et al. (2011a) investigated whether dogs would help a naïve human to find a hidden object. The object was hidden by a helper in one of four locations in a room. The dog witnessed the hiding process, but the human was absent and therefore unaware of the location of the object. Conditions varied presenting an object that was valuable either to the dog, the human, both, or neither. As in previous studies, dogs showed naïve humans the location of the hidden objects that were interesting for themselves (Miklósi et al., 2000; Virányi et al., 2006). However, dogs also sometimes indicated the location of objects in which they were not interested (i.e., a hole puncher, a vase). Moreover, when only the human was interested in the object, dogs often performed an informative showing behaviour, but indicating the *wrong* location of the object. They seemed to be motivated to help or at least willing to please the human, perhaps prompted by the human's utterances and search behaviour. But they were unable to infer the human goal in that situation.

In a second experiment, Kaminski et al. (2011a) investigated whether dogs could understand when the human needed helpful information to find a particular object—out of two—that she/he needed. Dogs were always presented with two objects. But only one of the objects was relevant for the owner because it was needed for a certain activity such as cutting or stapling paper. Then both the relevant and the irrelevant objects were hidden. Dogs did not differentiate between the object that the owner needed and the non-target object. They did not prefer to indicate the relevant one, probably because they did not understand what the human was looking for.

That is why Bräuer et al. (2013) raised the question whether dogs would help a human if the human's goal was made as obvious as possible. They used a setup in which a human tried to enter a target room in order to get a key. The tested dog could open the door to the target room by pushing a button. If the dogs were able to understand the human's goal and were motivated to help, they should open the door when the human tried to enter the target room. The help conditions in which the human expressed that she wanted to enter the target room were compared to a control condition in which the human did not try to enter the room.

Bräuer et al. (2013) found that dogs helped a human to open a door to a target room if the human explicitly communicated her goal to the dog. The results suggest that dogs are willing to help if they are able to recognise the human's goal. But the only effective way for a human to obtain help was to communicate her goal clearly by trying to open the door and giving ostensive cues to the dog in a natural way. How exactly the dogs perceived the human goal here is unclear. It is possible that they were instrumentally guided to the goal rather than really understanding what the human wanted by trying to open the door.

In conclusion, dogs have great difficulties in understanding humans' goals and intentions. It is unlikely that they interpret others' behaviour as goal-directed when they imitate and when they are deceived (Petter et al., 2009; Kaminski et al., 2011b; but see Range et al., 2007). The only clear evidence about recognition of intentions comes from communicative situations. Dogs understand humans' communicative intentions in a very flexible way (Kaminski et al., 2009b; Pettersson et al., 2011; Kaminski et al., 2012; Scheider et al., 2012). They are also able to recognise the goal of a human in a helping situation when it is expressed clearly by communicating with the dogs (Bräuer et al., 2013). In both cases, it is possible that dogs react to several ostensive cues. But no matter whether they are guided by these cues or have a more discerning understanding of humans' intentions, it enables them to react appropriately in most of these situations.

## 10.5 THEORY OF MIND IN DOMESTIC DOGS?

Some authors have suggested that due to the close evolutionary association between humans and dogs, the latter are likely candidates for finding evidence of theory of mind in a non-human animal (Topál et al., 2006; Virányi et al., 2006; Petter et al., 2009). Do dogs have a theory of mind? Are they able to attribute mental states—beliefs, intents, desires, pretending, knowledge to themselves and humans? The short answer is no. To date, there is no evidence that dogs understand that humans have beliefs, desires, and intentions that are different from their own.

But is it useful to raise the question whether dogs possess a full-fledged theory of mind like we do. Not only is it difficult to test, as many theory of mind experiments for non-verbal animals suffer from the drawback that there exist a number of other abilities that might account for the observed behaviour (Horowitz, 2011; Udell & Wynne, 2011; see also preceding description), but also for other species, namely chimpanzees, there is no clear answer to the question whether or not they have a theory of mind (Tomasello et al., 2003a, b; Call & Tomasello, 2008). Call, Tomasello, and colleagues have therefore argued that it makes no sense to answer this question with a 'yes' or 'no', but rather it makes more sense to ask which psychological states animals understand and to what extent. They also emphasise that there are many different ways in which an organism might understand the psychological functioning of others (Call & Tomasello, 2008; see also following text).

A comparison with other species can be helpful in order to find out what skills (if any) are possessed uniquely by dogs. Some skills discussed in the current chapter dogs do share with other social mammals and birds. However, in this book, we are especially interested in those skills that can be attributed to the selection pressures dogs were faced with during domestication as an adaptation to the human environment. Thus, the question is to what extent dogs' 'understanding of humans' has developed during domestication.

In their constant monitoring of humans, dogs clearly differ from other species, in particular from their closest living relatives—from wolves (Gacsi et al., 2005; see also Miklósi et al., 2003). It is surprising that they do not follow gaze into free space, a skill that is found in many primates and other mammals, birds, and even reptiles (Tomasello et al., 2001; Bugnyar et al., 2004; Bräuer et al., 2005; Kaminski et al., 2005; Wilkinson et al., 2010; see Shepherd, 2010, for a review). But if explanation by Teglas et al. (2012) is valid and dogs selectively follow human gaze when it is preceded by ostensive cues, then the fact that they do not follow gaze in some situations may be a result of their domestication history, as they were selected to be sensitive to human ostensive cues (see also Chapter 11, Topál).

Regarding visual perspective-taking, there is strong evidence that great apes, but also corvids and potentially goats, know what their conspecifics can see (Hare et al., 2000; Emery & Clayton, 2001; Dally et al., 2004; Kaminski et al., 2006; Bräuer et al., 2007). This is not surprising, as this skill is highly adaptive, both in communicative but also in competitive situations. Moreover, wolves are also sensitive to humans' attentional states under certain conditions (Udell et al., 2011; Udell & Wynne, 2011). It is possible that dogs' perspective-taking abilities reflect general mammalian skills rather than a special adaptation to the human environment. However, their sensitivity to the human eye—for example, their ability to distinguish between open and closed eyes—might be a unique adaptation to the human environment (Call et al., 2003).

Dogs do not know what humans have seen in the past and that seeing leads to knowing. But this seems to be a demanding task as, beside chimpanzees (Hare et al., 2001; Kaminski et al., 2008), there is only evidence for two corvid species that might understand that seeing leads to knowing when caching (Bugnyar & Heinrich, 2006; Dally et al., 2006).

Also regarding the understanding of others' intentions, dogs are outperformed only by great apes in the animal kingdom. This might be due to the fact that it is difficult for non-human animals to infer others' goals, or that up to now only primates have been tested in this task. Chimpanzees distinguish between an experimenter who is unwilling or unable to give them food (Call et al., 2004). They imitate selectively (Buttelmann et al., 2007) and understand others' goals in various helping situations (Warneken et al., 2007; Warneken & Tomasello, 2006, 2009). But in contrast to dogs, they do not understand humans' communicative intentions (Bräuer et al., 2006; Kaminski, 2011). The latter is most likely one of dogs' special adaptations to the human environment.

## 10.6 MIND READING OR BEHAVIOUR READING?

There is no evidence that dogs understand humans' knowledge about past events and beliefs, and they have problems understanding humans' intentions. But in many cases they are very skillful at solving social problems, leading many dog owners to conclude that 'my dog understands me'.

Indeed, dogs react appropriately in many social situations. The following 'toolkit' may help dogs to do so. First, they are extremely attentive and interested in what humans are doing. Second, they have excellent learning abilities: they are very flexible and quick to make associations and to generalise from known to similar situations in their human environment. Third, they are able to read subtle cues of human behaviour; and fourth, they have extensive experience with different communicative situations. Thus, such 'understanding' of the owner can be developed without any insightful recognition of others' subjective mind states (Whiten, 1997; Topál et al., 2006).

As Udell and Wynne (2011) have nicely stated, dogs are not readers of our minds; instead, they are exquisite readers of our behaviour. They solve these social problems successfully but probably use different strategies than we do. For example, dogs are very skilled in assessing humans' perspectives. But instead of really taking the other's visual perspective into account—i.e., imagining what she/he can see from her/his point of view—they might use some valid rules or assess what others can see on the basis of what they themselves can see. In the presence of humans, they avoid forbidden food when no opaque barrier is blocking it (Bräuer et al., 2004) and when it is illuminated (Kaminiski et al., 2013). In other situations, they seem to use the egocentric rule 'if I can see her/him, then she/he can see me'. Note that this rule is extremely successful unless they have to put their paw in a tunnel which the human has visual access to whilst they cannot see the human (Bräuer et al., 2013). Moreover, this rule is also valid to assess what humans can hear (Kundey et al., 2010; Bräuer et al., 2013).

Similarly, dogs solve the problem of reading human intentions under certain circumstances. Most likely they use the strategy to generalise past experiences in order to predict future human behaviour. They probably recognise that a particular sequence of events and/or actions precedes certain behaviour responses in humans. Dogs might have learned that if the human is turning her/his head in all directions and asking 'Where is it?' that she/he is looking for something (Kaminski et al., 2011a). But that is not enough to infer *what* she/he is looking for. Likewise, if she/he is moving towards a door accompanied by ostensive cues, dogs might have learned that she wants to open that door (Bräuer et al., 2013).

That also means that dogs might fail to predict humans' behaviour in completely novel situations, since in this case, interpreting another's behaviour as goal-directed and attending to the means by which others perform certain actions becomes necessary (Schwier et al., 2006). But in most situations, it is sufficient for dogs to simply monitor humans and 'predict' their behaviour on the basis of their past experience (Kaminski et al., 2011b).

Dogs' special talents lie in the understanding of humans' communicative intent. They not only know when communication is intended for them (Pettersson et al., 2011; Kaminski et al., 2012; see also Chapter 11, Topál), but they also understand the communicative intent in a new context (Kaminski et al., 2009b). It is possible that here they simply use a combination of different cues that they have learned during their extensive experience with communicative situations with humans. However, it is also possible that they have a deeper understanding of humans' communicative intents that has evolved as a special adaptation to the human environment.

In conclusion, although there is no evidence that domestic dogs possess a humanlike theory of mind, they 'understand' a lot about humans in their own way. They are very successful in solving social problems in their human environment, as they constantly monitor humans, learn valid associations, make adequate generalisations, and use egocentric strategies.

### Future Directions

We are left with these open questions:

- Under what circumstances do dogs follow human gaze?
- How exactly do dogs 'read' humans' intentions?
- Do dogs really understand humans' communicative intents?
- What experiences do dogs need to develop an ability to predict human behaviour?

## REFERENCES

Agnetta, B., Hare, B., Tomasello, M., 2000. Cues to food location that domestic dogs (*Canis familiaris*) of different ages do and do not use. Anim. Cogn. 3, 107–112.

Baron-Cohen, S., Cross, P., 1995. Reading the eyes: evidence for the role of perception in the development of a Theory of Mind. In: Davies, M., Stone, T. (Eds.), Folk psychology: the theory of mind debate. Blackwell Scientific Publications, Oxford, pp. 259–273.

Behne, T., Carpenter, M., Call, J., Tomasello, M., 2005. Unwilling versus unable: infants' understanding of intentional action. Dev. Psychol. 41, 328–337.

Bräuer, J., Call, J., Tomasello, M., 2004. Visual perspective taking in dogs (*Canis familiaris*) in the presence of barriers. Appl. Anim. Behav. Sci. 88, 299–317.

Bräuer, J., Call, J., Tomasello, M., 2005. All great ape species follow gaze to distant locations and around barriers. J. Comp. Psychol. 119, 145–154.

Bräuer, J., Call, J., Tomasello, M., 2007. Chimpanzees really know what others can see in a competitive situation. Anim. Cogn. 10, 439–448.

Bräuer, J., Kaminski, J., Riedel, J., Call, J., Tomasello, M., 2006. Making inferences about the location of hidden food: social dog, causal ape. J. Comp. Psychol. 120, 38–47.

Bräuer, J., Keckeisen, M., Pitsch, A., Kaminski, J., Call, J., Tomasello, M., 2013. Domestic dogs conceal auditory but not visual information from others. Anim. Cogn., 351–359.

Bräuer, J., Schönefeld, K., Call, J., 2013. When do dogs help humans? Appl. Anim. Behav. Sci. Published online.

Bugnyar, T., Heinrich, B., 2006. Pilfering ravens, *Corvus Corax*, adjust their behaviour to social context and identity of competitors. Anim. Cogn. 9, 369–376.

Bugnyar, T., Stöwe, M., Heinrich, B., 2004. Ravens, *corvus corax*, follow gaze direction of humans around obstacles. Proc. Royal Soc. London B Biol. Sci. 271, 1331–1336.

Buttelmann, D., Carpenter, M., Call, J., Tomasello, M., 2007. Enculturated chimpanzees imitate rationally. Dev. Sci. 10, F31–F38.

Butterworth, G., Jarrett, N., 1991. What minds have in common is space: spatial mechanisms serving joint visual attention in infancy. Br. J. Dev. Psychol. 9, 55–72.

Call, J., Bräuer, J., Kaminski, J., Tomasello, M., 2003. Domestic dogs (*canis familiaris*) are sensitive to the attentional state of humans. J. Comp. Psychol. 117, 257–263.

Call, J., Hare, B., Carpenter, M., Tomasello, M., 2004. 'Unwilling' versus 'unable': chimpanzees' understanding of human intentional action. Dev. Sci. 7, 488–498.

Call, J., Tomasello, M., 2008. Does the chimpanzee have a theory of mind? 30 years later. Trends Cogn. Sci. 12, 187–192.

Chance, M.R.A., 1967. Attention structure as the basis of primate rank orders. Man 2, 503–518.

Cooper, J.J., Ashton, C., Bishop, S., West, R., Mills, D.S., Young, R.J., 2003. Clever hounds: social cognition in the domestic dog (*Canis familiaris*). Appl. Anim. Behav. Sci. 81, 229–244.

Csibra, G., 2003. Teleological and referential understanding of action in infancy. Philos. Trans. Royal Soc. London B Biol. Sci. 358, 447–458.

Dally, J.M., Emery, N.J., Clayton, N.S., 2004. Cache protection strategies by western scrub-jays (*aphelocoma californica*): hiding food in the shade. Proc. Royal Soc. London B Biol. Sci. 271, S387–S390.

Dally, J.M., Emery, N.J., Clayton, N.S., 2006. Food-caching western scrub-jays keep track of who was watching when. Science 310, 1662–1665.

Emery, N.J., 2000. The eyes have it: the neuroethology, function and evolution of social gaze. Neurosci. Biobehav. Rev. 24, 581–604.

Emery, N.J., Clayton, N.S., 2001. Effects of experience and social context on prospective caching strategies by scrub jays. Nature 414, 443–446.

Gacsi, M., Gyori, B., Miklósi, A., Virányi, Z., Kubinyi, E., Topál, J., Csanyi, V., 2005. Species-specific differences and similarities in the behavior of hand-raised dog and wolf pups in social situations with humans. Dev. Psychobiol. 47, 111–122.

Gacsi, M., Miklósi, A., Varga, O., Topál, J., Csanyi, V., 2004. Are readers of our face readers of our minds? Dogs (*Canis familiaris*) show situation-dependent recognition of human's attention. Anim. Cogn. 7, 144–153.

Gomez, J.C., 1996. Nonhuman primate theories of (nonhuman primate) minds: some issues concerning the origins of mindreading. In: Carruthers, P., Smith, P.K. (Eds.), Theories of theories of mind. Cambridge University Press, Cambridge, pp. 330–343.

Hare, B., Brown, M., Williamson, C., Tomasello, M., 2002. The domestication of social cognition in dogs. Science 298, 1634–1636.

Hare, B., Call, J., Agnetta, B., Tomasello, M., 2000. Chimpanzees know what conspecifics do and do not see. Anim. Behav. 59, 771–785.

Hare, B., Call, J., Tomasello, M., 2001. Do chimpanzees know what conspecifics know? Anim. Behav. 61, 139–151.

Horowitz, A., 2011. Theory of mind in dogs? Examining method and concept. Learn. Behav. 39, 314–317.

Kaminski, J., 2011. Communicative cues among and between human and nonhuman primates: attending to specificity in triadic gestural interactions. In: Boos, M., Kolbe, M., Kappeler, P., Ellwart, T. (Eds.), Coordination in human and primate groups. Springer, Heidelberg, pp. 245–262.

Kaminski, J., Bräuer, J., Call, J., Tomasello, M., 2009a. Domestic dogs are sensitive to a human's perspective. Behaviour 146, 979–998.

Kaminski, J., Call, J., Tomasello, M., 2004. Body orientation and face orientation: two factors controlling apes' begging behavior from humans. Anim. Cogn. 7, 216–223.

Kaminski, J., Call, J., Tomasello, M., 2006. Goats' behaviour in a competitive food paradigm: evidence for perspective taking? Behaviour 143, 1341–1356.

Kaminski, J., Call, J., Tomasello, M., 2008. Chimpanzees know what others know, but not what they believe. Cognition 109, 224–234.

Kaminski, J., Neumann, M., Bräuer, J., Call, J., Tomasello, M., 2011a. Domestic dogs communicate to request and not to inform. Anim. Behav. 82, 651–658.

Kaminski, J., Nitzschner, M., Wobber, V., Tennie, C., Bräuer, J., Call, J., Tomasello, M., 2011b. Do dogs distinguish rational from irrational acts? Anim. Behav. 81, 195–203.

Kaminski, J., Pitsch, A., Tomasello, M., 2013. Dogs steal in the dark. Anim. Cogn. 16, 385–394.

Kaminski, J., Riedel, J., Call, J., Tomasello, M., 2005. Domestic goats, *Capra hircus*, follow gaze direction and use social cues in an object choice task. Anim. Behav. 69, 11–18.

Kaminski, J., Schulz, L., Tomasello, M., 2012. How dogs know when communication is intended for them. Dev. Sci. 15, 222–232.

Kaminski, J., Tempelmann, S., Call, J., Tomasello, M., 2009b. Domestic dogs comprehend human communication with iconic signs. Dev. Sci. 12, 831–837.

Kundey, S., De Los Reyes, A., Royer, E., Molina, S., Monnier, B., German, R., Coshun, A., 2011. Reputation-like inference in domestic dogs (*Canis familiaris*). Anim. Cogn. 14, 291–302.

Kundey, S.M.A., De Los Reyes, A., Taglang, C., Allen, R., Molina, S., Royer, E., German, R., 2010. Domesticated dogs (*Canis familiaris*) react to what others can and cannot hear. Appl. Anim. Behav. Sci. 126, 45–50.

Liebal, K., Pika, S., Call, J., Tomasello, M., 2004. To move or not to move: how great apes adjust to the attentional state of others. Interact. Stud. 5, 199–219.

Marshall-Pescini, S., Passalacqua, C., Ferrario, A., Valsecchi, P., Prato-Previde, E., 2011. Social eavesdropping in the domestic dog. Anim. Behav. 81, 1177–1183.

Marshall-Pescini, S., Colombo, E., Passalacqua, C., Merola, I., Prato-Previde, E., 2013. Gaze alternation in dogs and toddlers in an unsolvable task: evidence of an audience effect. Anim. Cogn. Published online.

Melis, A.P., Call, J., Tomasello, M., 2006. Chimpanzees (*Pan troglodytes*) conceal visual and auditory information from others. J. Comp. Psychol. 120, 154–162.

Merola, I., Prato-Previde, E., Marshall-Pescini, S., 2011. Social referencing in dog-owner dyads? Anim. Cogn. 15, 175–185.

Merola, I., Prato-Previde, E., Marshall-Pescini, S., 2012. Dogs' social referencing towards owners and strangers. PLoS One 7.

Miklósi, A., 2007. Dog behaviour, evolution, and cognition, first ed. Oxford University Press, Oxford.

Miklósi, A., Kubinyi, E., Gacsi, M., Virányi, Z., Csanyi, V., 2003. A simple reason for a big difference: wolves do not look back at humans but dogs do. Curr. Biol. 13, 763–766.

Miklósi, A., Polgardi, R., Topál, J., Csanyi, V., 2000. Intentional behavior in dog-human communication: an experimental analysis of 'showing' behaviour in the dog. Anim. Cogn. 3, 159–166.

Miklósi, A., Polgardi, R., Topál, J., Csanyi, V., 1998. Use of experimenter-given cues in dogs. Anim. Cogn. 1, 113–121.

Nitzschner, M., Melis, A.P., Kaminski, J., Tomasello, M., 2012. Dogs (*Canis familiaris*) evaluate humans on the basis of direct experiences only. PLoS One 7, e46880.

Peterson, R.O., Ciucci, P., 2003. The wolf as a carnivore. In: Mech, L.D., Boitani, L. (Eds.), Wolves—Behavior, ecology and conservation. University of Chicago Press, Chicago, IL, pp. 104–130.

Petter, M., Musolino, E., Roberts, W.A., Cole, M., 2009. Can dogs (*canis familiaris*) detect human deception? Behav. Proc. 82, 109–118.

Pettersson, H., Kaminski, J., Herrmann, E., Tomasello, M., 2011. Understanding of human communicative motives in domestic dogs. Appl. Anim. Behav. Sci. 133, 235–245.

Povinelli, D.J., 1994. What chimpanzees (might) know about the mind. In: Wrangham, R.W., McGrew, W.C., de Waal, F.B.M., Heltne, P.G. (Eds.), Chimpanzee cultures. Harvard University Press, Cambridge, MA, pp. 285–300.

Povinelli, D.J., Eddy, T.J., 1996. What young chimpanzees know about seeing. Monogr. Soc. Res. Child Dev. 61, 1–152.

Povinelli, D.J., Nelson, K.E., Boysen, S.T., 1990. Inferences about guessing and knowing by chimpanzees (*Pan troglodytes*). J. Comp. Psychol. 104, 203–210.

Range, F., Virányi, Z., 2011. Development of gaze following abilities in wolves (*Canis lupus*). PLoS One 6, e16888.

Range, F., Virányi, Z., Huber, L., 2007. Selective imitation in domestic dogs. Curr. Biol. 17, 868–872.

Roberts, W., Macpherson, K., 2011. Theory of mind in dogs: is the perspective-taking task a good test? Learn. Behav. 39, 303–305.

Santos, L.R., Nissen, A.G., Ferrugia, J.A., 2006. Rhesus monkeys, *Macaca mulatta*, know what others can and cannot hear. Anim. Behav. 71, 1175–1181.

Scheider, L., Kaminski, J., Call, J., Tomasello, M., 2012. Do domestic dogs interpret pointing as a command? Anim. Cogn. 361–372.

Schwab, C., Huber, L., 2006. Obey or not obey? Dogs (*canis familiaris*) behave differently in response to attentional states of their owners. J. Comp. Psychol. 120, 169–175.

Schwier, C., van Maanen, C., Carpenter, M., Tomasello, M., 2006. Rational imitation in 12-month-old infants. Infancy 10, 303–311.

Serpell, J.E., 1995. The domestic dog: Its evolution, behaviour and interactions with people. Cambridge University Press, Cambridge.

Shepherd, S.V., 2010. Following gaze: gaze-following behavior as a window into social cognition. Front. Integr. Neurosci. 4.

Soproni, K., Miklósi, A., Topál, J., Csanyi, V., 2001. Comprehension of human communicative signs in pet dogs (*Canis familiaris*). J. Comp. Psychol. 115, 27–34.

Svartberg, K., Forkman, B., 2002. Personality traits in the domestic dog (*Canis familiaris*). Appl. Anim. Behav. Sci. 79, 133–156.

Teglas, E., Gergely, A., Kupan, K., Miklósi, A., Topál, J., 2012. Dogs' gaze following is tuned to human communicative signals. Curr. Biol. 22, 209–212.

Tomasello, M., Call, J., Hare, B., 2003a. Chimpanzees understand psychological states—the question is which ones and to what extent. Trends Cogn. Sci. 7, 153–156.

Tomasello, M., Call, J., Hare, B., 2003b. Chimpanzees versus humans: it's not that simple. Trends Cogn. Sci. 7, 239–240.

Tomasello, M., Carpenter, M., Call, J., Behne, T., Moll, H., 2005. Understanding and sharing intentions: the origins of cultural cognition. Behav. Brain Sci. 28, 675–735.

Tomasello, M., Hare, B., Fogleman, T., 2001. The ontogeny of gaze following in chimpanzees, *Pan troglodytes*, and rhesus macaques. Macaca mulatta. Anim. Behav. 61, 335–343.

Topál, J., Erdõhegyi, Á., Mányik, R., Miklósi, Á., 2006. Mindreading in a dog: an adaptation of a primate 'mental attribution' study. Int. J. Psychol. Psychol. Ther. 6, 365–379.

Topál, J., Gacsi, M., Miklósi, A., Virányi, Z., Kubinyi, E., Csanyi, V., 2005. Attachment to humans: a comparative study on hand-reared wolves and differently socialized dog puppies. Anim. Behav. 70, 1367–1375.

Udell, M., Dorey, N., Wynne, C., 2011. Can your dog read your mind? Understanding the causes of canine perspective taking. Learn. Behav. 39, 289–302.

Udell, M., Wynne, C., 2011. Reevaluating canine perspective-taking behavior. Learn. Behav. 39, 318–323.

Virányi, Z., Range, F., 2011. Evaluating the logic of perspective-taking experiments. Learn. Behav. 39, 306–309.

Virányi, Z., Topál, J., Gacsi, M., Miklósi, A., Csanyi, V., 2004. Dogs respond appropriately to cues of humans' attentional focus. Behav. Proc. 66, 161–172.

Virányi, Z., Topál, J., Miklósi, Á., Csányi, V., 2006. A nonverbal test of knowledge attribution: a comparative study on dogs and children. Anim. Cogn. 9, 13–26.

Warneken, F., Hare, B., Melis, A.P., Hanus, D., Tomasello, M., 2007. Spontaneous altruism by chimpanzees and young children. PLoS Biol. 5, e184.

Warneken, F., Tomasello, M., 2006. Altruistic helping in human infants and young chimpanzees. Science 311, 1301–1303.

Warneken, F., Tomasello, M., 2009. Varieties of altruism in children and chimpanzees. Trends Cogn. Sci. 13, 397–402.

Whiten, A., 1997. The Machiavellian mindreader. In: Whiten, A., Byrne, R.W. (Eds.), Machiavellian intelligence II: extensions and evaluations. Cambridge University Press, New York, NY, pp. 144–173.

Whiten, A., 2000. Chimpanzee cognition and the question of mental re-representation. In: Sperber, D. (Ed.), Metarepresentation: a multidisciplinary perspective. Oxford University Press, Oxford, pp. 139–167.

Wilkinson, A., Mandl, I., Bugnyar, T., Huber, L., 2010. Gaze following in the red-footed tortoise (*Geochelone carbonaria*). Anim. Cogn. 13, 765–769.

Woodruff, G., Premack, D., 1979. Intentional communication in the chimpanzee: the development of deception. Cognition 7, 333–362.

# Chapter 11

# Dogs' Sensitivity to Human Ostensive Cues: A Unique Adaptation?

József Topál,[1] Anna Kis,[1,2] and Katalin Oláh[1,3]

[1]Institute of Cognitive Neuroscience and Psychology, Research Centre for Natural Sciences, Hungarian Academy of Sciences, Budapest, Hungary, [2]Department of Ethology, Eötvös University, Budapest, Hungary, [3]Department of Cognitive Psychology, Eötvös University, Budapest, Hungary

## 11.1 INTRODUCTION

One of the most striking trends in the social cognition literature is the rapid growth in the number of dog cognition papers published in the past few decades (see Bensky et al., 2013, for a comprehensive analysis). This can be interpreted as indicating that dogs have become, in a certain sense, the 'new chimpanzees' (cf. Bloom, 2004) for cognitive scientists, and raises a fundamental question: Why does this domestic species deserve such a privileged status in the study of social cognition?

The clue to the 'dog paper boom' lies in this species' *social competence*, a concept considered central to our understanding of the evolution of cognitive abilities. In fact, social competence (Oliveira, 2009; Taborsky & Oliveira, 2012), which is widely believed to be a key to answering the question 'What makes us human?', has gained increasing attention in the past few years. Although social competence, by default, manifests itself in conspecific interactions in a wide range of species, dogs can be seen as a unique case of how this system can manifest itself in interspecific (i.e., dog–human) interactions (Miklósi & Topál, 2013). It is increasingly assumed that dogs can be seen as animals displaying human competent social skills (Hare & Tomasello, 2005; Topál et al., 2009b), because their particular domestication history, including adaptive specialisation of social skills in the human social environment, paved the way for the emergence of an evolutionary novel, interspecific social competence. Humans are characterised by more pronounced collaborative (prosocial) attitudes than other apes (Hare & Tomasello, 2004; Rekers et al., 2011), and it has been suggested that this cooperative, prosocial bias may have been one of the key factors that made it possible for humans to develop higher-level cognitive skills (Richerson

The Social Dog. http://dx.doi.org/10.1016/B978-0-12-407818-5.00011-5

& Boyd, 1998; Moll & Tomasello, 2007). An increasing body of evidence suggests a similar trend towards collaboration with people in dogs (Miklósi et al., 2003; Gácsi et al., 2005). The specific differences in the social competence of dogs and apes (Wobber & Hare, 2009) raise the hypothesis that dogs have evolved interspecific social sensitivity (Bräuer et al., 2006), and domestication may have specifically enabled dogs to coordinate their own actions with humans in a collaborative manner (Horn et al., 2012; Ostojic & Clayton, 2013).

The idea that domestication enhanced the social skills of dogs in cooperative–communicative tasks involving humans gained some further support from studies in which dogs and wolves have been shown to differ (e.g., Gácsi et al., 2009a), while dogs and human infants display similarities with respect to their ability to react to challenges of the human social environment (e.g., Lakatos et al., 2009). In fact, in spite of their phylogenetic distance, dogs and human children often show comparable performance at the behavioural level in cooperative–communicative tasks involving humans. Dogs possess a wide variety of social-communication skills, and these skills are often manifested in an infant-analogue, sophisticated manner in interspecific interactions (towards people). Although the conclusions of these experimental results have often been challenged (e.g., Udell et al., 2009), leading to contrasting theories of the origins of dog social cognition (see Miklósi & Topál, 2013, for a recent review), the vivid debate on the interpretation of findings has led to the emergence of a new research direction in comparative cognition.

A core idea of this new wave of research is that in the course of domestication, human beings became an integral part of dogs' world, and the selective force entailing this interspecific social environment was one of the main factors of the adaptive specialisation of dogs' social cognition. In line with this assumption, Reid (2009) proposed that this social challenge could have been overcome by gaining skills that enable dogs to correctly interpret information provided by people, whereas others (Hare & Tomasello, 2005; Topál et al., 2009a) suggested functionally infant-analogue manifestations of social competence in dogs, including their responsiveness to human signals in social (learning) settings.

It stands to reason that domestic dogs and human infants are in a similar situation because of the enormous variations of human social behaviour, and the diversity in human communication poses an everyday challenge not only for young infants but also for dogs. When communicating, humans can use a potentially infinite number of communicative signals, leastways in the visual and the acoustic modalities. In addition, they apply certain signals for *initialising and maintaining* communication (e.g., addressing, eye contact) and rely on various behavioural cues to recognise human attention. It is also worth mentioning that people spontaneously interact with dogs in a very similar way to how they typically interact with infants (Mitchell, 2001), and they spontaneously use a specific intonation pattern both when they talk to infants (motherese) and dogs (doggerel) (Hirsh-Pasek & Treiman, 1982).

In this chapter, we focus on a key feature of dogs' interspecific social competence, the use of human ostensive referential communication as a source of information, arguing that this is more flexible than was formerly thought and shows functional similarities to human infants' corresponding skills (see Miklósi & Topál, 2012, for a review). To highlight these similarities (and differences), in the first part of the chapter, we give a short introduction to what human-like social competence entails and how communication contributes to forming socially competent behaviours (section 11.2). We also highlight the inferential nature of the human communication system and the role ostensive signals play in making the act (interaction) truly communicative (section 11.2.1). Then we provide a description of a human-specific system for fast and efficient transfer of knowledge (natural pedagogy) as compared to the characteristics of non-human types of social learning (section 11.2.2) and show that even young preverbal infants are able to learn efficiently from communication-guided interaction (section 11.2.3). In the next part of the chapter, we describe the infant-analogue manifestations of social competence in dogs as a set of skills reflecting evolutionary adaptation to the cognitively challenging human social environment (section 11.3). We also provide evidence about dogs' sensitivity to different types of human ostensive signals including eye contact and verbal addressing (section 11.3.1) and further discuss potential similarities between dogs' and preverbal infants' communication skills (referential expectation, genericity assumption—section 11.3.2). Finally, in section 11.4, 'Summary and Conclusions', we give a short overview of the infant-like characteristics of dogs' communication skills and share some thoughts regarding the possible adaptive function and the putative cognitive mechanisms that underpin these skills in dogs.

## 11.2  HUMAN COMMUNICATION SYSTEM: A UNIQUELY POWERFUL WAY OF KNOWLEDGE TRANSMISSION

One of the wonders of human evolution is the significant gap between humans' cognitive skills and that of any other animal—even primate—species. These human-specific capacities have enabled us to create unprecedentedly complex and rich cultures (Richerson & Boyd, 1998) that are characterised by a vast amount of accumulated and shared knowledge preserved over a great span of time. The accumulation of knowledge—and thus the preservation of cultures—could not be realised without uniquely effective mechanisms for acquiring, storing, and transferring knowledge, which poses, both on evolutionary and developmental time scales, a great adaptation challenge on the human cognitive system.

Although there is debate about the point at which communication exerts its effects in human evolution, the fact that humans possess a communicational system that is unique among species can hardly be challenged. Human communication—that is designed for acquiring, storing, and transferring relevant knowledge content—is clearly distinguishable from other communicational forms in that it consists of a highly complex system of linguistic symbols and grammatical structure. However, arguably a more striking feature of human communicative interactions is the *mode* with which information can be conveyed between individuals by the combination of linguistic, non-verbal, and meta-communicative cues. The content of the communicative intention is in great part conveyed by the social context, and a slight modification of one aspect may entirely change the meaning behind the message.

### 11.2.1 Recognising Communicative Intentions as a Key to 'Socially Competent' Interactions

The recognition that human communication is characterised by a number of factors that lie without the system of linguistic codes has prompted researchers to try to identify these additional aspects and preconditions for successful communication. Grice (1989), for example, argued that successful communication is dependent on the ability of perceiving others' communicative intentions and on a will to express our own. Grice's theory resonates with accounts that emphasise a role of cooperation in the evolutionary emergence of human cognitive skills, as his theory also claims that in inferring the meaning of an utterance, people rely on an expectation of the cooperative intent of the communicator (Co-operative Principle). Drawing on Grice's account, Sperber and Wilson (1986) put forward a theory, termed Relevance Theory, to explain how people come to successfully decode the intended meaning of a communication. According to this view, the mere realisation that a communication is intended for oneself is crucial in this process. This recognition in the recipient of the message is achieved by the *production of ostensive signals* (Sperber & Wilson, 1986). Ostensive signals consist of a set of verbal and non-verbal cues providing evidence of the communicator's intention to convey information.

Csibra (2010) further argues that—as he puts it—the *communicative intention* and the *informative intention* of a communication may be separated procedurally (but not conceptually), meaning that one might express or recognise one without the other (e.g., I may understand that someone is trying to say something to me without understanding what it is). In Csibra's view, ostensive cues serve as a means to identify the *communicative intention*. However, he rejects the possibility that ostensive signals work this way merely due to their ability to grasp attention. If this were true, anything that has the power to evoke the orientation reflex (e.g., any loud noise) may serve the function just as well. Instead, he claims that ostensive cues have the power to grab our attention *because* the human mind is wired to interpret them as a precursor for communication addressed to oneself.

### 11.2.2 Natural Pedagogy versus Social Learning: The Role of Communication in Humanly and Non-Humanly Acquired Knowledge

Reflecting in some aspects Sperber and Wilson's idea (1986), a recent theory, termed Natural Pedagogy (Csibra & Gergely, 2006, 2009), also emphasises how the realisation that a communication is addressed to oneself changes the processing of the content of communication. However, while the scope of the

Relevance Theory is narrowed down, as it comes from the field of linguistics, Natural Pedagogy, putting the issue in a broader evolutionary perspective, builds on the literature on social learning.

Csibra and Gergely (2006) argue for a communicational system that allows the efficient transmission of generic knowledge from one individual to another. According to the theory, human cultural products have become so complex that simple observational learning could not serve as a mechanism by which the accumulated knowledge is sustained and transmitted through generations. A key adaptation challenge for hominisation is, therefore, the necessity of sharing generic knowledge among group-mates in a fast and efficient manner. They identify the emergence of recursive tool use (making tools in order to make other tools with them) as a turning point at which observational learning ceases to be an efficient mechanism for cultural transmission. During recursive tool making, the goal and the benefit of the performed action cannot simply be inferred from the observable context, which makes a great part of human cultural habits opaque to the observer. In such opaque situations, distinguishing the necessary (and invariant) elements of an action sequence from the incidental ones becomes extremely hard, if not impossible. Imagine observing a person turn on a TV with a remote control by lifting the remote, pointing it at the TV, shaking it a couple of times, and then pressing a button. If one does not possess any prior knowledge about the functioning of remote controls, it would be impossible to decide which step contributed to the achievement of the desired outcome (the appearance of the picture on the screen) and which step was irrelevant. Human cultures are filled with complex phenomena such as this, which poses a great challenge on our cognitive system.

Csibra and Gergely (2009) propose that human cultural transmission is made possible by a species-specific adaptation that is responsible for disambiguating such situations via teaching. Evidence from archaeological studies also supports the idea that the evolution of our complex material culture would have been impossible without some forms of teaching (Tehrani & Riede, 2008), thereby strengthening the claims and evolutionary rationale of the Natural Pedagogy theory. With the help of this adaptation, people are able to (1) perceive others' communicative intentions and infer that the presented knowledge is relevant to them as well as to (2) express such communicative intentions by providing ostensive cues. This idea suggests that ostensive cues serve the function of calling the attention of the novice to the fact that the presented information is produced for their benefit and thus expresses the communicative intention. In other words, in addition to the addressing function (specifying the addressee), ostensive signals are indicators of the presence of a communicative intention, importantly however, without having the capacity to specify the knowledge to be transmitted. In such an ostensive setting, accompanying (or subsequently presented) referential signals direct the attention of the novice to the relevant aspects of a scenario, and this enables the addressee to decode the knowledge content of the communicative exchange.

This specialised communication system is highly interactive, and it results in three key properties of human social transmission. First, it enables the fast and efficient transfer of cultural knowledge even if the knowledge to be acquired is cognitively opaque (regarding its aim and/or the cause-effect relationships); and second, it ensures that naïve social learners can acquire knowledge even after one or a few observations. A further important claim of the theory is that ostensive-referential signals induce an expectation that the validity of the presented knowledge is not constrained to the given context, but that it is generalizable to other situations as well (e.g., all remote controls work this way and not just the one being used at the moment). While this account provides an explanation of how children acquire a great amount of cultural knowledge in a relatively short period of time (thus deals with the ontogenetic aspect), it also offers an account of the phylogeny of human cultures. Natural pedagogy may be the mechanism by which cognitively opaque cultural knowledge is sustained over generations (Csibra & Gergely, 2009).

In contrast, the non-human type of social/observational learning is a very slow and fortuitous process. The difficulties of observational learning lie in the fact that, without the guidance of ostensive signals, it cannot provide sufficient information regarding the following question: Who and when should be observed? How can the episodic and generalizable aspects of the presented knowledge content be separated? In the absence of natural pedagogy, these problems cannot be solved unless the naïve learner could gain (at least some) insight into the cause and effect relationships of an action.

Moreover, observational learning cannot answer the challenges of the generalisation problem. The naïve learner becomes informed about objects and events that are relevant in the particular context of 'here and now', but this mechanism provides no cues as to which pieces of information can be transferred to other contexts as well. The acquisition of nut cracking in wild chimpanzees, for example, clearly demonstrates the low efficiency of non-human types of social learning (i.e., knowledge transfer without being supported by natural pedagogy). Nut cracking has been described as a socially transmitted tradition of wild chimpanzees living in West Africa (Boesch et al., 1994). This can be regarded as a relatively complex behaviour pattern because in order to split the nut without pulverising it, chimps need to use stone tools (hammer, anvil) and they have to strike at the nut very precisely (applying about 1 kg force). A practiced chimpanzee mother can break open more than 100 nuts per day, providing extensive opportunity for her child to observe and learn the technique. Importantly, however, studying the acquisition of nut cracking in wild chimpanzee populations takes about 3–7 years (Inoue-Nakamura & Matsuzawa, 1997). However, others reported that chimpanzees living in a captive environment, after having reached the age of 3, can acquire the technique of nut cracking after a relatively brief exposure to a demonstration by an experienced conspecific (Marshall-Pescini & Whiten, 2008).

It seems that the learning process is time consuming and cognitively demanding because during the construction of generic knowledge, the naïve observer *could only rely on observation of individual events*—namely, the episodic manifestation of a particular nut-cracking action of the chimpanzee mother ('this particular nut can be split using this particular stone, in this particular manner'), which is in itself not very helpful for the naïve observer. To master the nut-cracking technique, the infant chimp *needs to extract generic knowledge* ('any nut-like object can be cracked by any "hammer-shaped" stone by implementing the observed motor pattern') by keeping in mind the invariant (relevant) and omitting the incidental (variable) components of the repeated action demonstrations. Accordingly, in lack of natural pedagogy, the utilisation of knowledge flexibly across a wide range of contexts can be acquired only by extensive 'trial-and-error' experience or by observing the tutor's behaviour many times under many different conditions.

In summary, it seems that introducing ostensive-referential guidance into social learning processes was a great invention of human evolution which allowed for more efficient transfer of knowledge. Qualitative and quantitative differences in the main characteristics of non-human and human social learning mechanisms are summarised in Table 11-1.

**TABLE 11-1** Main Characteristics of Non-Communicative Social-Observational Learning Mechanisms and Social Transmission of Knowledge Supported by Ostensive-Referential Signalling

|  | Non-Human Social-Observational Learning | Human Social Learning via Pedagogy |
| --- | --- | --- |
| Novice's role | passive (eavesdropper or passive recipient) | (inter)active |
| Focus of novice's attention | uncontrolled | guided by referential signals |
| Source of information | inflexible (based on dominance, kinship, social affiliation) | flexible (based on the expression of communicative intent) |
| Validity of information | context-embedded | generalizable to other contexts |
| Number of presentations necessary for successful knowledge transfer | several | one or few |
| Timing of knowledge transfer | based on present needs | flexible |
| Imitation of opaque elements | rarely (if any) | frequently |

## 11.2.3 Human Infants Are Socially Competent Participants of Pedagogical Knowledge Transfer

A specific aspect of communication-guided knowledge transfer is that natural pedagogy makes it possible to learn efficiently from communicative interactions well before a novice can acquire complex (adult-like) cognitive skills. In the past few years, ample evidence has been accumulated to support the claim that even preverbal infants readily interpret ostensive signals (such as eye contact or infant-directed speech) as indicating communicative intentions and are competent receivers of pedagogical knowledge manifestations (see Csibra, 2010, for a review). Ostensive signals not only grab and direct attention in infants from very early on (Farroni et al., 2002) but have the potential to unambiguously specify the addressee and trigger some inferential processes. In other words, ostensive-referential communication changes how we process what we see. A body of evidence supports the notion that, in observational learning situations, introducing information in an ostensive-referential manner *can modify what a novice learns* from a demonstration (Butler & Markman, 2010; Southgate et al., 2009). For example, 14-month-old children imitate a suboptimal and opaque means to achieve a goal (turning on a lamp with their forehead) when it is presented in an ostensive setting, but choose the prepotent action (using their hands) when the pedagogical setting is eliminated (Király et al., 2013).

This example clearly shows that ostension has a specific impact on infants' information processing and provides effective guidance for young infants on the available information content (message). The effect can be conceptualised as 'assumptions' or 'cognitive biases' affecting the interpretation of communicative acts. Recent evidence suggests that there are at least two scaffolding cognitive processes called 'referential expectation' and 'assumption of genericity'. These 'automatic' information processing dispositions prepare human infants to efficiently learn when they are addressed by ostensive signals. For example, it has been shown that infants below the age of one expect the direction of gaze to signal a referent at a specific location (Csibra & Volein, 2008; Senju et al., 2008). Importantly, this *referential expectation* is formed only when it is preceded by communicative (ostensive) signals (Senju & Csibra, 2008). These results suggest that infants appreciate the interactive nature of human communication and are prepared to benefit specifically from communication addressed to them. The fact that referential gestures in themselves (without accompanying ostensive signals) do not elicit attention to the subject of the communication further supports the idea of the aforementioned *dual nature of communication* (see Senju & Csibra, 2008), i.e., the claim that for successfully decoding a message it is important to recognise both the communicative and the informative intention (Csibra, 2010).

It has also been shown that in communicative situations preverbal infants tend to grab generic information even from a single observation and ignore episodic elements of the context. That is, when infants are presented with an object in an ostensive-referential context, they tend to pay attention to its generic (i.e., visual feature) rather than transient (i.e., location) properties. However, they remember

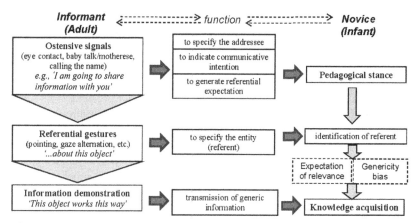

**FIGURE 11-1**   The interactional components of 'natural pedagogy' that serve for efficient knowledge transfer.

better the location information when the object is presented in a non-communicative context (Yoon et al., 2008). This *genericity bias* makes it possible to learn not only about the present situation but to extend the acquired knowledge to other contexts as well. For example, if one tries to teach a novice how to use a fork, it is essentially important for the novice to understand that the conveyed information refers to a kind of artefact and not just that particular object. If the novice did not have this comprehension, it would be necessary to demonstrate the function of an artefact before each instance of usage, which would be extremely costly. This aspect of human pedagogy contributes to the accumulation of cultural knowledge (Csibra & Gergely, 2009) since the ability to learn generalizable knowledge of artefact functions saves us the trouble of having to invent everything ourselves.

On the basis of these findings, *natural pedagogy* can be conceptualised as a step-by-step social interaction process based on mutual skills in informant and recipient (Figure 11-1). An important feature of human communication that arguably cannot be found in any other animal species (Csibra, 2007) is that with its help, the information conveyed is not stalled between the boundaries of the current, but it endows the recipient with knowledge that can be efficiently used in other situations as well. As described previously, this is the crucial factor that arguably allowed for the evolvement of the complex cultural niche we live in today and the sophisticated cognitive architecture the human race is equipped with.

## 11.3  THE INTERACTIVE NATURE OF DOG–HUMAN COMMUNICATION AS A HALLMARK OF DOGS' SOCIAL COMPETENCE

One of the important conclusions of the previous description of human-specific communication is that natural pedagogy opens the possibility for cognitively immature individuals (young infants) to profit from this highly interactive system and to acquire new, relevant, and generalizable knowledge. This conclusion and recent theoretical and experimental works on dog domestication (e.g., Miklósi & Topál, 2013) raise an intriguing question: Do dogs demonstrate a human-tunedness in their communicative skills that would enable them to participate in 'pedagogical' knowledge transfer as recipients?

In the following sections, we focus on the notion of the existence of infant-like social competence in dogs, examining the ways in which a dog can engage in communicative interactions with humans. We argue that dogs' interspecific social competence, which allows for the establishment of a wide range of social relationships with humans, can be characterised by some preverbal infant-like features, including the sensitivity to human eye contact and other ostensive signals as well as some understanding of the referential nature of human directional gestures.

## 11.3.1 Sensitivity to Human Ostensive Signals

A characteristic feature of human communication is that any action that gives evidence about the content of the corresponding informative intention of the communicator (in the form of referential gestures and the manifestation of specific knowledge with regard to the referred object/subject/event) can acquire communicative value if these behaviours are preceded or accompanied by ostensive signals. That is, ostensive signals are 'designed' to give evidence about the *communicative intention* of the communicator and are generally applied for initialising and maintaining communicative interactions.

### 11.3.1.1 Eye Contact

When communicating with a (preverbal) infant or a dog, face-to-face interactions in general and making eye contact in particular could potentially be communicatively meaningful signals. In fact, looking at each other's eyes in a face-to-face setting triggers enhanced attention to the human's face in infants and dogs alike. This is supported by recent investigations on 'left gaze bias' in dogs and human infants, which raised the possibility that the human face has a special importance for both (Guo et al., 2009). Left gaze bias, when the left hemiface is inspected first and/or for longer periods, is widely believed to be associated with the processing of facial-emotional information because this perceptual function is located in the right hemisphere, which receives visual input from the left visual field. Adult dogs demonstrate a significant left gaze bias only towards neutral human faces but not towards objects or other hetero-specific faces (Guo et al., 2009). Moreover, dogs, like humans, show a left gaze bias towards human faces expressing positive or negative emotions, but they show a similar gaze bias towards dog images only if such an image displays a threatening (negative) expression (Racca et al., 2012).

The eyes within the face are of great importance in human communication because one of the key 'ostensive' components of the signal patterns displayed during communicative interactions is eye contact. Eye contact specifically confirms for both the communicator and the recipient that the partner is 'on line' and, in addition, it is an obvious signal for the recipient that she/he is the intended addressee of the communication. The tendency to look into each other's eyes is based on face preference: ample evidence suggests that in human infants face and gaze perceptions are closely linked together from birth. Human infants innately detect eye contact (Bátki et al., 2000) and the fact that new-borns show selective preference for upright faces looking at the baby (Farroni et al., 2002) suggests that eye contact has some communicative function for infants from very early on (Csibra, 2010). It is worth mentioning that in non-human primates eye contact seems to play a role in affiliative processes (social bond) and has no communicative significance. This is supported by the finding that although eye contact and face-to-face interaction can also be observed in the mother–infant interactions of rhesus macaques (Ferrari et al., 2009), these behaviour patterns gradually diminish a few weeks after birth.

Needless to say, staring at the eyes of another is a strong attention-getter for adult individuals in all (social) species (Emery, 2000) because it can signal either coercive (threatening) or prosocial (information sharing) attitudes towards the group-mate. Concerning these two contradictory functions, we should note that the prosocial (information sharing) role of eye contact has become predominant during hominisation, whereas for the 'rest of the world', eye contact usually induces fear or aggression and this signal has little (if any) ostensive-communicative property. Recent evidence suggests that there is at least one notable exception to this: the domestic dog.

Interestingly, although dogs use eye contact in intraspecific situations in a wolf-like manner, this signal can carry different meaning for dogs and wolves when used in interspecific interactions (towards humans). That is, direct eye contact between dogs has little (if any) ostensive-communicative property. This signal is commonly used to indicate dominance (submissive wolves/dogs avoid eye contact with dominant members of the pack) and as a form of ritualised aggression (Schenkel, 1967). The same holds true, of course, for the use of eye contact between dogs and humans in many cases. Eye contact with an unfamiliar human in a 'threatening approach' task has the potential to evoke fear (Vas et al., 2005) and to increase heart rate (Gácsi et al., 2013) in dogs. Moreover, in a 'forced eye contact' task, dogs show less tolerance of eye contact with an unfamiliar than with a familiar human (Hernádi et al., 2012). Additionally, however, dogs often establish eye contact with humans for the same purpose people do: demanding attention, initialising communicative interaction, or just keeping in contact.

Findings from dog-wolf comparative research indicate an early emerging and permanent preference for face-to-face interactions and eye contact with humans in dogs, but not in hand-reared wolves, and support the idea that this behaviour may reflect the dogs' adaptation to the human social environment. For example, in a reinforced eye-contact task, in which food reinforcement always followed the eye contact, dog puppies showed a significantly stronger tendency to look in the human's eye from 4 weeks of age onward (Gácsi et al., 2005). In a food search task, while the subject was waiting for a human's pointing signal, 8-week-old dogs showed significantly shorter latency to gaze at the human's face than wolves, and the difference still existed in juveniles and adults (Gácsi et al., 2009a). Moreover, Gácsi et al. (2009a) found that willingness to form eye contact increases with age in dogs but not in hand-reared wolves. Virányi et al. (2008) also reported that 4-month-old dogs and wolves differ in their readiness to establish and maintain eye contact with a human while waiting to be informed in a food search task. They concluded that although wolves *can be trained* to prefer to make eye contact with a human (after extensive formal training; see Virányi et al., 2008), dogs have a *spontaneous tendency* to gaze at a human's face from very early on.

There is now ample evidence to show that dogs are willing to initiate eye contact with humans in a wide range of situations (e.g., begging: Gácsi et al., 2004; out-of-reach food task: Miklósi et al., 2000). This tendency is often studied in 'unsolvable task' paradigms, in which an experimenter baits a container with an unobtainable reward to prevent the dog from solving the task. In these conflict situations, human-directed gazing is a characteristic behavioural trait of dogs and is widely interpreted as 'requesting help' (Miklósi et al., 2000). Although the flexible use of human-directed gazing behaviour in such situations increases with age (Passalacqua et al., 2011; Hori et al., 2013) and learning experiences during ontogeny can evidently modulate communicative behaviour (Barrera et al., 2011), available results concerning the prominent role learning may play in dogs' looking at a human's face are not yet conclusive (Bentosela et al., 2008; Gaunet, 2008; Yamamoto et al., 2011). In addition to the striking individual variation (Marshall-Pescini et al., 2009; Jakovcevic et al., 2010), breed differences have also been shown to exist in dogs' tendency to make eye contact with humans in communicative situations. Interestingly, dog breeds selected for cooperative working roles (e.g., retrievers) and 'short-nosed' dog breeds (in which the ganglion cells occur more centrally in the retina, enabling them to perform much more focused attention on the human, e.g., bulldogs) are better in visually guided cooperative interactions involving eye contact with humans (Passalacqua et al., 2011: Gácsi et al., 2009b). A more recent study further confirmed the role of innate factors in the individual differences in dogs' gazing to the human face. In an unsolvable task, the latency, frequency, and duration of human-directed gazing showed associations with dopamine receptor gene polymorphism (Hori et al., 2013), a gene that has been shown to be associated with different aspects of social behaviour in dogs (social impulsivity, aggressiveness: Héjjas et al., 2007, 2009) as well as in human infants (temperament, attachment: Lakatos et al., 2000; Holmboe et al., 2011).

In summary, converging evidence indicates that dogs' preference for human faces (eye contact) can be interpreted as an innate predisposition, an adaptive orientation mechanism which ensures that dogs will fixate on and learn about the most relevant social stimuli in their environment. This 'infant-like' mechanism of preferential orientation may help dogs overcome their 'wolf-like' predisposition to seeing face-to-face interactions with humans as an aversive/threatening situation. More importantly, this ensures comprehensive learning experience for young dogs about the communicative-collaborative nature of eye contact. If so, communicative skills may be a by-product of preferential attention to faces, and especially to the eye region, which may carry important communication signals (Miklósi et al., 2003).

### 11.3.1.2 Verbal Addressing in an Ostensive Manner

In addition to eye contact, verbal signals can also have an ostensive function because these cues, among others, can indicate that the communication is directed to a specific addressee. Importantly, however, there is a difference between eye contact and verbal signals regarding their addressing function: without being able to decode the content of the speech, one cannot decide whether he is the one being addressed. Therefore, when someone initiates communicative interaction with a preverbal infant or a dog, the best way to reduce the inherent ambiguity of verbal addressing is to use a specific intonation pattern, 'motherese' (higher average pitch and exaggerated intonation contour).

In human communicative interactions, motherese is an efficient tool for disambiguating the addressee from very early on: the acoustic characteristics of infant-directed verbal signals effectively grab even few-day-old infants' attention (Cooper & Aslin, 1994). It has also been raised that the use of such characteristic intonation and rhythmic pattern has the potential to grab the dogs' attention while interacting with people (doggerel: Hirsh-Pasek & Treiman, 1982).

Empirical findings suggest that acoustic features of human verbal signals modulate dogs' responsiveness both when commanded to do an act (Fukuzawa et al., 2005) and when informed about the location of a reward (Petterson et al., 2011; Scheider et al., 2011). In this latter study, dogs have been shown to be more responsive to human directional gestures when they were addressed in a high-pitched, 'informing' voice compared to addressing in a lower-pitch, imperative manner.

It is also worth mentioning that, as naming and eye contact are usually provided in a highly contingent manner, mentioning the name becomes 'decodable' for even preverbal infants (Newman, 2005) as a reliable signal assuring that the communicator is talking to the named subject. Although intuitively very reasonable, there is no solid evidence that the dog's own name would acquire an ostensive function via strong association with other potentially ostensive signals (eye contact, dog-directed speech). In a recent study, Kaminski et al. (2012) found that dogs react similarly to a human's directional gestures both when their name or an unfamiliar name is called in an ostensive-addressing manner.

These results suggest that for dogs, independently of the lexical meaning of the vocal communication, high-pitched dog-directed intonation pattern (like eye contact—see preceding text) is a reliable indicator of ostension. The finding that dogs do not recognise their names as specific signals indicating humans' communicative intention is further supported by Virányi et al. (2008), who reported that in an ambiguous situation, calling the dog's name in an imperative mode without accompanying eye contact is insufficient to evoke obedience.

Although these studies give some support for the conclusion that dog-directed verbal addressing attracts the dog's attention towards the speaker efficiently, current evidence is insufficient to assess the exact role specific intonation pattern and name calling play in making manifest the actor's communicative intention for the dog.

### 11.3.1.3 Combined Use of Verbal and Non-Verbal Ostensive Signals

Needless to say, verbal signals in human-to-dog communication are usually accompanied by further non-verbal signals (e.g., eye contact) that give evidence about the communicative intention of the communicator, and this can help to overcome the difficulties in interpreting the situation. In fact, in most studies of dogs' interspecific communication skills, the human communicator uses verbal addressing signals in combination with eye contact (e.g., Ittyerah & Gaunet, 2009). Dogs are highly sensitive to such combined signals, and these cues, in comparison with non-social attention-getters (e.g., sound signal or a visual marker), are not simply more salient than non-social cues, but ostensive signals are *more effective* in terms of signalling the location of a reward in an object choice context (e.g., Agnetta et al., 2000; Udell et al., 2008) and can also be crucial for overcoming difficulties in observational learning tasks (e.g., Pongrácz et al., 2004, but see Range et al., 2009, for a more sophisticated picture).

Interestingly, dogs show a strong initial bias towards maladaptive response in object choice tasks if the inefficient or mistaken solution is cued ostensively by a human. Increasing evidence suggests that adult pet dogs readily follow deceptive pointing despite having been informed of the correct location of the reward (Szetei et al., 2003); they prefer to select an empty container despite having seen the correct one being baited (Erdőhegyi et al., 2007; Topál et al., 2009a) and adopt inefficient responses in object manipulation tasks as a result of repeated observations of human action demonstrations (Kupán et al., 2011). In a quantity discrimination task, it has also been shown that after having seen a human's ostensively cued preference for a smaller amount of food, dogs change their 'natural' preference for the larger quantity and show a selection bias towards the smaller one (Prato-Previde et al., 2008; Marshall-Pescini et al., 2011). Task performance in human children can also be biased by ostensive signals even when these cues serve to highlight an inefficient or mistaken solution (Moriguchi & Itakura, 2005; Topál et al., 2008). It is also worth noting that in some cases learning experience can help dogs to overcome their tendency to rely 'blindly' on ostensive-communicative signals because they are able to learn to stop responding to ostensively cued 'deceptive' pointing (e.g., Elgier et al., 2009).

More importantly, there are three experimental paradigms: the A-not-B error task (Topál et al., 2009a, 2010; Marshall-Pescini et al., 2010; Kis et al., 2012; Sümegi et al., 2013), the quantity discrimination task (Marshall-Pescini et al., 2012), and the two-way object choice task (Kaminski et al., 2012), in which the role different human ostensive signals and non-ostensive attention-getters play in dogs' communicative responsiveness was more or less systematically tested.

The findings from these experiments provide a somewhat coherent picture, and although the details are not fully understood, the main conclusions regarding how dogs perceive different manifestations of human communicative intentions can be summarised in a few points:

1. Dogs treat specific human behaviours not simply as discriminative cues: for dogs, human ostensive signals indicate that they are being instructed, and this is an important and often indispensable element of their readiness to respond to human referential gestures (pointing, gaze shift, etc.).
2. Eye contact is the 'strongest' human ostensive signal; it has a primary role in inducing specific responsiveness. High-pitched vocalisation has a secondary but still an important role in this respect, whereas calling the dog by its name is of minor importance.
3. Ostensively cued human behaviours can often act as imperatives for the dog, inducing a 'ready-to-obey' attitude that may result from the domestication of dogs and/or from their extensive experience with humans. There is no solid evidence of a dog-like responsiveness to eye contact and verbal addressing in hand-reared wolves.
4. While interacting with humans, non-ostensive and/or non-social attention-getting signals may also affect dogs' response; however, dogs' specific responsiveness to ostensive signals is based on the perception of the communicative intention behind the action.

## 11.3.2 Interpretation Biases in Information Processing

It is increasingly accepted that dogs can learn a lot by simply observing humans' actions, preferences, or object manipulations, especially if the human behaviour is embedded in an ostensive-communicative context. The results concerning dogs' sensitivity to ostensive signals (see preceding text) raise the possibility that they interpret these cues as indicating that further information is to be expected. Concerning the dogs' ability to profit from human communication, the intriguing question is whether they interpret and respond to human actions differently depending on whether these displays are preceded or accompanied by ostensive signals.

As pointed out previously, ostensive signals have an important modulatory role in human communicative interactions, as these signals can specifically change how (even a preverbal) human infant processes the communicator's actions (see section 11.2.3). We also mentioned that the impact of ostensive signals on young infants' information processing can be conceptualised as two kinds of cognitive biases: *referential expectation* and *genericity assumption*. In the next section, we discuss whether dogs, in comparison with infants, show an ability to understand the referential character of human cuing and to learn something generic in an ostensive communicative context.

### 11.3.2.1 Referential Expectation

Understanding the designating function of directional gestures, or in other words, the referential interpretation of pointing or looking at an object or event, is an inherent component of triadic communication (which includes the signaler, the recipient, and a target object). Such triadic or 'shared' engagement has been shown to occur in preverbal infant–adult as well as in adult dog–human interactions (e.g., Parise et al., 2008; Kaminski et al., 2009a, 2009b).

Gaze following is one of the key indicators of a subject's involvement in triadic communication situations. Spontaneous gaze following into distant 'empty' space (i.e., even if the partner's attention is not guided by an external stimulus) has been shown to occur in young human infants as well as in many non-human species (for a review, see Shepherd, 2010). Gaze following may also occur in interspecific (human–animal) interactions: accumulating evidence suggests that corvids (Schloegl et al., 2008), great apes (Bräuer et al., 2005), and canids (Range & Virányi, 2011) also show an ability to follow humans' gaze.

Gaze following in non-human subjects and preverbal infants is standardly interpreted either as a 'cognitively blind', socially facilitated orientation response that supports associative learning (Povinelli & Eddy, 1996) or as evidence of their understanding of perception and attention (Baron-Cohen, 1991). Moreover, the ability to follow the gaze of others to spaces behind barriers is often interpreted as an indicator of visual perspective taking (e.g., Kaminski et al., 2009a). All these explanations presume some understanding of the referential nature of looking; importantly however, according to these accounts, gaze following has no explicit communicative value because the response does not necessarily depend on whether or not the source of the gaze has a communicative intention.

If the communicative intention behind an object-directed gaze matters to the social partner, then gazing in a particular direction is more likely to be followed by the partner if the signal is presented in an ostensive context indicating the actor's addressee-directed communicative intention. It seems that human infants gradually acquire the ability to establish a link between referentiality and ostensive addressing: while 1–6-month-olds follow others' gaze 'reflexively' (Hood et al., 1998), older ones form strong referential expectations about the gazing behaviour of others only in ostensive-communicative contexts (e.g., Okumura et al., 2013). For preverbal infants, in addition to verbal addressing, eye contact has been shown to be a crucial factor in detecting the association between gaze direction and the target object (Senju & Csibra, 2008).

As mentioned previously, the preference to look at humans' faces and to make eye contact with humans (see preceding text) might have led to enhanced skills in reading human visual attention in dogs in comparison with wolves. In fact, hand-reared wolves (who received intensive obedience training and eye contact on a daily basis), unlike pet dogs, follow a human's gaze shifts into distant space 'reflexively' (Agnetta et al., 2000; Range & Virányi, 2011).

Dogs, however, are very skillful in using human gaze in object choice situations in which they both follow and direct human gaze (for a review, see Kaminski, 2009). Ample evidence suggest that dogs are not only sensitive to whether or not a human is watching them (e.g., Kaminski et al., 2013) but are sensitive to the direction of human visual attention in various contexts; e.g., when they beg (Gácsi et al., 2004), perform forbidden actions (Call et al., 2003), or are asked to fetch an object (Kaminski et al., 2009a). Moreover, relying on the gaze direction of humans in an ostensive-addressing context, dogs are able infer who is commanded by the human experimenter (Virányi et al., 2008). These findings raise the possibility that, relying on the direction of a human's gaze, dogs generate expectations about the potential referents (i.e., who/what is the addressee/target object).

Dogs themselves also use eye contact and gazing as a component of communication in situations in which they have to direct the attention of someone to something. For example, in unsolvable problem-solving tasks, dogs, like toddlers, show an increased tendency to make eye contact and gaze alternation between the target object and the human participant (potential helper), and this supports the intentional and communicative-referential nature of gaze alternation in both dogs and children (Marshall-Pescini et al., 2013).

A few studies provide more direct evidence for the claim that dogs selectively shape referential expectations about human gaze cues. Soproni et al. (2001), for example, investigated the dog's behaviour in a two-way object choice task in which making eye contact with the dog (ostensive addressing) preceded shifts of a human's gaze either to the baited container (looking at target) or to the 'empty' space (looking above target). The finding that dogs reliably chose the baited container in the 'looking at target' condition but had difficulty doing so in the 'looking above target' condition supports the notion that in ostensive contexts they tend to interpret humans' gaze cues as referential signals. Others also concluded that the ostensive context specifically enhances the referential understanding of directional signals. Kaminski et al. (2012) reported that human gaze cues can acquire a communicative-referential meaning by preceding or accompanying ostensive signals. In another recent study using eye-tracking methods, Téglás et al. (2012) also found that dogs tend to follow a human's gaze towards a target object only if it is preceded by eye contact and high-pitched addressing. It is also noteworthy to mention that although a wide range of non-human animals (apes: Brauer et al., 2005; horses: McKinley & Sambrook, 2000; ravens: Schloegl et al., 2007) follow the gaze of humans, there is no evidence to support that any of these species (except the dog) would do this preferentially in ostensive-communicative situations.

In addition to gaze shift, human pointing gestures also seem to induce referential expectations in dogs. Ample evidence suggest that dogs, unlike wolves (Virányi et al., 2008) or apes (Povinelli et al., 1997), reliably utilise human pointing even from early puppyhood (Riedel et al., 2008), and like children, they show some capacity to generalise this knowledge to relatively unfamiliar directional gestures (Lakatos et al., 2009). Comparative evidence on the dogs' and children's performance in pointing tasks suggest that although there is considerable individual variation in both infants' and dogs' ability to understand the referential character of pointing gestures, adult pet dogs show a performance similar to 16–18-month-old toddlers (Lakatos et al., 2009; Pfandler et al., 2013).

Some have challenged these findings and propose that pointing following in dogs is based on associative learning mechanisms without any specific, 'infant-like' understanding of the human's communicative-referential intention (e.g., Wynne et al., 2008). Converging evidence, however, indicates that dogs are not simply skillful at using human pointing gestures (as a discriminative stimulus), but their response depends on contextual information including the human's communicative motive (e.g., Petterson et al., 2011; Scheider et al., 2011).

To date, there is no consensus regarding the question whether dogs perceive human pointing in an ostensive context as an informative gesture (e.g., informing them about the location of the reward) or as an imperative order (i.e., a strong command that has to be obeyed). There are indications that dogs often tend to interpret pointing and other referential gestures as more or less strong imperatives (Topál et al., 2009a; Kaminski et al., 2012; Kirchhofer et al., 2012; Kis et al., 2012), whereas others suggest that for dogs human pointing falls somewhere between informing signal and imperative command, a suggestion, which, in some situations, can be ignored (Scheider et al., 2013). The imperative-referential nature of pointing is also supported by the fact that human pointing can even outweigh dogs' spatial bias to search for an object where they saw it disappear (Plourde & Fiset, 2013). It is also worth mentioning that, unlike in human children, referential understanding of human directional signals in dogs seems to be limited to cooperative contexts (Wobber & Hare, 2009; Petterson et al., 2011).

### 11.3.2.2 Genericity Assumption

As mentioned previously, even preverbal infants can acquire generalizable information by communication. This is so because in addition to referential expectation, there is another interpretation bias in social information processing: in ostensive contexts, infants tend to assume that 'they are supposed to learn something generic' (Csibra, 2010). In contrast, dogs do not seem to show a similar bias to receive generic information about objects or events in ostensive contexts; they tend to pick up information from the communication that is restricted to the 'here and now'.

The difference between dogs and preverbal infants is clearly shown by those studies in which Topál et al. (2008, 2009b) investigated how ostensive communication affects the perseverative search error in the A-not-B object search tasks. These studies reported that the A-not-B search error—the phenomenon that subjects repeatedly look for a hidden object at an initial hiding place (A) even after the object has been displaced in full view of them—is strongly tied to the ostensive-communicative character of the object hiding action in both dogs and preverbal infants. These results raise the possibility that if one location is misleadingly indicated by the human's ostensive signals, this may be interpreted by the subject either as communicating information about some generalizable property of the referent kind (e.g., 'this type of object is usually found in container A') or as 'imperatives' with the function of performing the observed action (e.g., 'produce search behaviour at location A').

As one of the crucial components of the A-not-B task is the identity of the person that dogs/infants interact with, Topál et al. (2009b) also investigated how they react if, after the 'A' trials, the identity of the hiding person is changed and a new experimenter continues hiding the object at the new location. They hypothesised that if the ostensive hiding action were interpreted as an imperative order associated with a specific 'instructor', then the perseverative search bias might be expected to diminish during the 'B' trials. In contrast, if the ostensive hiding action is (mis-)interpreted as conveying some generalizable information about the type of the hidden object or the function of the hiding location that is not related to the identity of the particular demonstrator, switching the experimenter should not reduce the tendency to commit the A-not-B error.

The finding that dogs did not perseverate after switching the experimenter confirms the hypothesis that they did not extend the scope of the learned imperative to the new context but anchored communication to the episodic situation and to the specific communicator. Infants, however, tended to generalise their erroneously learnt object-finding action to the new person context probably due to their bias to expect generic information that could lead to a misinterpretation of communication and to such characteristic errors (Topál et al., 2009a).

## 11.4 SUMMARY AND CONCLUSIONS ON THE 'INFANT-LIKENESS' OF DOGS' COMMUNICATION SKILLS

The experimental findings presented in the preceding sections provide strong support for the idea that social competence in dogs has been affected by the challenges of the human social environment. We propose that as a consequence of these adaptation requirements (i) dogs have evolved some special skills for interacting and communicating with humans, (ii) these skills provide the basis for the emergence of a more 'infant-like' and less 'wolf-like' social competence in the dog, which (iii) makes this species remarkably adept at learning from humans in communicative contexts.

More specifically, recent studies provide corroborating evidence that similarly to preverbal human infants, dogs fulfil at least two out of three operational criteria for being a recipient in 'pedagogical' knowledge transfer.

First, dogs show special sensitivity to ostensive cues that signal the human's communicative intention, and they also show some evidence of recognising the information transferring nature of communicative contexts. As in infants, ostensive-communicative cues have the potential to guide the dogs' attention and influence their inferences and interpretations in human–dog interactions.

Second, dogs seem to comprehend the referential character of human cuing in a way similar to human infants. Dogs show differential sensitivity to human directional signals (gaze-shift, pointing) in communicative contexts as compared to non-ostensive situations. The finding that ostensive-communicative addressing signals facilitate, for example, gaze-following behaviour in dogs suggests that for dogs human ostensive signals indicate that they are being instructed (i.e., the communicator has an informative intention).

Importantly, however, human ostensive-communicative referential signals seem insufficient to create an 'infant-like' expectation to receive generic information about objects or events in dogs. That is, whereas in infants ostensive signals induce a specific receptive attitude towards learning something generic, it seems that in dogs, an ostensive context induces a 'ready-to-learn' or 'ready-to-obey' attitude that is largely tied to the here and now. However, regardless of its generic or episodic character, an important consequence of this receptive attitude is that both dogs and infants are willing to reproduce those cognitively 'opaque' actions that they have seen in a communicative referential context even if the action is unusual or represents a counterproductive solution to the problem. Thus, the adaptive function of dogs' interspecific communicative skills can be conceptualised as not (only) knowledge acquisition from human partners but as facilitating behavioural synchronisation in order to avoid conflicts and/or to co-act in terms of common actions without necessarily comprehending the causal structure of the collaborative interaction. There is no denying that communication always involves some kind of coordination among partners. But it seems that in dog–human ostensive-communicative interactions the essential coordination is not about the interpretation of contextual cues (what is mutually assumed) but simply on the manner in which the recipient (dog) will respond to the human's communicative act.

Infants' receptivity to ostensive signals may also have the social function of achieving a higher level of interactional synchrony. In fact, synchronising activities have a general facilitating effect on the interaction of human partners, and such activities are often guided by ostensive signals. Synchronisation is essential for group cohesion (Engel & Lamprecht, 1997), and in humans, it can be achieved by the employment of different means, such as the tendency to follow social rules (de Waal, 1996), the ability for emotional contagion (Hatfield et al., 1993) and imitative learning skills, assuring a high level of behavioural conformity between partners (Meltzoff, 1999).

Dogs are also equipped with abilities that are crucial in establishing behaviour synchrony, and these abilities are often tied to ostensive contexts (see Figure 11-2). For example, recent studies provide evidence of social referencing in dog–human interactions: dogs readily use the emotional information provided by a human about a novel object to guide their own behaviour towards it (Merola et al., 2012). There is also empirical evidence suggesting empathy-based emotionally connected yawn contagion in dogs (Silva et al., 2013). Moreover, dogs can efficiently use human behaviour as a cue for selecting functionally similar behaviour (Fugazza & Miklósi, 2013), and it has also been reported that dogs are inclined to follow social rules of the group in both the short (Topál et al., 2005) and the long term (Kubinyi et al., 2003).

It is also noteworthy to mention that dogs, like preverbal infants, are extremely proficient in using signals that reliably indicate human communicative or informative intentions, even if they cannot fully understand the 'message' (i.e., the content of this intention). Although ostensive-referential communicative interactions are often described as a process in which partners intentionally try to influence each other's mental state while viewing the partner's ostensive signals as the offering of evidence for communicative intentions (Sperber & Wilson, 1986), this is just one way in which more sophisticated agents may interact, and there are other, cognitively less demanding ways to interact communicatively. In fact, a basic act of communication involves one agent drawing another agent's attention to a situation. As the communicative situation is a shared situation, it is a potential source of shared knowledge and intentions. Importantly, however, an agent (dog or preverbal infant) can recognise the shared nature of the ostensively guided situation without being able to recognise that the content of the communication (the manifested knowledge) is also shared between the partners. That is, dogs as well as infants can grasp the explicit or overt dimension of

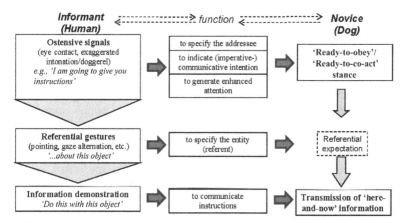

**FIGURE 11-2**   The interactional components of dog–human communicative interaction.

communication without any sophisticated mind-reading abilities or conceptual thinking skills.

Thus, we propose that those motivational processes and cognitive skills that enable dogs (and preverbal infants) to (1) monitor, (2) gain, (3) maintain, and (4) shift attention while interacting with humans and to (5) share a common focus on something with the human partner can be regarded as the 'minimum requirements' for socially competent responding in ostensive-communicative situations.

Accordingly, dogs' and preverbal infants' receptivity to ostensive-communicative signals are not necessarily tied to complex mental processing, and it is still unclear whether a particular type of self-knowing consciousness is involved in initiating communicative interactions and responding to the manifestation of communicative intention in others. Despite the fact that dogs' reactions and expectations help them to act in line with the presumed content of the communicative interaction, we are not necessarily entitled to call this phenomenon 'recognition' of intention. Although dogs' sensitivity to human communication is unprecedented among non-human animals, for them, human ostensive signals may simply act as reliable signs for anticipating further informative stimuli, and we have no reason to suppose a recognition of the 'conceptual bond' between the communicative and the informative intent of the human.

In summary, although we may say that recent research effort in the field of dogs' social cognition raises more research questions than it answers, the generally consistent pattern of findings that emerged from the previously cited studies provides an increasingly firm basis for understanding what is 'infant-like' about dogs' interspecific communication skills.

### Future Directions

- Dogs seem to be especially attuned to human ostensive communication. The idea of infant likeness of dogs' social competence, however, requires further systematic comparative investigations (between wolves, dogs, and children).
- There is also a compelling need for more systematic investigation of the relative importance of different types of human ostensive signals (e.g., eye contact, name calling, specific intonation) in dogs' communicative responsiveness.
- It is also unknown whether or not dogs can recognise the ostensive nature of those turn-taking situations in which a human partner initiates contingent responding (an essential structural property of human communicative exchanges).
- The role of associative reinforcement learning in dogs' communicative responsiveness is strongly debated. Results are still largely inconsistent due to a lack of systematic data collection and properly controlled experiments. Further comparative (and longitudinal) studies between dogs and human infants would be necessary to clarify the exact role of individual experience (associative learning) in both species.

## ACKNOWLEDGEMENTS

This work was supported by the Hungarian Scientific Research Fund (OTKA K100695).

## REFERENCES

Agnetta, B., Hare, B.A., Tomasello, M., 2000. Cues to food location that domestic dogs *(Canis familiaris)* of different ages do and do not use. Anim. Cogn. 3, 107–112.

Baron-Cohen, S., 1991. Precursors to a theory of mind: understanding attention in others. In: Whiten, A. (Ed.), Natural theories of mind: evolution, development, and simulation of everyday mindreading. Blackwell, Oxford, pp. 39–50.

Barrera, G., Mustaca, A., Bentosela, M., 2011. Communication between domestic dogs and humans: effects of shelter housing upon the gaze to the human. Anim. Cogn. 14, 727–734.

Bátki, A., Baron-Cohen, S., Wheelwright, S., Connelan, J., Ahluwalia, J., 2000. Is there an innate gaze module? Evidence from human neonates. Infant Behav. Dev. 23, 223–229.

Bensky, M.K., Gosling, S.D., Sinn, D.L., 2013. The world from a dog's point of view: a review and synthesis of dog cognition research. Adv. S. Behav. 45, 209–406.

Bentosela, M., Barrera, G., Jakovcevic, A., Elgier, A.M., Mustaca, A.E., 2008. Effect of reinforcement, reinforcer omission and extinction on a communicative response in domestic dogs *(Canis familiaris)*. Behav. Proc. 78, 464–469.

Bloom, P., 2004. Can a dog learn a word? Science 304, 1605–1606.

Boesch, C., Marchesi, P., Marchesi, N., Fruth, B., Joulian, F., 1994. Is nut cracking in wild chimpanzees a cultural behaviour? J. Hum. Evol. 26, 325–338.

Bräuer, J., Call, J., Tomasello, M., 2005. All great ape species follow gaze to distant locations and around barriers. J. Comp. Psychol. 119, 145–154.

Bräuer, J., Kaminsky, J., Riedel, J., Call, J., Tomasello, M., 2006. Making inferences about the location of hidden food: social dog, causal ape. J. Comp. Psychol. 120, 38–47.

Butler, L.P., Markman, E.M., 2010. Pedagogical cues influence children's inductive inference and exploratory play. In: Ohlsson, S., Catrambone, R. (Eds.), Proceedings of the 32nd annual meeting of the Cognitive Science Society. Cognitive Science Society, Austin, TX, pp. 1417–1422.

Call, J., Bräuer, J., Kaminski, J., Tomasello, M., 2003. Domestic dogs are sensitive to the attentional state of humans. J. Comp. Psychol. 117, 257–263.

Cooper, R.P., Aslin, R.N., 1994. Developmental differences in infant attention to the spectral properties of infant-directed speech. Child Dev. 65, 1663–1677.

Csibra, G., 2007. Teachers in the wild. Trends Cogn. Sci. 11, 95–96.

Csibra, G., 2010. Recognizing communicative intentions in infancy. Mind Lang. 25, 141–168.

Csibra, G., Gergely, G., 2006. Social learning and social cognition: The case for pedagogy. Processes of change in brain and cognitive development. Atten. and Perform. XXI, 249–274.

Csibra, G., Gergely, G., 2009. Natural pedagogy. Trends Cogn. Sci. 13, 148–153.

Csibra, G., Volein, A., 2008. Infants can infer the presence of hidden objects from referential gaze information. Br. J. Dev. Psychol. 26, 1–11.

Elgier, A.M., Jakovcevic, A., Barrera, G., Mustaca, A.E., Bentosela, M., 2009. Learning and owner-stranger effects on interspecific communication in domestic dogs *(Canis familiaris)*. Behav. Proc. 81, 44–49.

Emery, N.J., 2000. The eyes have it: the neuroethology, function and evolution of social gaze. Neurosci. Biobehav. Rev. 24, 581–604.

Engel, J., Lamprecht, J., 1997. Doing what everybody does? A procedure for investigating behavioural synchronization. J. Theor. Biol. 185, 255–262.

Erdőhegyi, Á., Topál, J., Virányi, Z., Miklósi, Á., 2007. Dogs use inferential reasoning in a two-way choice task—only if they cannot choose on the basis of human-given cues. Anim. Behav. 74, 725–737.

Farroni, T., Csibra, G., Simion, F., Johnson, M.H., 2002. Eye contact detection in humans from birth. Proc. Natl. Acad. Sci. USA 99, 9602–9605.

Ferrari, P.F., Paukner, A., Ionica, C., Suomi, S.J., 2009. Reciprocal face-to-face communication between rhesus macaque mothers and their newborn infants. Curr. Biol. 19, 1768–1772.

Fugazza, C., Miklósi, Á., 2013. Deferred imitation and declarative memory in domestic dogs. Anim. Cogn. http://dx.doi.org/10.1007/s10071-013-0656-5. in press.

Fukuzawa, M., Mills, D.S., Cooper, J.J., 2005. More than just a word: non-semantic command variables affecting obedience in the domestic dog (Canis familiaris). Appl. Anim. Behav. Sci. 91, 129–141.

Gácsi, M., Győri, B., Miklósi, Á., Virányi, Z.S., Kubinyi, E., Topál, J., Csányi, V., 2005. Species-specific differences and similarities in the behavior of hand raised dog and wolf puppies in social situations with humans. Dev. Psychobiol. 47, 111–122.

Gácsi, M., Győri, B., Virányi, Z., Kubinyi, E., Range, F., Belényi, B., Miklósi, Á., 2009a. Explaining dog wolf differences in utilizing human pointing gestures: selection for synergistic shifts in the development of some social skills. PLoS One 4, e6584.

Gácsi, M., Maros, K., Sernkvist, S., Faragó, T., Miklósi, Á., 2013. Human analogue safe haven effect of the owner: behavioural and heart rate response to stressful social stimuli in dogs. PLoS One 8, e58475.

Gácsi, M., McGreevy, P., Kara, E., Miklósi, Á., 2009b. Effects of selection for cooperation and attention in dogs. Behav. Brain Func. 5, 31.

Gácsi, M., Miklósi, Á., Varga, O., Topál, J., Csányi, V., 2004. Are readers of our face readers of our minds? Dogs (Canis familiaris) show situation-dependent recognition of human's attention. Anim. Cogn. 7, 144–153.

Gaunet, F., 2008. How do guide dogs of blind owners and pet dogs of sighted owners (Canis familiaris) ask their owners for food? Anim. Cogn. 11, 475–483.

Grice, H.P., 1989. Studies in the Way of Words. Harvard University Press, Cambridge, MA, p. 406.

Guo, K., Meints, K., Hall, C., Hall, S., Mills, D., 2009. Left gaze bias in humans, rhesus monkeys and domestic dogs. Anim. Cogn. 12, 409–418.

Hare, B., Tomasello, M., 2004. Chimpanzees are more skilful in competitive than in cooperative cognitive tasks. Anim. Behav. 68, 571–581.

Hare, B., Tomasello, M., 2005. One way social intelligence can evolve: the case of domestic dogs. Trends Cogn. Sci. 9, 439–444.

Hatfield, E., Cacioppo, J.T., Rapson, R.L., 1993. Emotional contagion. Curr. Dir. Psychol. Sci. 2, 96–99.

Héjjas, K., Kubinyi, E., Rónai, Z., Székely, A., Vas, J., Miklósi, Á., Sasvári-Székely, M., Kereszturi, E., 2009. Molecular and behavioral analysis of the intron 2 repeat polymorphism in the canine dopamine D4 receptor gene. Genes Brain Behav. 8, 330–336.

Héjjas, K., Vas, J., Topál, J., Szántai, E., Rónai, Z., Székely, A., Kubinyi, E., Horváth, Z., Sasvári-Székely, M., Miklósi, Á., 2007. Association of polymorphisms in the dopamine D4 receptor gene and the activity-impulsivity endophenotype in dogs. Anim. Gen. 38, 629–633.

Hernádi, A., Kis, A., Turcsán, B., Topál, J., 2012. Man's underground best friend: domestic ferrets, unlike the wild forms, show evidence of dog-like social-cognitive skills. PloS One 7, e73267.

Hirsh-Pasek, K., Treiman, R., 1982. Doggerel: motherese in a new context. J. Child Lang. 9, 229–237.

Holmboe, K., Nemoda, Z.S., Fearon, R.M.P., Sasvári-Székely, M., Johnson, M.H., 2011. Dopamine D4 receptor and serotonin transporter gene effects on the longitudinal development of infant temperament. Genes Brain Behav. 10, 513–522.

Hood, B.M., Willen, J.D., Driver, J., 1998. Adult's eyes trigger shifts of visual attention in human infants. Psychol. Sci. 9, 131–134.

Hori, Y., Kishi, H., Inoue-Murayama, M., Fujita, K., 2013. Dopamine receptor D4 gene (DRD4) is associated with gazing toward humans in domestic dogs *(Canis familiaris)*. Open J. Anim. Sci. 3, 54–58.

Horn, L., Virányi, Z., Miklósi, Á., Huber, L., Range, F., 2012. Domestic dogs *(Canis familiaris)* flexibly adjust their human-directed behavior to the actions of their human partners in a problem situation. Anim. Cogn. 15, 57–71.

Inoue-Nakamura, N., Matsuzawa, T., 1997. Development of stone tool use by wild chimpanzees *(Pan troglodytes)*. J. Comp. Psychol. 111 (2), 159.

Ittyerah, M., Gaunet, F., 2009. The response of guide dogs and pet dogs *(Canis familiaris)* to cues of human referential communication (pointing and gaze). Anim. Cogn. 12, 257–265.

Jakovcevic, A., Elgier, A.M., Mustaca, A.E., Bentosela, M., 2010. Breed differences in dogs' *(Canis familiaris)* gaze to the human face. Behav. Proc. 84, 602–607.

Kaminski, J., 2009. Dogs *(Canis familiaris)* are adapted to receive human communication. In: Berthoz, A., Christen, Y. (Eds.), Neurobiology of Umwelt: how living beings perceive the world. Springer Verlag, Berlin, pp. 103–107.

Kaminski, J., Bräuer, J., Call, J., Tomasello, M., 2009a. Domestic dogs are sensitive to a human's perspective. Behaviour 146, 979–998.

Kaminski, J., Pitsch, A., Tomasello, M., 2013. Dogs steal in the dark. Anim. Cogn. 16, 385–394.

Kaminski, J., Schulz, L., Tomasello, M., 2012. How dogs know when communication is intended for them. Dev. Sci. 15, 222–232.

Kaminski, J., Tempelmann, S., Call, J., Tomasello, M., 2009b. Domestic dogs comprehend human communication with iconic signs. Dev. Sci. 12, 831–837.

Király, I., Csibra, G., Gergely, G., 2013. Beyond rational imitation: learning arbitrary means actions from communicative demonstrations. J. Exp. Child Psychol. in press. DOI 0.1016/j.jecp.2012.12.003.

Kirchhofer, K.K.C., Zimmermann, F., Kaminski, J., Tomasello, M., 2012. Dogs *(Canis familiaris)*, but not chimpanzees *(Pan troglodytes)*, understand imperative pointing. PloS One 7, e30913.

Kis, A., Topál, J., Gácsi, M., Range, F., Huber, L., Miklósi, Á., Virányi, Z., 2012. Does the A-not-B error in adult pet dogs indicate sensitivity to human communication? Anim. Cogn. 15, 737–743.

Kubinyi, E., Miklósi, Á., Topál, J., Csányi, V., 2003. Social mimetic behaviour and social anticipation in dogs: preliminary results. Anim. Cogn 6, 57–63.

Kupán, K., Miklósi, Á., Gergely, G., Topál, J., 2011. Why do dogs *(Canis familiaris)* select the empty container in an observational learning task? Anim. Cogn. 14, 259–268.

Lakatos, G., Soproni, K., Dóka, A., Miklósi, Á., 2009. A comparative approach to dogs' *(Canis familiaris)* and human infants' comprehension of various forms of pointing gestures. Anim. Cogn. 12, 621–631.

Lakatos, K., Tóth, I., Nemoda, Z., Ney, K., Sasváry-Székely, M., Gervai, J., 2000. Dopamine D4 receptor (DRD4) gene polymorphism is associated with attachment disorganization in infants. Mol. Psychiatry 5, 633–637.

Marshall-Pescini, S., Colombo, E., Passalacqua, C., Merola, I., Prato-Previde, E., 2013. Gaze alternation in dogs and toddlers in an unsolvable task: evidence of an audience effect. Anim. Cogn. 16, 933–943.

Marshall-Pescini, S., Passalacqua, C., Barnard, S., Valsecchi, P., Prato-Previde, E., 2009. Agility and search and rescue training differently affects pet dogs' behaviour in socio-cognitive tasks. Behav. Proc. 81, 416–422.

Marshall-Pescini, S., Passalacqua, C., Miletto Petrazzini, M.E., Valsecchi, P., Prato-Previde, E., 2012. Do dogs *(Canis lupus familiaris)* make counterproductive choices because they are sensitive to human ostensive cues? PLoS One 7, e35437.

Marshall-Pescini, S., Passalacqua, C., Valsecchi, P., Prato-Previde, E., 2010. Comment on "Differential sensitivity to human communication in dogs". Science 329, 142–c.

Marshall-Pescini, S., Prato-Previde, E., Valsecchi, P., 2011. Are dogs *(Canis familiaris)* misled more by their owners than by strangers in a food choice task? Anim. Cogn. 14, 137–142.

Marshall-Pescini, S., Whiten, A., 2008. Social learning of nut-cracking behavior in East African sanctuary-living chimpanzees *(Pan troglodytes schweinfurthii)*. J. Comp. Psychol 122, 186–194.

McKinley, J., Sambrook, T.D., 2000. Use of human-given cues by domestic dogs *(Canis familiaris)* and horses *(Equus caballus)*. Anim. Cogn 3, 13–22.

Meltzoff, A.N., 1999. Born to learn: what infants learn from watching us. In: Fox, N., Worhol, J.G. (Eds.), The role of early experience in infant development. Pediatric Institute Publications, Skillman, NJ, pp. 1–10.

Merola, I., Prato-Previde, E., Marshall-Pescini, M., 2012. Dogs' social referencing towards owners and strangers. PLoS One 7 (10), e47653. http://dx.doi.org/10.1371/ journal. pone.0047653.

Miklósi, Á., Kubinyi, E., Topál, J., Gácsi, M., Virányi, Z., Csányi, V., 2003. A simple reason for a big difference: wolves do not look back at humans but dogs do. Curr. Biol. 13, 763–766.

Miklósi, Á., Polgárdi, R., Topál, J., Csányi, V., 2000. Intentional behaviour in dog–human communication: an experimental analysis of 'showing' behaviour in the dog. Anim. Cogn. 3, 159–166.

Miklósi, Á., Topál, J., 2012. The evolution of canine cognition. In: Vonk, J., Shackelford, T. (Eds.), The Oxford handbook of comparative evolutionary psychology. Oxford University Press, Oxford, pp. 513–568.

Miklósi, Á., Topál, J., 2013. What does it take to become "best friends"? Evolutionary changes in canine social competence. Trends Cogn. Sci. 17, 287–294.

Mitchell, R.W., 2001. Americans' talk to dogs: similarities and differences with talk to infants. Res. Lang. Soc. Int. 34, 183–210.

Moll, H., Tomasello, M., 2007. Cooperation and human cognition: the Vygotskian intelligence hypothesis. Phil. Trans. Roy. Soc. B: Biol. Sci. 362, 639–648.

Moriguchi, Y., Itakura, S., 2005. Does pointing comprehension disturb controlling action? Evidence from 2-year-old children. Proc. 4th IEEE Int. Conf. Dev. Learn., 102–105.

Newman, R.S., 2005. The cocktail party effect in infants revisited: listening to one's name in noise. Dev. Psychol. 41, 352–362.

Okumura, Y., Kanakogi, Y., Kanda, T., Ishiguro, H., Itakura, S., 2013. Infants understand the referential nature of human gaze but not robot gaze. J. Exp. Child Psychol. 116, 86–95.

Oliveira, R.F., 2009. Social behavior in context: hormonal modulation of behavioral plasticity and social competence. Int. Comp. Biol. 49, 423–440.

Ostojic, L., Clayton, N.S., 2013. Behavioural coordination of dogs in a cooperative problem-solving task with a conspecific and a human partner. Anim. Cogn. 16, 765–772.

Passalacqua, C., Marshall-Pescini, S., Barnard, S., Lakatos, G., Valsecchi, P., Prato-Previde, E., 2011. Human-directed gazing behaviour in puppies and adult dogs, *Canis lupus familiaris*. Anim. Behav. 82, 1043–1050.

Pettersson, H., Kaminski, J., Herrmann, E., Tomasello, M., Kaminski, J., 2011. Understanding of human communicative motives in domestic dogs. Appl. Anim. Behav. Sci. 133, 235–245.

Pfandler, E., Lakatos, G., Miklósi, Á., 2013. Eighteen-month-old human infants show intensive development in comprehension of different types of pointing gestures. Anim. Cogn. 16, 711–716.

Pongrácz, P., Miklósi, Á., Timar-Geng, K., Csányi, V., 2004. Verbal attention getting as a key factor in social learning between dog *(Canis familiaris)* and human. J. Comp. Psychol. 118, 375–383.

Povinelli, D.J., Reaux, J.E., Bierschwale, D.T., Allain, A.D., Simon, B.B., 1997. Exploitation of pointing as a referential gesture in young children, but not adolescent chimpanzees. Cogn. Dev. 12, 423–461.

Povinelli, D.J., Eddy, T.J., 1996. Factors influencing young chimpanzees' *(Pan troglodytes)* recognition of attention. J. Comp. Psychol 110, 336–345.

Parise, E., Reid, V.M., Stets, M., Striano, T., 2008. Direct eye contact influences the neural processing of objects in 5-month-old infants. Soc. Neurosci. 3, 141–150.

Plourde, V., Fiset, S., 2013. Pointing gestures modulate domestic dogs' search behavior for hidden objects in a spatial rotation problem. Learning and Motivation 44, 282–293.

Prato-Previde, E., Marshall-Pescini, S., Valsecchi, P., 2008. Is your choice my choice? The owners' effect on pet dogs' *(Canis lupus familiaris)* performance in a food choice. Anim. Cogn. 11, 167–174.

Racca, A., Guo, K., Meints, K., Mills, D.S., 2012. Reading faces: differential lateral gaze bias in processing canine and human facial expressions in dogs and 4-year-old children. PLoS One 7, e36076.

Range, F., Virányi, Z., 2011. Development of gaze following abilities in wolves *(Canis lupus)*. PLoS One 6, e16888.

Range, F., Heucke, S.L., Gruber, C., Konz, A., Huber, L., Virányi, Z., 2009. The effect of ostensive cues on dogs' performance in a manipulative social learning task. Appl. Anim. Behav. Sci. 120, 170–178.

Reid, P.J., 2009. Adapting to the human world: dogs' responsiveness to our social cues. Behav. Proc. 80, 325–333.

Rekers, Y., Haunsend, D.B.M., Tomasello, M., 2011. Children, but not chimpanzees, prefer to collaborate. Curr. Biol. 21, 1756–1758.

Richerson, P.J., Boyd, R., 1998. The evolution of human ultra-sociality. In: Eibl-Eibisfeldt, I., Salter, F. (Eds.), Ideology, warfare, and indoctrinability. Berghan Books, Oxford, New York, pp. 71–95.

Riedel, J., Schumann, K., Kaminski, J., Call, J., Tomasello, M., 2008. The early ontogeny of human–dog communication. Anim. Behav 73, 1003–1014.

Scheider, L., Grassmann, S., Kaminski, J., Tomasello, M., 2011. Domestic dogs use contextual information and tone of voice when following a human pointing gesture. PLoS One 6, e21676.

Scheider, L., Kaminski, J., Call, J., Tomasello, M., 2013. Do domestic dogs interpret pointing as a command? Anim. Cogn. 16, 361–372.

Schenkel, R., 1967. Submission: Its features and function in wolf and dog. Am. Zool. 7, 319–329.

Schloegl, C., Kotrschal, K., Bugnyar, T., Lorenz, K., Gru, F., 2007. Gaze following in common ravens, *Corvus corax*: ontogeny and habituation. Anim. Behav 74, 769–778.

Schloegl, C., Schmidt, J., Scheid, C., Kotrschal, K., Bugnyar, T., 2008. Gaze following in non-human animals: the corvid example. In: Columbus, F. (Ed.), Animal behaviour: new research. Nova Science Publishers, New York, pp. 73–92.

Senju, A., Csibra, G., 2008. Gaze following in human infants depends on communicative signals. Curr. Biol. 18, 668–671.

Senju, A., Csibra, G., Johnson, M.H., 2008. Understanding the referential nature of looking: infants' preference for object-directed gaze. Cognition 108, 303–319.

Shepherd, S.V., 2010. Following gaze: gaze-following behavior as a window into social cognition. Front. Integr. Neurosci. 4 (5). http://dx.doi.org/10.3389/fnint.2010.00005.

Silva, K., Bessa, J., Sousa, L., 2013. Familiarity connected or stress-based contagious yawning in domestic dogs *(Canis familiaris)*? Some additional data. Anim. Cogn. 16, 1007–1009.

Soproni, K., Miklósi, Á., Topál, J., Csányi, V., 2001. Comprehension of human communicative signs in pet dogs. J. Comp. Psychol. 115, 122–126.

Southgate, V., Chevallier, C., Csibra, G., 2009. Sensitivity to communicative relevance tells young children what to imitate. Dev. Sci. 12, 1013–1019.

Sperber, D., Wilson, D., 1986. Relevance: communication and cognition. Harvard University Press, Cambridge, MA, p. 142.

Sümegi, Z., Kis, A., Miklósi, Á., Topál, J., 2013. Why do adult dogs *(Canis familiaris)* commit the A-not-B search error? J. Comp. Psychol. in press.

Szetei, V., Miklósi, Á., Topál, J., Csányi, V., 2003. When dogs seem to lose their nose: an investigation on the use of visual and olfactory cues in communicative context between dog and owner. Appl. Anim. Behav. Sci. 83, 141–152.

Taborsky, B., Oliveira, R.F., 2012. Social competence: an evolutionary approach. Trends Cogn. Sci. 27, 679–688.

Téglás, E., Gergely, A., Kupán, K., Miklósi, Á., Topál, J., 2012. Dogs' gaze following is tuned to human communicative signals. Curr. Biol. 22, 209–212.

Tehrani, J.J., Riede, F., 2008. Towards an archaeology of pedagogy: learning, teaching and the generation of material culture traditions. World Archaeology 40 (3), 316–331.

Topál, J., Gergely, G., Erdőhegyi, Á., Csibra, G., Miklósi, Á., 2009a. Differential sensitivity to human communication in dogs, wolves, and human infants. Science 325, 1269–1272.

Topál, J., Gergely, G., Miklósi, Á., Erdőhegyi, Á., Csibra, G., 2008. Infants' perseverative search errors are induced by pragmatic misinterpretation. Science 321 (5897), 1831–1834.

Topál, J., Kubinyi, E., Gácsi, M., Miklósi, Á., 2005. Obeying social rules: a comparative study on dogs and humans. J. Cult. Evol. Psychol 3, 213–238.

Topál, J., Miklósi, Á., Gácsi, M., Dóka, A., Pongrácz, P., Kubinyi, E., Virányi, Z., Miklósi, Á., Csányi, V., 2009b. The dog as a model for understanding human social behavior. Adv. S. Behav. 39, 71–116.

Topál, J., Miklósi, Á., Sümegi, Z., Kis, A., 2010. Response to comments on "Differential sensitivity to human communication in dogs, wolves and human infants." Science 329, 142d, 1624.

Udell, M.A.R., Dorey, N.R., Wynne, C.D.L., 2008. Wolves outperform dogs in following human social cues. Anim. Behav. 76, 1767–1773.

Udell, M.A.R., Dorey, N.R., Wynne, C.D.L., 2009. What did domestication do to dogs? A new account of dogs' sensitivity to human actions. Biol. Rev. 85, 327–345.

Vas, J., Topál, J., Gácsi, M., Miklósi, Á., Csányi, V., 2005. A friend or an enemy? Dogs' reaction to an unfamiliar person showing behavioural cues of threat and friendliness at different times. Appl. Anim. Behav. Sci. 94, 99–115.

Virányi, Z., Gácsi, M., Kubinyi, E., Topál, J., Belényi, B., Ujfalussy, D., Miklósi, Á., 2008. Comprehension of human pointing gestures in young human-reared wolves and dogs. Anim. Cogn. 11, 373–387.

de Waal, F.B.M., 1996. Good natured: The origins of right and wrong in humans and other animals. Harvard University Press, Cambridge, MA.

Wobber, V., Hare, B., 2009. Testing the social dog hypothesis: are dogs also more skilled than chimpanzees in non-communicative social tasks? Behav. Proc. 81, 423–428.

Wynne, C.D.L., Udell, M.A.R., Lord, K., 2008. Ontogeny's impacts on human–dog communication. Anim. Behav. 76, e1–e4.

Yamamoto, M., Ohtani, N., Ohta, M., 2011. The response of dogs to attentional focus of human beings: a comparison between guide dog candidates and other dogs. J. Vet. Behav. Clin. Appl. Res. 6, 4–11.

Yoon, J.M., Johnson, M.H., Csibra, G., 2008. Communication-induced memory biases in preverbal infants. Proc. Natl. Acad. Sci. USA 105, 13690–13695.

# Do Dogs Show an Optimistic or Pessimistic Attitude to Life?

## A Review of Studies Using the 'Cognitive Bias' Paradigm to Assess Dog Welfare

Oliver Burman

*University of Lincoln, Riseholme Park, Lincoln, UK*

## 12.1 INTRODUCTION

There is an increasing acceptance that the study of affective states (i.e., emotions and moods) in animals is a critical component in our understanding of animal welfare (Dawkins, 2000). The capacity for animals to experience affective states such as fear or pleasure is central to the study of animal welfare (Dawkins, 2006), and there is increasing interest in the scientific study of animal affect across a range of research disciplines reflecting considerable public interest. The development of methodologies to more accurately assess affective state and thus welfare is therefore of great interest to researchers and members of the public alike, given the likely benefit for those associated with captive animals.

### 12.1.1 Why Use Cognition to Assess Animal Welfare?

In the absence of any direct way of measuring subjective affective states, indirect indicators of affect have been developed, typically measures of physiology or behaviour (Broom, 1991) that are assumed to correlate with associated subjective states, but these indicators are not without significant limitations. Physiological measures (e.g., heart rate) reflect arousal but may not differ according to valence (i.e., positive or negative)—although there may be exceptions to this [e.g., salivary IgA (Skandakumar et al., 1995)]. This makes the interpretation of such measures in terms of affective state and welfare problematic—especially when considered in isolation. For example, an increase in heart rate was observed in dogs following the appearance of their owner and also following the appearance of a stranger (Palestrini et al., 2005). This could have been because the dogs were excited by the arrival of both individuals, but could also be explained by excitement at the arrival of the owner and distress at the arrival of an unfamiliar person.

Behavioural measures can also be difficult to interpret, preventing the clear predictions necessary to allow the assessment of animal welfare in a more sensitive and targeted manner (Mendl et al., 2010b). For example, self-grooming in dogs can be interpreted as an indicator of relaxation and appropriate self-maintenance, but can also develop into an abnormal self-mutilating behaviour ('over-grooming') thought to be associated with an attempt to relieve stress and/or anxiety (Rooney et al., 2009). There is therefore a clear requirement for the development and refinement of alternative indicators of affective state that are less susceptible to these limitations.

Affective states in both humans and animals incorporate cognitive, behavioural, and physiological components in addition to the associated conscious subjective experience (Winkielman et al., 2007), and it is the cognitive components of affect that have been the focus of recent attention (Paul et al., 2005). This is based on robust findings from human psychology that affective state can influence the way we process information and our decision making (Bishop, 2007). For example, anxious people (in a negative affective state) bias their attention to threatening stimuli (Mogg & Bradley, 1998). They also make more negative interpretations of ambiguous stimuli compared to individuals in positive affective states. For instance, Eysenck et al. (1991) demonstrated that clinically anxious human subjects were more likely to interpret ambiguous sentences as being threatening compared to either recovered clinically anxious or normal control subjects, suggesting that their biased interpretation of ambiguity reflected their anxious mood state. Measuring cognitive outputs such as decision making can therefore be used to reveal what affective states are being experienced.

This differentiation between individuals experiencing differently valenced affective states (i.e., negative versus positive) means that cognitive indicators of animal affect and welfare are able to discriminate between affective states of different valence (e.g., anxiety, pleasure)—an advance on many physiological measures of welfare. Cognitive approaches may even allow us to differentiate between affective states of the same valence (e.g., anxiety, depression) (Mendl et al., 2009). Cognitive measures therefore allow clear predictions to be made regarding the expected response of an animal to particular conditions (e.g., its housing environment) or events (e.g., social isolation) that it experiences that are assumed to influence affective state.

However, whilst there are benefits to a cognitive approach, there are also limitations (see later), and so it is important to emphasise that this does not mean that a cognitive approach should be considered as a replacement for existing measures. Indeed, a combined approach using complementary cognitive, behavioural, and physiological measures—each selected specifically for its suitability for the research question under investigation—is likely to result in the most comprehensive and accurate assessment of animal welfare.

## 12.1.2 Why Assess Dog Welfare?

Large numbers of dogs are kept as companion animals, e.g., an estimated 10 million owned dogs in the UK (Murray et al., 2010), as well as working animals (e.g., guide dogs, military dogs) and laboratory animals. Whilst there is clearly a need for additional, accurate, and reliable measures of affect in animal welfare research in general, the dog is a very suitable 'model' species for such development because there are surprisingly few reliable welfare indicators for this species (Rooney et al., 2007) given the long historical relationship between humans and dogs.

Whilst improving our ability to assess the welfare of dogs will help to inform the general public as to how to best care for their pets, there are numerous more specific examples of the sorts of situations in which better assessment of dog welfare might have an impact, including (1) behavioural problems, e.g., separation-related behaviour (Bradshaw et al., 2002); (2) rescue/shelter environments and recuperation/rehoming (e.g., Wells, 2004); (3) training techniques, e.g., electronic collars (University of Lincoln et al., 2013); (4) veterinary treatment, e.g., corticosteroids (Notari & Mills, 2011); (5) working dogs, e.g., military (Rooney et al., 2007); (6) laboratory-housed dogs (Hubrecht, 1993); and (7) obesity (German, 2006). One topic of considerable recent public interest is the effect of artificial selection and domestication on dog welfare, with major health and welfare problems associated with the breeding of pedigree dogs (Rooney & Sargan, 2009; Bateson, 2010). For example, syringomyelia is a condition particularly associated with Cavalier King Charles spaniels in which artificial selection for the retention of puppyish facial features into adulthood has resulted in dogs with underdeveloped dome-shaped skulls that can lead to the formation of fluid-containing cavities in their cervical spinal cords. Over time, these cavities can damage the dogs' spinal cords, resulting in considerable pain as well as muscular weakness in the back and limbs (Rusbridge, 2005). Refinement of the welfare assessment of dogs will therefore have an input into the process of addressing this problem by investigating the link between genotypic and phenotypic abnormalities in dogs and their relation to affective state.

Owned dogs also offer specific advantages as subjects for cognition-related research on affect compared to other species. For example, they have high-level learning and memory capabilities (Kaminski et al., 2004) and excellent multimodal perception (Pretterer et al., 2004). It is no surprise that dogs are being increasingly considered as potential models for human psychiatric disorders (Cyranoski, 2010) and have been identified as being valuable models for understanding human social cognition (Miklósi et al., 2007). As such, research on owned dogs, as well as having a direct benefit to dogs themselves in terms of improving our ability to determine their welfare in a range of contexts, will inform us about biological concepts likely to have a significant benefit common to mammalian species, including improving our understanding of affective state in humans (Paul et al., 2005).

### 12.1.3 Cognitive Approaches to Assessing Affective State

There are several different types of what might be described as cognitive approaches to the assessment of affective state. It is worth clarifying that, although these approaches are labelled 'cognitive', they rely on the recording of a behavioural and/or physiological response for the interpretation of underlying cognitive processes. One such approach, incentive contrast, is based around sensitivity to reward change—in which an animal responds to unexpected changes in reward quantity/quality in a way that is consistent with experiencing an emotional state (Flaherty, 1999). For example, a decrease in responding following an unexpected reduction in reward quality/quantity is thought to reflect the subject experiencing a state akin to 'disappointment', whereas behaviour reflecting 'elation' is exhibited following an unexpected increase in reward quality/quantity. Although incentive contrast has been demonstrated in a wide range of species, and was first shown in dogs by Bentosela et al. (2009), it is only relatively recently that the potential of this paradigm to assess affective state has been explored.

Burman et al. (2008a) showed that laboratory rats that experienced an unexpected decrease in reward quantity, from high to low (12 to 1 food pellet), exhibited a significant reduction in response (an increased latency to feed) compared to those animals that only ever experienced a reward of low quantity (1 pellet). However, the crucial finding was that this sensitivity to reward loss was significantly prolonged for animals housed without enrichment (and therefore in a putative negative affective state) compared to those housed with enrichment (and in a more positive affective state), indicating enhanced sensitivity to reward loss. These findings indicated for the first time that an animal's sensitivity to reward change can be influenced by affective state, and so this paradigm could be used as a potential indicator of animal welfare. Further investigation of this paradigm is currently under way, using owned dogs and those housed in rescue shelters.

For the rest of this chapter, I focus on the cognitive approach that has been most frequently studied in dogs, the cognitive affective bias paradigm (Paul et al., 2005) (henceforth referred to as cognitive bias), and review the recent studies that have used this paradigm to assess dog welfare before proposing directions for future research.

## 12.2 COGNITIVE BIAS

### 12.2.1 What Is Cognitive Bias?

The cognitive bias paradigm is based on theoretical and empirical findings, obtained from human studies (e.g., Mogg & Bradley, 1998), that affective state can influence cognitive processes including judgement, memory, and/or attention. For example, people in negative affective states attend more to threatening stimuli (*attention bias*), better remember negative events (*memory bias*), and judge ambiguous stimuli more negatively (are relatively 'pessimistic') (*judgement bias*) compared to individuals in a positive affective state (see Paul et al., 2005). Translation from human to (non-human) animal studies was first demonstrated by Harding et al. (2004) in a seminal study investigating judgement bias in laboratory rats, with the hypothesis that, as shown for humans, animals will judge the same ambiguous stimulus differently according to their affective state.

Harding et al. (2004) trained rats to press a lever ('go') when an auditory tone ('A') was sounded in order to gain a food reward, but to refrain from pressing the lever ('no-go') when a different tone ('B') was played in order to avoid an aversive burst of white noise. Once the rats could discriminate accurately between, and respond appropriately to, the two tones, the rats were split into two treatment groups. One group experienced a standard (predictable) laboratory husbandry regime, and the other experienced an unpredictable husbandry regime known to be mildly aversive (Willner, 1997). After 9 days, both groups were then tested for judgement bias by observing how they responded to non-reinforced ambiguous 'probe' tones intermediate between the trained 'reference' tones.

Based on human findings in judgement bias studies, the prediction was that those rats exposed to the unpredictable husbandry regime, because they would be in a negative affective state, would judge the ambiguous tones negatively compared to the rats housed in predictable conditions, and this would be demonstrated by lower (and slower) levels of lever pressing, i.e., as though they anticipated a negative outcome. In contrast, it was predicted that rats housed in standard conditions, being in a more positive affective state, would judge the same ambiguous tones more positively, reflected by increased (and faster) levels of lever pressing, i.e., as though they expected a positive outcome. As predicted, the authors found that rats exposed to the aversive unpredictable husbandry regime were slower to lever press and tended to lever press less in response to the ambiguous tones compared to rats housed under predictable conditions, what could be considered a 'pessimistic' response, and this was the first evidence that the cognitive bias paradigm could be used to identify affective states in animals.

The findings of Harding et al. (2004) resulted in a considerable surge of interest in the use of the cognitive bias paradigm as a means of assessing affective state, with a subsequent proliferation of studies using a range of methodologies and affective treatments in a variety of different animal species, including rats (e.g., Burman et al., 2008b; Brydges et al., 2011); mice (e.g., Boleij et al., 2012); starlings (e.g., Matheson et al., 2008; Brilot et al., 2010); sheep (e.g., Doyle et al., 2010a); bees (Bateson et al., 2011); chickens (Wichman et al., 2012); macaques (Bethell et al., 2012); goats (Briefer & McElligott, 2013); and dogs (e.g., Mendl et al., 2010a).

When assessing cognitive bias in non-human animals, the majority of tasks have used judgement bias rather than either attention or memory bias—although an exception is Bethell et al. (2012), who used an attention bias paradigm to investigate the impact of a potentially stressful veterinary health check on the affective state of captive rhesus macaques. Despite focusing almost uniformly on judgement bias, there is considerable variation between studies in aspects including, amongst other things, (1) the subject species; (2) the treatment; (3) the type of discriminative stimulus (e.g., spatial, auditory); (4) the contingencies (e.g., reward versus no reward, reward versus punisher); (5) the dependent variable (e.g., lever presses, latency/speed); (6) the number of intermediate probe stimuli; (7) the number of presentations of probe stimuli; (8) the approach to statistical analysis; and (9) the experimental design (e.g., correlational, within-subjects, between-subjects). For example, in terms of the type of discriminative stimulus, researchers have used auditory tones (Harding et al., 2004), spatial cues (Burman et al., 2008b), grey-scale hues (Bateson & Matheson, 2007), tactile cues (Brydges et al., 2011), and olfactory cues (Bateson et al., 2011) (see Mendl et al., 2009, for a review).

One area where there has been some consistency in methodology—not least in the use of the same subject species—has been in studies using the cognitive bias paradigm to assess welfare in dogs. Whilst reviewing the methodological differences between the various studies allows the identification of areas for future research as well as potential limitations, the similarities in approach between different studies also provides an excellent opportunity to reflect on the utility and efficacy of the cognitive bias paradigm thus far, and a suitable time point at which to assess the potential of this testing paradigm for the assessment of dog welfare.

## 12.2.2 Cognitive Bias in Dogs

In this section I review, in chronological order, the current studies on dogs, inevitably focusing on those studies that have already been published in peer-reviewed journals or refereed reports.

### 12.2.2.1 Mendl et al. (2010a). Dogs Showing Separation-Related Behaviour Exhibit a 'Pessimistic' Cognitive Bias

Dogs exhibiting separation-related behaviour (SRB) during periods when they have been left alone are considered to be anxious (Bradshaw et al., 2002), but it was previously unclear whether or not experience of this negative affective state was confined to instances of social separation or if the dogs' overall background affective state (mood) was negative. For this study, the authors recruited dogs currently housed in rescue shelters. Shortly after admission to the shelter, they were given a previously validated SRB test (Blackwell et al., 2003) to determine their behavioural response to being left alone, and their behaviour during this 5-minute test was used to generate an SRB score. This SRB score was calculated by summing the total amount of time (seconds) during the 5-minute test that the dog spent exhibiting the predefined separation-related behaviours: vocalisation (i.e., barking, whining, and howling); scratching at the door; destructive behaviour; and toileting. A few days later, the dogs went on to receive a cognitive judgement bias test adapted for dogs (see later) from a design first developed in rats (Burman et al., 2008b). This test used spatial locations as discriminative stimuli. When a food bowl was placed in one location (Positive 'P'), it contained a food reward; when placed in a second location (Negative 'N'), it contained no food reward. The P and N locations (balanced between dogs) were equidistant from the dogs' starting position (4 m). Dogs were kept behind a visual barrier during the 'baiting' of the different locations, and a trial was ended if the bowl had not been visited within 30 seconds of the dogs being released from the start position. The dogs were trained to discriminate between the two reference locations (P and N), with the premise that they should quickly learn that when the bowl is placed at the rewarded location (P), then they should go to it to get the food, but that when the bowl is in the non-rewarded location (N), then it should be either ignored or visited without urgency. Successful discrimination was based on individual performance, with a criterion that for six consecutive trials (three P trials and three N trials), the slowest run to the P location was faster than the quickest run to the N location. This study therefore used a correlational approach to investigate the relationship between performance in a cognitive bias test (individual dogs tested just once) and in an SRB test.

Once criterion had been achieved (average 29.42 trials; range 21–61), testing followed immediately. During testing, the bowl (non-rewarded) was placed in one of three ambiguous probe locations intermediate between the P and N locations: NP (near P); M (middle); and NN (near N). Each probe was presented three times in a balanced order, separated by four standard P/N trials, with each dog receiving a total of 41 test trials. The authors also included an additional 'olfactory control' in which an empty bowl was placed in the P location to confirm that dogs were not using olfactory cues to decide whether or not to approach the bowl, regardless of location/spatial cues. Raw latency data was adjusted using the following calculation, in order to take into account differences in running speeds between individual dogs:

$$\text{adjusted score} = \frac{\left( \begin{array}{c} \text{mean latency to probe location} \\ - \text{ mean latency to positive location} \end{array} \right)}{\left( \begin{array}{c} \text{mean latency to negative location} \\ - \text{ mean latency to positive location} \end{array} \right)} \times 100$$

Dogs showed a clear generalisation across the five different locations, running fastest to P and slowest to N, with all five locations (P, NP, M, NN, N) significantly different from one another except P and NP. Dogs with high SRB scores showed significantly increased latencies for the M probe (and approaching significance for NN), suggesting a 'pessimistic' judgement of ambiguity; i.e., they behaved as though they expected a negative outcome. Separation-related behaviour scores did not correlate with the dogs' latency to approach either of the reference locations (P and N), suggesting that there was no underlying difference in general speed related to SRB, either due to high SRB dogs simply being slower or having a decreased food motivation. The authors concluded that those dogs that exhibited high levels of SRB responded in a 'pessimistic' way to ambiguous spatial stimuli, suggesting an underlying negative affective state. This study therefore successfully modified a cognitive judgement bias task for use in dogs and demonstrated for the first time that this approach could be used to determine the affective state of dogs in relation to commonly experienced welfare problems.

### 12.2.2.2 Burman et al. (2011). Using Judgement Bias to Measure Positive Affective State in Dogs

The aim of the Burman et al. (2011) study was to assess the impact of experiencing a positive 'search & forage task' (i.e., searching for hidden food treats within a familiar maze) immediately prior to cognitive bias testing, with the prediction that the positive experience of the 'search & forage task' would result in a post-consummatory positive (i.e., 'optimistic') judgement of ambiguous stimuli compared to when tested without experiencing the task. This therefore contrasts with many studies in which the focus is on identifying negative affective states, instead focusing on investigating whether a cognitive judgement bias task could be used to identify positive affective states in dogs. There has been a recent rise in interest regarding the assessment of positive affective states (e.g., Boissy et al., 2007; Yeates & Main, 2008), but there still remains a relative lack of research in this area, relating both to the circumstances when positive affective states might occur and to how they might be measured.

In contrast to Mendl et al. (2010a), this study used laboratory-housed dogs of a single breed (beagles). This allowed the authors to control many of the aspects of variation, such as breed/size, age, sex, and experience of the dogs, not possible with the majority of 'owned' or rescue dog populations. The cognitive bias paradigm used by Burman et al. (2011) also differed from that used in Mendl et al. (2010a), and was adapted from a visual grey-scale task that was first developed in starlings (Bateson & Matheson, 2007). This allowed the researchers to determine whether different types of cognitive judgement bias tasks are effective in assessing affective state in dogs.

The reference stimuli were boxes covered with two different shades of grey, with three intermediate shades acting as ambiguous probe stimuli during testing. One grey shade contained food (P), and the other (balanced between dogs) contained no food (N). Boxes were 'baited' out of sight of the dogs and always placed in the same position 6 m away from the dogs' starting point. The dogs were trained to discriminate between the two reference stimuli, with the premise that they should quickly learn that when the box is the rewarded shade of grey, then they should go to it to get the food, but that when the box is the non-rewarded shade of grey, then it should be either ignored or visited without urgency.

Once criterion was achieved, with dogs having to run faster to the rewarded box for six consecutive trials, with at least half a second difference between the slowest run to the rewarded box and the fastest run to the non-rewarded box, then the dogs were tested. Criterion was achieved with a mean of 93.2 trials (range 30–200). Training occurred 2 weeks prior to testing, and so additional 'refresher' training was carried out to ensure that the dogs could still achieve criterion. Each dog received a total of 90 test trials. Raw latency data were adjusted using the same adjustment as Mendl et al. (2010a).

Dogs tested after experiencing the 'search & forage task' did not differ in their response to either P or N shades compared to when tested without experiencing the 'search & forage task', suggesting that there were no clear differences in food motivation or fatigue. There was a general 'optimistic tendency' to approach all three probes quickly, possibly due to the minimal cost associated with a 'no reward' contingency (see Mendl et al., 2009). The authors found a significant difference between the treatments in the dogs' latency to approach the M ambiguous probe stimulus. However, contrary to what was expected, dogs responded more positively to the ambiguous shade (i.e., an 'optimistic' bias) when they had *not* experienced the 'search & forage task'.

The authors speculated that this result might have been due to testing the dogs after the apparently rewarding 'search & forage task' had ended, with the consequence that the affective state of the dogs at the point of testing may have been negative due to the removal/termination of a perceived rewarding event, e.g., akin to 'disappointment' (Burman et al., 2008a). It therefore appears that the affective state at the point of cognitive bias testing can be different to that experienced during the treatment. For example, Doyle et al. (2010a), when investigating restraint stress in sheep, observed an apparent 'optimistic' bias in sheep following release from aversive restraint. For long-term treatments, this may be less of a problem because the affective (mood) state induced by the treatment is expected to carry over into the testing period. It is therefore important to consider over what time period we expect the treatment to be investigated to affect the animals. Is it something that affects them generally, and so testing at any time will reflect their affective state, or is it something that will directly influence them only at that point in time and so may not carry over into the testing period?

An alternative interpretation of the results of this study is that laboratory-housed dogs tested without experiencing the 'search & forage task' might have been more 'optimistic' due to the excitement of human interaction during CB testing. The suggestion is that, for those dogs for whom human interaction may be less regular, whether they are owned dogs that are left alone for long periods of time, laboratory dogs, or dogs housed at rescue shelters, the increase in arousal generated by interacting with the researcher during testing influences the results, overshadowing any effect of treatment.

So, although this study demonstrated that a cognitive judgement bias task based on a visual grey scale could be used to measure affective state in dogs—extending the number of different tasks available to people working with dogs—it also identified some specific difficulties that should be taken into consideration for future studies.

### 12.2.2.3 Müller et al. (2012). Brief Owner Absence Does Not Induce Negative Judgement Bias in Pet Dogs

The Müller et al. (2012) study investigated whether dogs' cognitive bias tested in the brief absence of their owners would judge ambiguous stimuli negatively, due to being in a negative affective state of moderate anxiety, compared to when tested in the comforting presence of their owner. As well as improving our understanding of how dogs cope with periods of separation from their owners and the impact of that separation on affective state, this study also relates to issues of social support. The presence of their dog during the experience of a stressful event (e.g., carrying out an arithmetic task) has been shown to reduce physiological measures of stress in female dog owners (Allen et al., 1991), and it is therefore of interest to see if the presence of an owner has a similarly supportive influence on their dog.

In this study, Müller et al. (2012) therefore avoided the previously mentioned issue whereby a short-term treatment (and its potential influence on affective state) 'stops' before testing commences, because the treatment that they selected (owner presence/absence) occurred during testing itself. They predicted that, if brief separation from the owner results in moderate anxiety, then dogs tested in the absence of their owners would show a more 'pessimistic' cognitive bias compared to when they were tested with their owners present. They used the previously described spatial judgement bias task (Burman et al., 2008b; Mendl et al., 2010a) with some modifications, and with the subjects tested in their home environment. Criterion was reached on average after 42 trials (range 30–90). Dogs received a total of 104 test trials. Dogs were able to observe the researcher positioning the food bowl, and when placing the bowl at the M probe location, half the time the researcher approached from the P side and half from the N side. Latency data were adjusted as for Mendl et al. (2010a).

No effect of owner presence/absence was observed on the dogs' response to the ambiguous probe locations. The authors suggest that either the dogs were not anxious during their owner's absence or that the CB test was not sufficiently sensitive to identify such a short-term 'treatment'—although short-term treatments have been shown to influence cognitive bias in other animal species. For example, Burman et al. (2009) found that altering light levels during testing influenced anxiety levels in rats as reflected in a more 'pessimistic' response to ambiguity for those rats tested under bright lights.

Müller et al. (2012) did find that latencies to the M probe were shorter when the researcher placed the bowl approaching from the 'positive' side, suggesting an additional spatial generalisation based on the direction of human approach. Latencies were also found to be significantly longer on the second testing day compared to the first, potentially reflecting the dogs' learning that the probes were non-rewarded (e.g., Doyle et al., 2010b). There were large individual differences in how dogs responded to the probe locations. Some showed a typical generalisation (latency: P < NP < M < NN < N), whilst others showed a similar response to two of the three probes, and a differential response for the third, a response pattern that the authors attributed to potential personality differences.

## 12.2.2.4 Titulaer et al. (2013). Cross Sectional Study Comparing Behavioural, Cognitive and Physiological Indicators of Welfare Between Short and Long Term Kennelled Domestic Dogs

Despite the best efforts of staff, dogs housed in kennels at rescue shelters can appear to have compromised welfare (e.g., Hiby et al., 2006). The aim of the Titulaer et al. (2013) study was to compare behavioural, cognitive, and physiological indicators of welfare between those dogs housed in a rescue shelter for short periods (ST: 1 week–3 months) compared to those kennelled for longer periods (LT: > 6 months). The prediction was that dogs kennelled for longer periods—and therefore presumably exposed to stressors for longer—would show more indications of poor welfare, including a pessimistic cognitive bias compared to more recent arrivals at the shelter.

The authors adopted a 'multiple indicator approach' to assessing dog welfare, using a combination of behavioural observations (e.g., repetitive behaviour), physiological measures (e.g., urinary cortisol), cognitive measures (cognitive bias), and shelter staff personal assessment. Subjects were matched for breed, sex, age, and neuter status. For the cognitive bias testing, they used the spatial judgement bias task in the same way as established by Mendl et al. (2010a), with testing occurring immediately after each dog reached criterion—and so no refresher trials were needed—with each dog tested only once.

No difference was found between treatment groups (short- versus long-term kennelling) in response to probe stimuli. Generalisation between the five spatial stimuli was observed from the rewarded location (shortest latency) to the unrewarded location (longest latency), with differences between all pairs except the rewarded location and the probe closest too it. The authors found few other differences between the treatment groups for the non-cognitive measures. There was no difference in urinary cortisol, a result that would suggest no differences in physiological stress response, and the only behavioural difference was that LT dogs spent longer resting during undisturbed observation periods—although there were some differences as determined by shelter staff assessment (e.g., LT dogs were considered more likely to growl or bark at other dogs). Taking these results together, the authors speculated that perhaps the length of time that a dog is kennelled in a rescue shelter is not a major influence on welfare state, or, alternatively, at the time points at which the dogs were assessed, the dogs had already become habituated to the environment.

## 12.2.2.5 Defra Report (a)—University of Bristol, University of Lincoln, FERA. (2013). Studies to Assess the Effect of Pet Training Aids, Specifically Remote Static Pulse Systems, on the Welfare of Domestic Dogs

The use of 'e-collars' (remote static pulse collar systems) as training aides for dogs is controversial. Although banned in some European countries, e-collars are considered to be beneficial tools for training by many dog trainers and owners. The commercially available devices investigated in this University of

Bristol et al. (2013) study worked by delivering an electrical stimulus to the dog via their collar, controlled by a remote hand-held device. There is currently little scientific evidence about the impact of e-collars on dog welfare and so, for this reason, researchers were commissioned by Defra (Department for Environment, Food and Rural Affairs, UK) to assess the welfare of dogs trained with e-collars.

In a large-scale field study, owned dogs with prior experience of e-collar training were compared in a cognitive bias test to matched control dogs (no e-collar experience), alongside several other commonly used measures of dog welfare (e.g., salivary cortisol). The authors used the modified spatial judgement bias task (Burman et al., 2008b; Mendl et al., 2010a). However, because dogs were tested in their owners' homes, there was not always sufficient space to have the same-sized test area for all dogs, and so, in order to allow testing in the familiar home environment, the majority of dogs were tested using two different sizes of test area and so speed, rather than latency, was recorded as the dependent variable.

Although the authors found no significant differences in speed to approach the ambiguous probe locations between the 'e-collar' and control group dogs—suggesting that there was no apparent effect of e-collar training experience on dog affective state as assessed by a cognitive bias task—they did note a potentially confounding effect of using different-sized test areas, with dogs tested in the smaller-sized area showing faster running speeds to all probe locations—an important consideration for the design of future studies where consistency of test environment is hard to ensure. The authors did not speculate on what might have caused this effect, but it is likely to be related to overall distance travelled, with the cumulative extra distance covered by those dogs using the larger-sized area over the course of training, potentially resulting in increased levels of fatigue and therefore slower running speeds when it came to testing— particularly given that dogs were tested on the same day as training.

### 12.2.2.6  Defra Report (b)—University of Lincoln. (2013). Studies to Assess the Effect of Pet Training Aids, Specifically Remote Static Pulse Systems, on the Welfare of Domestic Dogs; Field Study of Dogs in Training

In this University of Lincoln (2013) study, subjects were owned dogs referred for problems (e.g., poor recall to owner) often addressed by the use of e-collars, and allocated to three treatment groups: either (1) dogs with professional trainers experienced with using e-collars where the use of e-collars was incorporated into training; (2) dogs with professional trainers experienced with using e-collars where the use of e-collars was *not* incorporated into training; or (3) dogs with professional trainers who did not typically use e-collars where the use of e-collars was *not* incorporated into training. Dogs were matched for age, breed, and sex where possible. The authors used the modified spatial judgement bias task (Burman et al., 2008b; Mendl et al., 2010a), and individual dogs were cognitive bias tested on three separate occasions. Cognitive bias tests were carried out (1) prior to training; (2) immediately after training; and (3) after a further 3 months, in order to determine 'baseline' responses and any subsequent differences in affective state in response to the different training regimes. There were no differences in running speed to any of the three ambiguous probes between dogs experiencing the different treatment regimes when tested at any of the three time points, suggesting that there was no apparent effect of incorporating the use of e-collars into training on dog affective state, at least as assessed by a cognitive bias task.

### 12.2.3 Conclusions

For the studies with dogs as subjects that have been described here, only two of the six studies found significant effects of their chosen treatments, and one of these was in the opposite direction to what was predicted. Yet, if we consider all cognitive bias studies on non-human animals published thus far, then the results initially appear far more encouraging—although we should take into account a potential publication bias in favour of 'positive' results. This raises the question as to whether or not this apparent discrepancy in results is a consequence of factors related to dogs per se, or just a consequence of those particular studies involving dogs that have been carried out so far. Further research is therefore still clearly needed to determine the efficacy of the cognitive bias paradigm as an indicator of welfare in dogs. Fortunately, studies with dogs provide an excellent opportunity for the necessary further validation, standardisation, and refinement of the paradigm, and some of the key points of consideration that appear to need addressing will now be discussed, focusing on those likely to be most relevant to dogs.

### 12.3 POINTS OF CONSIDERATION

### 12.3.1 Repeat Testing

One of the potential limitations of the cognitive judgement bias paradigm is the issue of repeatedly presenting ambiguous 'probe' stimuli to the subjects—whether these be visual, olfactory, tactile, or auditory-based. This is the critical moment in cognitive bias testing when, on the basis of how it responds to the presentation of an ambiguous 'probe' stimulus with which it has not previously experienced a reward outcome, an animal is 'asked' what outcome it anticipates. In the dog studies reviewed here, responses to the probe stimuli showed a generalisation effect, with probes nearer to the rewarded stimulus being approached faster than probes nearer to the unrewarded stimulus, and the middle probe being intermediate. This suggests that the probe stimuli were perceived as ambiguous rather than novel—with the training to the reference stimuli establishing a framework based on which subjects could then judge how to respond to the probe stimuli. However, it is not always clear for how long these stimuli are perceived as ambiguous to the subjects. For example, given that the probe stimuli are generally unrewarded, how many presentations does it take before the subjects learn this contingency? Doyle et al. (2010b) observed that sheep appeared able to learn that probe stimuli were unrewarded when the number of presentations was increased, as reflected in fewer approaches to the ambiguous positions over time (associated with no expectation of reward).

This issue has typically been dealt with by restricting the total number of probe trials presented during testing, and by presenting probe trials interspersed with standard 'reference' trials in order both to reduce the number of probe trials as a proportion of total test trials and to increase the interval between probe presentations. All of the six dog studies reviewed here have used this approach to some extent. For three of the studies (Mendl et al., 2010a; Titulaer et al., 2013; University of Lincoln et al., 2013), probe trials made up just 9 out of 41 test trials, with only 3 probe trials for each ambiguous probe. It could be argued that, given that it takes the subjects considerably more than 9 trials to learn the initial discrimination (e.g., 21–61 trials; Mendl et al., 2010a), then it is unlikely that the outcome of the probe trials would be learned so fast. However, exposure to the probe trials comes after criterion has already been achieved, and so the speed of learning about new, closely related stimuli is likely to be faster at this point.

Alternative ways to avoid this issue include using variable reinforcement when training the initial discrimination (Parker, 2008), or presenting each probe stimulus just once, although both these alternatives come with their own limitations, such as the need for an extended training period. A further possibility is to use just one probe stimulus rather than the three stimuli (NP, M, NN) used by all of the dog studies reviewed here, positioned at intermediate distances (whether spatial or visual) between the two learned reference stimuli. But which probe should be selected? The middle probe positioned midway between the two reference stimuli might be considered the most ambiguous, but even the calculation of a perceptually 'middle' probe can be complicated because stimuli may operate on a logarithmic scale rather than a standard linear scale. For example, because auditory stimuli operate on a logarithmic scale, Enkel et al. (2009) selected three tones of 3 kHz, 5 kHz, and 7 kHz that were roughly equidistant between their reference stimuli of 2 kHz and 9 kHz rather than 3.75 kHz, 5.5 kHz, and 7.25 kHz—those frequencies that appear intermediate according to a linear scale. And, whilst treatment differences in dog studies have been observed for the middle probe only (e.g., Mendl et al., 2010a; Burman et al., 2011), studies using other species have found differences at other probes (e.g., NP: Bateson & Matheson, 2007; NN: Burman et al., 2008b). So, including just a single probe stimulus may reduce the chances of identifying a potential treatment effect. It is also a matter of debate as to whether different affective states might result in differences at specific probes, i.e., anxiety at the NN probe and depression at the NP probe (Burman et al., 2008b; Mendl et al., 2009; Salmeto et al., 2011), or whether such findings are coincidental.

If the use of the cognitive judgement bias paradigm is not to be limited, then the issue of repeat testing becomes most problematic for those studies that require the use of a within-subjects/repeated measures experimental design. Between-subjects designs may not always be possible for the research question under consideration, particularly when dogs are being used as subjects. For example, there may not always be sufficient dogs available for a between-subjects design due to limitations on recruitment, it could be that a study is required to assess treatment efficacy (i.e., testing dogs before and after the implementation of a treatment), or it could be a longitudinal study during which the same dogs are tested multiple times. Three of the six dog studies reviewed here used a within-subjects design, and this inevitably increased the total number of probe stimuli to which the individual subjects were exposed (see Table 12-1). But, by testing over a greater period of time (i.e., not just on the same day as training) and maintaining a low proportion of probe presentations during testing, then the level of probe exposure may be acceptable. Müller et al. (2012) compared approach latencies between their 2 test days (separated by an interval of 4–9 days) and found that latencies were significantly longer on the second day of testing—suggesting that the dogs had learned about the ambiguous probes being unrewarded by the time of the second testing.

However, this area needs to be directly investigated in more detail, particularly with dogs as subjects, in order to determine the constraints of repeat testing. For instance, if we were to attempt to assess the welfare of dogs with chronic pain conditions, could a cognitive judgement bias test be used more than once for each individual and, if so, how long should the interval between tests be and how many probe presentations within each test? Could we train dogs to respond to more than one type of stimulus (e.g., auditory and visual cues) such that the presentation of ambiguous probes could be extended by having more than one probe type? At present, such questions have not been answered satisfactorily and so a degree of complexity is added to the interpretation of studies that include repeat testing (both between- and within-subjects experimental designs).

## 12.3.2 Individual Variation

The most obvious similarity between the six studies reviewed in this chapter is that all used dogs as subjects, but this still leaves ample opportunity for influence of individual variation in the form of the breed of dogs and their previous experience/background history. For instance, one study used laboratory-housed beagles (Burman et al., 2011), two used dogs (mixed breed) housed in rescue shelter kennels (Mendl et al., 2010a; Titulaer et al., 2013), and three used owned 'pet' dogs (mixed breed) (Müller et al., 2012; University of Lincoln et al., 2013; University of Lincoln, 2013). Whilst it is encouraging that the cognitive bias paradigm appears to be trainable/testable for dogs of a variety of breeds with diverse backgrounds and housed in contrasting environments, it is still unclear what effect specific differences might have on their performance. These are

**TABLE 12-1** Overview of Recent Studies of Cognitive Judgement Bias as a Measure of Affective State and Welfare in Dogs

| Dog Breed and Study | Housing Environment | Stimulus Type | Response | Positive Reinforcer | Negative Reinforcer | Treatment | Design | Repeated Testing? | Trained and Tested on the Same Day? | Total Number of Probe Test Trials (Probe Frequency) | Additional Welfare Measures | Predicted Bias? |
|---|---|---|---|---|---|---|---|---|---|---|---|---|
| Beagle[1] | Laboratory | Visual (grey scale) | Go/no-go (locomotion) | Food | No food | Search & forage task | Within-subjects | Yes | No | 18 (0.2) | No | Opposite: at middle probe only |
| Mixed[2] | Rescue shelter | Spatial location | Go/no-go (locomotion) | Food | No food | Separation-related behaviour (SRB) | Correlational | No | Yes | 9 (0.22) | No | Yes: at middle probe only |
| Mixed[3] | Home (pet) | Spatial location | Go/no-go (locomotion) | Food | No food | Owner presence/absence | Within-subjects | Yes | No | 24 (0.23) | No | No |
| Mixed[4] | Home (pet) | Spatial location | Go/no-go (locomotion) | Food | No food | E-collar vs. no e-collar experience | Between-subjects | No | Yes | 9 (0.22) | Yes | No |

| | | | | | | | | | | | |
|---|---|---|---|---|---|---|---|---|---|---|---|
| Mixed[5] | Home (pet) | Spatial location | Go/no-go (locomotion) | Food | No food | E-collar during training vs. no e-collar during training | Between-subjects | Yes | Yes | 27 (0.22) | Yes | No |
| Mixed[6] | Rescue shelter | Spatial location | Go/no-go (locomotion) | Food | No food | Length of stay in rescue shelter | Between-subjects | No | Yes | 9 (0.22) | Yes | No |

[1]Burman et al. (2011).
[2]Mendl et al. (2010).
[3]Müller et al. (2012).
[4]University of Lincoln et al. (2013).
[5]University of Lincoln (2013).
[6]Titulaer et al. (2013).

areas worthy of future investigation given that there are likely to be both genetic and environmental influences on the way that animals respond in the cognitive bias paradigm. For example, congenitally helpless rats, a strain of rat used as a genetic animal model of depression, showed an increased negative response to ambiguity compared to a strain of non-helpless rats, and rats treated with drugs to simulate a stress response showed a reduction in positive responding compared to sham-treated rats (Enkel et al., 2009).

Where possible, between-subjects studies (e.g., University of Lincoln et al., 2013; University of Lincoln, 2013) have matched dogs for age/breed/sex and neuter status when allocating individuals to treatment groups to take such factors into consideration, and within-subjects studies (e.g., Müller et al., 2012) have not identified significant breed or sex differences on performance. However, more targeted research specifically investigating, for example, potential sex, breed, and age effects would be informative for our understanding of how individual dog performance in cognitive bias tests might be influenced when considering both affect-related and non-affect-related (e.g., motivation) factors. Future research is also needed to identify the influence of dog background/experience on results by, for instance, comparing the same treatment in both shelter and owned dogs. For example, Mendl et al. (2010a) used dogs from rescue shelters to investigate separation-related behaviour, and so it would be interesting to know if similar results would be obtained if owned dogs tested in their home environment were used as subjects.

Although Müller et al. (2012) observed no significant sex and breed differences in their study, they did find a strong effect of individual differences revealing that individuals, even when other factors are taken into consideration, differed markedly in their response to ambiguous probes during cognitive bias testing. It is notable that, despite this variation, significant treatment effects have been observed (Mendl et al., 2010a; Burman et al., 2011), but it is a potential limiting factor. Variation between individuals is likely to influence performance in a number of ways. For example, in many studies, some dogs do not achieve the predetermined criterion at all (resulting in a potential inclusion bias) (e.g., Müller et al., 2012); there is a considerable range when considering how many trials individuals take to reach criterion (e.g., 30–200 trials: Burman et al., 2011), as well as variation in response to the ambiguous probes. Whether or not this individual variation is attributable to differences in genetic background and/or environmental experience or to stable behavioural characteristics (i.e., temperament/personality) remains open to question and further investigation.

### 12.3.3 Treatments

A range of different affect-related treatments have been studied in dogs thus far (see Table 12-1), with more likely to be investigated in the future. The use of a range of different treatments is helpful, not only because each study investigates a new topic of interest to the welfare assessment of dogs, but because we can assess the effectiveness of the cognitive bias approach to identify changes in affective state as a result of different treatments that may vary in a number of ways, including long- versus short-term; positive versus negative; and severe versus mild treatments.

There are many topics worthy of further investigation within the context of dog welfare. Some of these topics are associated with particular breeds/types of dogs. For example, brachycephalic obstructive airway syndrome (BOAS) is a disorder that is especially common in dogs with shortened muzzles in which dogs experience symptoms of respiratory stress, such as snoring, shortness of breath, and wheezing (Packer et al., 2012), and syringomyelia (see earlier) is a condition particularly associated with Cavalier King Charles spaniels. The risk factors for other dog welfare issues appear to be multifactorial. For instance, obesity is increasing in the pet population and is a consequence of dogs consuming more energy than they expend, and can be influenced by a number of factors including genetics (i.e., breed), neutering, and disease (German, 2006).

Additional areas of welfare concern in dogs that are worthy of further investigation include tail chasing, a behaviour that can be observed in a range of dog breeds and is often viewed as a 'normal' behaviour by dog owners who may not appreciate that it can also be a sign of distress for dogs (Burn, 2011); and dog–dog aggression. For all these topics/treatments, cognitive bias may be a useful technique for determining their impact on the affective state of dogs, helping us to identify appropriate and effective interventions. Indeed, the cognitive bias paradigm may also be suitable in some instances to investigate the efficacy of interventions by testing before, during, and/or after receiving pharmacological and/or behavioural guidance.

Thus far, cognitive bias studies in dogs have tended to focus on negative affective treatments, e.g., the experience of electronic collars (e.g., University of Lincoln et al., 2013; University of Lincoln, 2013), with only one study (Burman et al., 2011) investigating the effect of a putative positive treatment (i.e., the experience of a 'search & forage' task). It is worth remembering that we should investigate the impact of 'positive' treatments, such as providing dogs with additional environmental enrichment (Wells, 2004). This therefore remains an area for future research, although there is an understandable priority towards identifying negative affective states so that these can be minimised.

An important consideration when deciding what treatment to investigate is whether or not it would be of value to have additional tests/measures running in conjunction with cognitive bias testing to help with the interpretation of null results. In the absence of any other measure of welfare, differentiating between an affective treatment that has had no effect, versus a cognitive bias test that has failed to identify an effect when it has actually occurred, is difficult. For this reason, the inclusion of suitable additional measures of welfare is likely to be of benefit. For example, in their study of short- versus long-term kennelling of dogs, Titulaer et al. (2013) included additional measures of welfare (e.g., general behaviour, urinary cortisol:creatinine ratio) alongside a cognitive judgement bias test and, being able to consider all the welfare measures together, gave extra weight to their conclusion that the length of time that a dog is kennelled in a rescue shelter may not have a major influence on welfare state—at least at the time points assessed in their study.

## 12.3.4 Human Influence

One of the benefits of working with dogs is that their familiarity and socialisation with humans means that handling during experiments, for instance, does not require the levels of habituation required for other species (e.g., rodents). However, this close relationship with humans means that there is also the potential for human presence to exert a confounding influence on dog behaviour because of dogs' subtle understanding of human cues (see Chapter 11 by Topál). This possible confound is therefore an important consideration to be taken into account when carrying out a cognitive bias test with dogs. While this issue has yet to be studied directly, there are some suggestions from the current literature that human behaviour may well be influencing performance.

When one is considering tests of cognitive judgement bias, then human influence could come either from the handler who releases the dog at the start of each trial or from the researcher who 'baits' the food bowls/goal boxes. In all of the reviewed dog studies of cognitive bias, the handler was blind to the stimulus presented (i.e., reference or probe), either by having his/her eyes closed (e.g., Burman et al., 2011), by being positioned behind a visual barrier (Mendl et al., 2010a; Titulaer et al., 2013; University of Lincoln et al., 2013; University of Lincoln, 2013), or by being blindfolded (Müller et al., 2012). Being positioned behind a visual barrier (Titulaer et al., 2013; University of Lincoln et al., 2013; University of Lincoln, 2013) meant that dogs were also unable to see the food bowls being 'baited' or positioned. In contrast, for both Burman et al. (2011) and Müller et al. (2012), the dogs were able to see the positioning of the food bowl/goal box. Interestingly, Müller et al. (2012) found that the direction from which the researcher came when placing the middle probe bowl appeared to influence dog perception of outcome: if the researcher approached from the side of the 'rewarded' location, then the dogs' latency to approach was shorter than if the researcher approached from the side of the 'unrewarded' location. It therefore appears as though the researcher influenced the generalisation process in some way—as if the researcher was himself/herself associated with the value of the reference stimulus to which he/she was closest when placing the bowl. This highlights the potential for human influence on behavioural performance and decision making.

Further human influence relates to the previous social (non-conspecific) experience of the dogs being tested. Dogs housed in rescue shelters may react quite differently to human contact during cognitive bias testing due to their relative lack of contact with people during kennel housing. This could mean that without frequent contact they are more wary of humans than owned dogs and so require additional habituation/familiarisation with the research team or, if housed in the shelter due to human–dog aggression, then they may have to be excluded from the studies resulting in a selection bias. In contrast, dogs experiencing infrequent human interaction may become so excited by contact with researchers during testing that their emotional reaction to the human presence might outweigh experimental treatments (e.g., Burman et al., 2011).

### 12.3.5 Test Design

Thus far, all published dog cognitive bias studies have used go/no-go tasks and latency/speed measures. There are a number of limitations with the use of this approach that have been the focus of discussion (see Mendl et al., 2009). For instance, interpreting when an animal does not approach a stimulus (no go) is difficult because it may have decided not to approach but, alternatively, it may not approach for some other reason (e.g., distraction, fatigue, satiation). Latency measures are also problematic because, unless you want the testing to last indefinitely, a predetermined 'cut-off' is needed so that if a subject has not responded to the stimulus within a set period time, then the researcher knows to end the trial. But, if 'cut-off' values, e.g., 30 seconds (Mendl et al., 2010a), are allocated to individuals that fail to respond to the stimulus, this results in a number of problems. First, although they have been allocated a value of 30 seconds, this does not reflect their actual response, a value that is, of course, unknown and could feasibly have ranged from 30 seconds upwards; i.e., had the trial been longer, then they could have approached after 31 seconds, but they could also not have approached for many hours, if ever. Second, the presence of lots of identical 'cut-off' values in the data influences the choice of statistical analysis.

Go/go 'active choice' tasks (e.g., Matheson et al., 2008) in which the animal has to make an active response for all stimuli avoids the issues mentioned previously but can be difficult to train. For instance, training an animal to respond actively to gain no reward is likely to be difficult if not impossible, so, when using go/go tasks, reward contingencies are typically high-value reward versus lower-value reward (e.g., reward size: Parker, 2008; delayed reward: Matheson et al., 2008) or reward versus punisher (with the individual actively responding to avoid experiencing the punisher (e.g., electric foot-shock: Enkel et al., 2009). Given the trainability of dogs, then this might be an avenue worthy of further investigation; however, the use of a punisher may not be considered ethically appropriate for studies with owned or rescue dogs, or for studies investigating animal welfare.

The choice of reward contingency is another issue that has been much deliberated (see Mendl et al., 2009). Studies of cognitive judgement bias have used rewarded versus unrewarded (Burman et al., 2008b); high reward versus lower reward (Parker, 2008; Matheson et al., 2008); and reward versus punisher (Doyle et al., 2010a), and it may be that the choice of contingency influences the sensitivity of the cognitive bias test to different affective states (e.g., Mendl et al., 2009), although this has yet to be studied in a systematic way. In the dog studies reviewed here, all used reward versus no reward. This raises the additional question of how a lack of reward is perceived by the dogs. Again, this is something that requires further study, but, based on a dimensional framework of affect (Mendl et al., 2010b), the absence/removal of an expected reward is likely to be negatively valenced but low arousal in comparison to the presence and/or expectation of a punisher (e.g., electric shock: Enkel et al., 2009; presence of a dog: Doyle et al., 2010a) that would be negatively valenced but high arousal.

All except one of the dog studies reviewed here used the spatial task first developed in rats (Burman et al., 2008b) and subsequently modified for dogs (Mendl et al., 2010a). This consistency in task/discrimination should allow us to make comparisons between the studies—although almost inevitably there are other differences between studies, which means such comparisons must still be viewed cautiously. The number of trials required to reach criterion, where given, showed similar ranges (30–90 trials: Müller et al., 2012; and 21–61 trials: Mendl et al., 2010a), although it should be noted that the criterion was slightly different for Müller et al. (2012). It is interesting to note that Burman et al. (2011), who used a grey-scale visual discrimination task and not a spatial task, did observe an apparent decrement in performance in terms of the number of trials to reach criterion (30–200 trials: Burman et al., 2011), suggesting that that particular visual task could have been more difficult for the dogs to learn. However, this could also be related to other factors that differed between the studies, such as the use of one breed of dogs (beagles) housed under laboratory conditions. It would therefore be interesting to assess the efficacy of other types of tasks that do not require the physical demands of the spatial task—a potential limiting factor if researchers are interested in assessing the welfare of dogs with physical constraints (e.g., obesity, pain)—especially if other factors could be kept as constant as possible. For example, Fernandes (2012) attempted to train dogs to discriminate between human facial expressions as an alternative approach to assessing judgement of ambiguity—an approach that avoids the need for the dog to engage in strenuous activity. The author used images of 'angry' and 'happy' human faces as reference stimuli, before presenting faces of intermediate emotional expression (i.e., morphed faces containing elements of both 'happy' and 'angry' expressions) as the ambiguous probe stimuli.

All the dog studies reviewed thus far have relied on using food as a reward, and this has a number of potential implications for task performance in addition to the complex relationship between food motivation and affect discussed earlier. Effects on task performance include which, and how many, dogs reach criterion—because not all dogs will be equally food motivated (individual variation)—and how long training/testing can last and the size of the food rewards, given that training/testing completion should occur well before satiation is reached, especially if latency/speed is measured. Techniques such as training/testing using valued food treats as rewards, training/testing several hours after the subject has last been fed, and/or restricting the number of trials in any one session are typically used to address these concerns. Alternatively, studies using non-food (e.g., social) rewards could be investigated in the future, although one might expect similar potential confounds to occur (e.g., individual variation in social motivation, social 'fatigue/satiation', affective influence on levels of social interaction).

## 12.4 CONCLUSIONS

The cognitive bias paradigm has been the focus of interest from increasing numbers of researchers in a variety of fields, extending beyond its initial grounding in the assessment of animal welfare, and has been successfully adapted for a range of species, including dogs. There are appreciable benefits associated with this approach: the potential to determine affective valence in addition to arousal; the ability to generate closely defined predictions; as well as the undeniable appeal of an approach that appears to offer something new and different to other welfare measures. Directions for future research using dogs are likely to include the refinement of the paradigm and its development to allow the investigation of a range of issues relating to dog welfare (see 'Future Directions') as well as the development of different types of cognitive bias, e.g., memory and attention bias, for assessing affective states in dogs.

With a new measure such as this, interest is likely to continue for the foreseeable future. However, before immediately progressing with the cognitive bias paradigm to assess aspects of dog welfare, researchers should always consider all available welfare measure/indicators, in order to determine which are likely to be the most appropriate for the particular situation/treatment that is being investigated. These measures may include other cognitive approaches such as incentive contrast—a paradigm that initially appears more robust to repeat testing than cognitive bias—or the use of behavioural and physiological indicators of welfare. For example, in rescue shelters, not all dogs are easily handled, and their behaviour can be strongly influenced by the presence of humans. In this specific instance, welfare measures that can be carried out remotely (e.g., behavioural observation via video camera), and do not require the handling and training that are, thus far, requirements of the cognitive bias paradigm, are vital.

On those occasions when it has been decided that cognitive bias is the most suitable approach, it is also helpful, where possible, to include additional measures of welfare alongside cognitive bias. The reason is that, if, for example, there is found to be no effect of the treatment on cognitive bias, the inclusion of the other measures may clarify whether or not the treatment had no effect or if the cognitive bias test was not able to identify it. In conclusion, provided that there is an awareness of some of the limitations surrounding the paradigm, and that caution is taken with both experimental design and the interpretation of results, then the cognitive bias paradigm represents an exciting opportunity for researchers to contribute to an improvement in our ability to assess the welfare of dogs.

**Future Directions**

- Refinement of the cognitive bias paradigm: Improving the efficacy of the cognitive bias paradigm and increasing its versatility in addressing issues related to dog welfare. Developing attention and memory bias paradigms for dogs.
- Individual variation: Investigating the influence of genetic and environmental factors on differing cognitive bias performance between individuals.
- Human/animal interactions: Use of the cognitive bias paradigm to investigate the human/dog bond, e.g., social support.
- Genetic-based problems in dog welfare: Using the cognitive bias paradigm to investigate problems, e.g., chronic pain, arising from the selection of breeds/types.
- Behaviour-based problems in dog welfare: Use of the cognitive bias paradigm to investigate behavioural problems in dogs, e.g., separation-related behaviour, and their effective treatment.

# REFERENCES

Allen, K.M., Blascovich, J., Tomaka, J., Kelsey, R.M., 1991. Presence of human friends and pet dogs as moderators of autonomic responses to stress in women. J. Pers. Soc. Psychol. 61, 582.

Bateson, M., Desire, S., Gartside, S.E., Wright, G.A., 2011. Agitated honeybees exhibit pessimistic cognitive biases. Curr. Biol. 21, 1070–1073.

Bateson, M., Matheson, S., 2007. Performance on a categorisation task suggests that removal of environmental enrichment induces pessimism' in captive European starlings (*Sturnus vulgaris*). Anim. Welfare-Potters Bar Then Wheathampstead 16, 33.

Bateson, P., 2010. Independent inquiry into dog breeding. Dogs Trust, London.

Bentosela, M., Jakovcevic, A., Elgier, A.M., Mustaca, A.E., Papini, M.R., 2009. Incentive contrast in domestic dogs (*Canis familiaris*). J. Comp. Psychol. 123, 125–130.

Bethell, E.J., Holmes, A., MacLarnon, A., Semple, S., 2012. Evidence that emotion mediates social attention in rhesus macaques. PloS One 7, e44387.

Bishop, S.J., 2007. Neurocognitive mechanisms of anxiety: an integrative account. Trends Cogn. Sci. (Regul. Ed.) 11, 307–316.

Blackwell, E., Casey, R., Bradshaw, J., 2003. The assessment of shelter dogs to predict separation-related behaviour and the validation of advice to reduce its incidence post-homing. Report to RSPCA. RSPCA, Horsham, UK.

Boissy, A., Manteuffel, G., Jensen, M.B., Moe, R.O., Spruijt, B., Keeling, L.J., Winckler, C., Forkman, B., Dimitrov, I., Langbein, J., 2007. Assessment of positive emotions in animals to improve their welfare. Physiol. Behav. 92, 375–397.

Boleij, H., Klooster, J.V., Lavrijsen, M., Kirchhoff, S., Arndt, S.S., Ohl, F., 2012. A test to identify judgement bias in mice. Behav. Brain Res. 233, 45–54.

Bradshaw, J., McPherson, J., Casey, R., Larter, I., 2002. Aetiology of separation-related behaviour in domestic dogs. Vet. Rec. 151, 43–46.

Briefer, E.F., McElligott, A.G., 2013. Rescued goats at a sanctuary display positive mood after former neglect. Appl. Anim. Behav. Sci. 146 (1–4), 45–55.

Brilot, B.O., Asher, L., Bateson, M., 2010. Stereotyping starlings are more 'pessimistic'. Anim. Cogn. 13, 721–731.

Broom, D.M., 1991. Animal welfare: concepts and measurement. J. Anim. Sci. 69, 4167–4175.

Brydges, N.M., Leach, M., Nicol, K., Wright, R., Bateson, M., 2011. Environmental enrichment induces optimistic cognitive bias in rats. Anim. Behav. 81, 169–175.

Burman, O., McGowan, R., Mendl, M., Norling, Y., Paul, E., Rehn, T., Keeling, L., 2011. Using judgement bias to measure positive affective state in dogs. Appl. Anim. Behav. Sci. 132, 160–168.

Burman, O.H., Parker, R., Paul, E.S., Mendl, M., 2008a. Sensitivity to reward loss as an indicator of animal emotion and welfare. Biol. Lett. 4, 330–333.

Burman, O.H., Parker, R., Paul, E.S., Mendl, M., 2008b. A spatial judgement task to determine background emotional state in laboratory rats, *Rattus norvegicus*. Anim. Behav. 76, 801–809.

Burman, O.H., Parker, R., Paul, E.S., Mendl, M.T., 2009. Anxiety-induced cognitive bias in non-human animals. Physiol. Behav. 98, 345–350.

Burn, C.C., 2011. A vicious cycle: a cross-sectional study of canine tail-chasing and human responses to it, using a free video-sharing website. PloS One 6, e26553.

Cyranoski, D., 2010. Genetics: pet project. Nature 466, 1036–1038.

Dawkins, M.S., 2000. Animal minds and animal emotions. Am. Zool. 40, 883–888.

Dawkins, M.S., 2006. A user's guide to animal welfare science. Trends Ecol. Evol. 21, 77–82.

Doyle, R.E., Fisher, A.D., Hinch, G.N., Boissy, A., Lee, C., 2010a. Release from restraint generates a positive judgement bias in sheep. Appl. Anim. Behav. Sci. 122, 28–34.

Doyle, R.E., Vidal, S., Hinch, G.N., Fisher, A.D., Boissy, A., Lee, C., 2010b. The effect of repeated testing on judgement biases in sheep. Behav. Processes 83, 349–352.

Enkel, T., Gholizadeh, D., von Bohlen und Halbach, O., Sanchis-Segura, C., Hurlemann, R., Spanagel, R., Gass, P., Vollmayr, B., 2009. Ambiguous-cue interpretation is biased under stress-and depression-like states in rats. Neuropsychopharmacology 35, 1008–1015.

Eysenck, M.W., Mogg, K., May, J., Richards, A., Mathews, A., 1991. Bias in interpretation of ambiguous sentences related to threat in anxiety. J. Abnorm. Psychol. 100, 144.

Fernandes, J.M.G.M., 2012. Developing visual discrimination tasks for dogs. Master's thesis. University of Lisbon, Portugal.

Flaherty, C.F., 1999. Incentive relativity. Cambridge University Press, Cambridge.

German, A.J., 2006. The growing problem of obesity in dogs and cats. J. Nutr. 136, 1940S–1946S.

Harding, E.J., Paul, E.S., Mendl, M., 2004. Animal behaviour: cognitive bias and affective state. Nature 427, 312.

Hiby, E.F., Rooney, N.J., Bradshaw, J.W., 2006. Behavioural and physiological responses of dogs entering re-homing kennels. Physiol. Behav. 89, 385–391.

Hubrecht, R.C., 1993. A comparison of social and environmental enrichment methods for laboratory housed dogs. Appl. Anim. Behav. Sci. 37, 345–361.

Kaminski, J., Call, J., Fischer, J., 2004. Word learning in a domestic dog: evidence for "fast mapping". Science 304, 1682–1683.

Matheson, S.M., Asher, L., Bateson, M., 2008. Larger, enriched cages are associated with 'optimistic' response biases in captive European starlings (*Sturnus vulgaris*). Appl. Anim. Behav. Sci. 109, 374–383.

Mendl, M., Brooks, J., Basse, C., Burman, O., Paul, E., Blackwell, E., Casey, R., 2010a. Dogs showing separation-related behaviour exhibit a 'pessimistic' cognitive bias. Curr. Biol. 20, R839–R840.

Mendl, M., Burman, O.H., Paul, E.S., 2010b. An integrative and functional framework for the study of animal emotion and mood. Proc. Royal Soc. B. Biol. Sci. 277, 2895–2904.

Mendl, M., Burman, O.H., Parker, R., Paul, E.S., 2009. Cognitive bias as an indicator of animal emotion and welfare: emerging evidence and underlying mechanisms. Appl. Anim. Behav. Sci. 118, 161–181.

Miklósi, A., Topál, J., Csanyi, V., 2007. Big thoughts in small brains? Dogs as a model for understanding human social cognition. Neuroreport 18, 467–471.

Mogg, K., Bradley, B.P., 1998. A cognitive-motivational analysis of anxiety. Behav. Res. Ther. 36, 809–848.

Müller, C.A., Riemer, S., Rosam, C.M., Schößwender, J., Range, F., Huber, L., 2012. Brief owner absence does not induce negative judgement bias in pet dogs. Anim. Cogn. 15, 1031–1035.

Murray, J., Browne, W., Roberts, M., Whitmarsh, A., Gruffydd-Jones, T., 2010. Number and ownership profiles of cats and dogs in the UK. Vet. Rec. 166, 163–168.

Notari, L., Mills, D., 2011. Possible behavioral effects of exogenous corticosteroids on dog behavior: a preliminary investigation. J. Vet. Behav. Clin. Appl. Res. 6, 321–327.

Packer, R., Hendricks, A., Burn, C., 2012. Do dog owners perceive the clinical signs related to conformational inherited disorders as 'normal' for the breed? A potential constraint to improving canine welfare. Anim. Welfare—UFAW J. 21, 81.

Palestrini, C., Previde, E.P., Spiezio, C., Verga, M., 2005. Heart rate and behavioural responses of dogs in the Ainsworth's strange situation: a pilot study. Appl. Anim. Behav. Sci. 94, 75–88.

Parker, R.M., 2008. Cognitive bias as an indicator of emotional state in animals. Ph.D. dissertation. University of Bristol, UK.

Paul, E.S., Harding, E.J., Mendl, M., 2005. Measuring emotional processes in animals: the utility of a cognitive approach. Neurosci. Biobehav. Rev. 29, 469–491.

Pretterer, G., Bubna-Littitz, H., Windischbauer, G., Gabler, C., Griebel, U., 2004. Brightness discrimination in the dog. J. Vis. 4 (3), 241–249.

Rooney, N., Gaines, S., Hiby, E., 2009. A practitioner's guide to working dog welfare. J Vet. Behav. Clin. Appl. Res. 4, 127–134.

Rooney, N.J., Gaines, S.A., Bradshaw, J.W., 2007. Behavioural and glucocorticoid responses of dogs (Canis familiaris) to kennelling: investigating mitigation of stress by prior habituation. Physiol. Behav. 92, 847–854.

Rooney, N.J., Sargan, D., 2009. Pedigree dog-breeding in the UK: a major welfare concern? An Independent Scientific Report Commissioned by the RSPCA.

Rusbridge, C., 2005. Neurological diseases of the Cavalier King Charles spaniel. J. Small Anim. Pract. 46, 265–272.

Salmeto, A.L., Hymel, K.A., Carpenter, E.C., Brilot, B.O., Bateson, M., Sufka, K.J., 2011. Cognitive bias in the chick anxiety–depression model. Brain Res. 1373, 124–130.

Skandakumar, S., Stodulski, G., Hau, J., 1995. Salivary IgA: a possible stress marker in dogs. Anim. Welfare 4, 339–350.

Titulaer, M., Blackwell, E.J., Mendl, M., Casey, R.A., 2013. Cross sectional study comparing behavioural, cognitive and physiological indicators of welfare between short and long term kennelled domestic dogs. Appl. Anim. Behav. Sci. 147 (1), 149–158.

University of Lincoln, 2013. Studies to assess the effect of pet training aids, specifically remote static pulse systems, on the welfare of domestic dogs; field study of dogs in training. Defra, Lincoln, UK.

University of Lincoln, University of Bristol, Food & Environment Research Agency, 2013. Studies to assess the effect of pet training aids, specifically remote static pulse systems, on the welfare of domestic dogs. Defra, Lincoln, UK.

Wells, D.L., 2004. A review of environmental enrichment for kennelled dogs, Canis familiaris. Appl. Anim. Behav. Sci. 85, 307–317.

Wichman, A., Keeling, L.J., Forkman, B., 2012. Cognitive bias and anticipatory behaviour of laying hens housed in basic and enriched pens. Appl. Anim. Behav. Sci. 140, 62–69.

Willner, P., 1997. Validity, reliability and utility of the chronic mild stress model of depression: a 10-year review and evaluation. Psychopharmacology (Berl) 134, 319–329.

Winkielman, P., Knutson, B., Paulus, M., Trujillo, J.L., 2007. Affective influence on judgments and decisions: moving towards core mechanisms. Rev. Gen. Psychol. 11, 179.

Yeates, J., Main, D., 2008. Assessment of positive welfare: a review. Vet. J. 175, 293–300.

# Chapter 13

# Wagging to the Right or to the Left: Lateralisation and What It Tells of the Dog's Social Brain

Marcello Siniscalchi and Angelo Quaranta

*Department of Veterinary Medicine, Section of Behavioural Sciences and Animal Bioethics, University of Bari 'Aldo Moro', Italy*

## 13.1 INTRODUCTION

In dogs, behavioural asymmetries, which directly reflect different activities of the two brain hemispheres, are evident at both structural and functional levels, including paw preferences (Wells, 2003; Quaranta et al., 2004) and asymmetric tail wagging (Quaranta et al., 2007).

For more than a century, the study of differential information processing by the brain hemispheres was confined to the human species. Now we know that lateralisation of brain functions is a widespread characteristic of both vertebrates (Bradshaw & Rogers, 1993; Rogers & Andrew, 2002; Rogers et al., 2013) and invertebrates (Letzkus et al., 2006; Rogers & Vallortigara, 2008; Frasnelli et al., 2012).

Interestingly, the basic pattern of lateralisation is similar across non-human animals and fundamentally the same as that in humans (MacNeilage et al., 2009).

In general, it could be claimed that the activity of the left hemisphere involves use of learned templates or rules, categorising stimuli by focusing on relevant cues and responding to features that are invariant and repeated (MacNeilage et al., 2009; Rogers et al., 2013): this is referred to as 'instruction-driven' processing (Toppino & Long, 2005).

In contrast, the activity of the right cerebral hemisphere has been associated with response to novelty (it specialises in picking up on details) and the expression of intense emotions, such as aggression, escape behaviour, and fear (summarised by Rogers & Andrew, 2002; Rogers et al., 2013); this is referred to as 'stimulus-driven' processing (Toppino & Long, 2005).

Moreover, right hemisphere activation strongly influences the fight-or-flight response throughout the dominant control of the pituitary-adrenal axis (Wittling, 1990; Wittling et al., 1995).

The Social Dog. http://dx.doi.org/10.1016/B978-0-12-407818-5.00013-9

Although investigations have mainly been carried out on conventional animal model systems (e.g., pigeons, mice, monkeys), in the past few years, evidence for functional lateralisation in domesticated and pet species has also started to appear.

Understanding the impact of cerebral lateralisation on dog behaviour is particularly interesting mainly for two reasons:

1. Several studies have shown an association between lateralisation and different functions relevant to successful social communication in both humans (e.g., human processing of language including comprehension of emotional prosody; see Mitchell & Crow, 2005) and dogs (e.g., dogs' ability to process conspecific vocalisations; see Siniscalchi et al., 2008; and to recognise the facial expressions of different emotions; see Racca et al., 2012).
2. Dogs play a number of significant roles within the human community.

## 13.2 BRAIN LATERALISATION AND DOGS' EMOTIONS: THE TAIL-WAGGING EXPERIMENT

The complementary specialisation of the two sides of a dog's brain is clearly apparent in tail-wagging behaviour. To test this idea, researchers measured dogs' asymmetric tail-wagging responses to different emotive stimuli at the University of Bari, Italy, in collaboration with the B.R.A.I.N. Centre for Neuroscience, University of Trieste, also in Italy.

To conduct the study, researchers gathered family pet dogs of mixed breed at the Department of Veterinary Medicine of Bari. The dogs were placed in a large rectangular wooden box with an opening at the front to allow them to view the different stimuli. An opaque plastic panel was used to cover the opening between stimuli presentations. The stimuli presented were as follows: the dog's owner; an unknown person; an unfamiliar dog with agonistic approach behaviour; and a cat. The dogs' behaviour was recorded using a video camera placed on the roof of the 'testing box', and asymmetrical tail wagging was scored by measuring angles associated with the maximum excursions of the tail from the midline to the left and to the right side of the dog's body (see Figure 13-1).

Our results clearly showed that both direction and amplitude of tail wagging were strictly related to the emotion elicited by different stimuli (see Figure 13-2): a tail-wagging bias towards the dog's right side (left hemisphere activation) was found when the dog saw stimuli eliciting approach responses such as the owner (presumably associated with positive emotions), whereas a bias towards the dog's left side (right hemisphere activation) was found when the dog saw stimuli from which it wished to withdraw, such as the unfamiliar dog displaying an agonistic approach (presumably associated with negative emotions). It is interesting to note that the amplitude of wagging behaviour is a crucial cue in dogs' perception of emotions because it is a determinant for estimating 'quantitatively' the type of emotional arousal elicited by stimuli: a significant bias of tail wagging to the dog's right side was also observed when dogs were

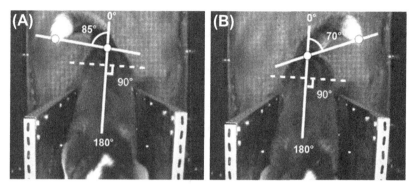

**FIGURE 13-1**   Methods to estimate angles to score amplitude of wagging behaviour of the dogs. Angle A: wagging to the right; angle B: wagging to the left. *(From Quaranta et al., 2007; with permission from Elsevier.)*

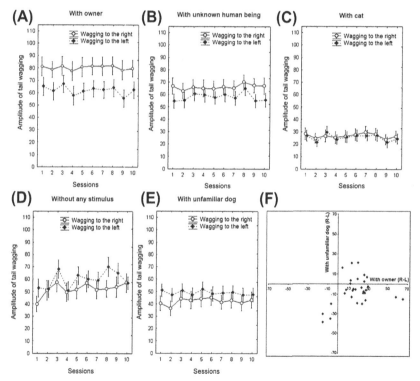

**FIGURE 13-2**   Amplitudes of tail wagging (degrees) to the left and right side when dogs were looking at different stimuli. A significant right bias in wagging behaviour was observed with the owner (A), the unfamiliar human being (B), and the cat (C); a significant left bias was observed in the absence of stimuli (D) and with the unfamiliar dominant dog (E). A sample of data from individual animals is given in (F), showing a scattergram of a right–left (difference) score for each dog, with the 'With owner' condition on one axis and the 'With unfamiliar dog' condition on the other axis. *(From Quaranta et al., 2007; with permission from Elsevier.)*

shown an unfamiliar human being, but with somewhat less enthusiasm (i.e., amplitude) than towards the owner, whereas the presentation of a cat again elicited a right side wag, although with less intensity than towards the unknown human being (in the latter case, the wagging behaviour would probably reflect the tendency of dogs to approach cats 'with malice aforethought').

Neuroanatomical evidence exists regarding the fact that asymmetries in tail wagging reflect different activation of the two hemispheres, since the rubrospinal tract, which is the predominantly volitional pathway from the brain to the spinal cord in dogs, decussates just caudal of its origin from the red nucleus and descends into the contralateral funiculus, terminating on interneurons at all levels of the spinal cord (see Buxton & Goodman, 1967).

Based on these results, it could be said that the complementary specialisation of dog brain hemispheres is consistent with Davidson's *laterality-valence* hypothesis, which suggests left hemisphere specialisation with regard to processing positive emotions and right hemisphere specialisation for processing negative emotions (Davidson, 1995). Similar results were reported in adult dogs when presented with pictures of expressive dog faces with a left gaze bias (right hemisphere activation) while looking at negative conspecific facial expressions and a right gaze bias (left hemisphere activation) towards positive ones (Racca et al., 2012). Although the *laterality-valence* hypothesis was based on neuropsychological studies in humans (Canli et al., 1998), approach-withdrawal complementary specialisation is a common feature of several animal species (Schneirla, 1959; Hanbury et al., 2013).

Evidence of an asymmetric control of a medial organ such as the tail has attracted considerable interest even among the media (see, e.g., Blakeslee, 2007; McNeilage, 2007), since research on behavioural asymmetries associated with specialisation of the left and right sides of the brain has usually focused on asymmetric use of paired organs, such as forelimbs; in addition, the example of measuring the side bias of tail wagging in domestic animals offers a simple solution to the problem of quantitatively estimating positive–negative emotions using a simple, non-invasive method with direct implications for both animal welfare and veterinary behavioural medicine.

Nevertheless, in the light of evidence from the tail-wagging experiment, it cannot be assumed that asymmetries of tail movements have direct implications for social communication between dogs: it simply reflects a different activation of the two hemispheres of the dog's brain in response to stimuli with different emotional valence.

What about the possibility that asymmetric tail wagging could have a role in social communication? To evaluate this possibility, Canadian researchers investigated the approach behaviour of free-ranging dogs to the asymmetric tail wagging of a life-size robotic dog replica (Artelle et al., 2011). Unexpectedly, their results showed that there was a preference to approach the model (i.e., the 'robodog' was approached continuously without stopping) when its tail was wagging to the left side. Although these results show that dogs actually perceive and respond to asymmetries in tail wagging, dogs would be expected to be more likely to approach a right-wagging dog replica (left hemisphere activation → expression of positive emotion). Authors reported that a possible explanation for the stop response during the approach to the model with a right-biased tail wag may originate when subjects are presented with a signal that would otherwise be positive (right wag) yet is not accompanied by additional reciprocal visual or acoustical responses by the robotic model. In addition, another variable that should be taken into account when analysing these results is that in animals, humans included, there is striking evidence that visual attention is greater for stimuli that are located in the left side of the visual field (Butler et al., 2005). Curiously, in dogs, this pattern appears when they look at neutral human faces but not when they look at neutral dog faces (Guo et al., 2009).

We feel that the more appropriate strategy for studying the communicative function of asymmetric tail wagging is to present subjects with tail movements by real dogs, because even in the presence of a good dog replica, we believe that robotic movements are not properly biological: to do so, we are currently investigating the possible lateralised effects of presenting moving video images of wagging dogs, digitalised in such a way as to allow us to produce controlled mirror images of dogs showing prevalent left- or right-asymmetric tail-wagging behaviour. We expect these stimuli to induce different approach-withdrawal tendencies and/or in general affect social responses in tested subjects. Moreover, to test accurately the different emotional responses of dogs to right-left wag presentations, we are also measuring the cardiac activity of tested subjects during the experiment (manuscript in preparation). Even though it is not easy for human eyes to detect asymmetric tail-wagging movements (in the tail-wagging experiment described previously, we analysed video footage using a frame-by-frame technique), this would probably not be the case for dogs because the flicker fusion rate (the rate at which successive flashes from a stationary light source become undetectable by the retina and the sensation of the light becomes 'steady') of their eyes is higher than that of humans (Miller & Murphy, 1995) and, as a consequence, canine species would perform better than us at detecting fast movements such as tail wagging.

## 13.3   RIGHT AND LEFT IN THE CANINE WORLD: COMPLEMENTARY SPECIALISATIONS OF THE TWO SIDES OF THE DOG'S BRAIN

In dogs, right hemisphere advantage in processing negative emotions associated with the sight of threatening stimuli is also manifested in their asymmetrical head-turning responses to 2D visual stimuli (Siniscalchi et al., 2010).

Asymmetries in the use of the eyes have been shown in animals as different as chickens, fishes, and cetaceans (Rogers & Andrew, 2002; MacNeilage et al., 2009; Siniscalchi et al., 2012b), and the overall evidence available suggests that these animals are able to activate the hemisphere more appropriate to particular conditions by using lateral fixation with the contralateral eye (Vallortigara et al., 1999). Over-all, this pattern of evidence fits nicely with what is reported on dog vision: although dogs show considerable variation between breeds in terms of the size of their visual fields, depending on the shape of their heads, the proportion of crossed fibres in the optic nerve is 75% (Fogle, 1992); thus, direct monocular input is processed primar-ily in the opposite side of the brain (i.e., neural structures located on the right side of a dog's brain are mainly fed by input from the left visual hemifield and vice versa). In the experimental situation, during feeding behaviour, silhouette drawings of different animal models were presented to dogs simultaneously to the left and right visual hemifields through two retro-illuminated panels (Figure 13-3).

When threatening stimuli such as the silhouette of a snake were presented, dogs preferentially turned their heads to the left, thus confirming the dominant control of the right hemisphere in processing fear (the silhouette of a snake seems to be an alarming stimulus for most mammal species; see Lobue & DeLoache, 2008; see Figure 13-4). Interestingly, the specialisation of the right hemisphere for expressing intense emotion also included aggression, since an evident left bias in their head-turning behaviour was revealed in response to the presentation of a cat silhouette displaying a defensive threat posture (arched lateral displayed body with the tail erected).

When stimuli were presented simultaneously in the two visual hemifields (bin-ocular presentations), dogs displayed shorter latencies to react and longer latencies to resume feeding when they turned their head to the left than when they turned their head to the right (i.e., when their right hemisphere was in charge; see Figure 13-5).

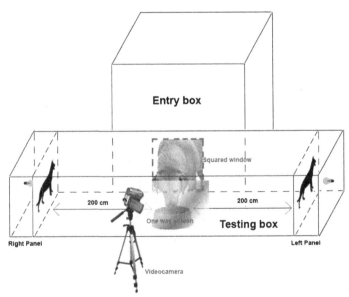

**FIGURE 13-3**   Schematic representation of the testing apparatus. *(From Siniscalchi et al., 2010; with permission from Elsevier.)*

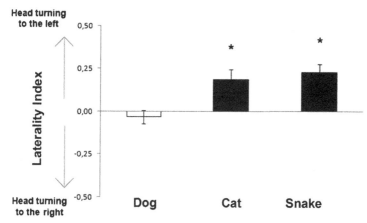

**FIGURE 13-4**   Laterality index (LI) of head-turning responses to 2D visual stimuli (group means with SEM are shown). Asterisks indicate significant biases (P < 0.05). *(From Siniscalchi et al., 2010; with permission from Elsevier.)*

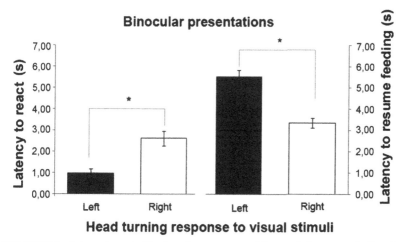

**FIGURE 13-5**   Latency to react of head-turning responses and time needed to resume feeding after stimulus presentation (group means with SEM are shown; * = P < 0.001). *(From Siniscalchi et al., 2010; with permission from Elsevier.)*

Moreover, when stimuli were presented to only one visual hemifield (monocular presentation), higher overall responsiveness to stimuli shown in the left hemifield (right hemisphere) was observed, irrespective of the type of stimulus. Taken together, these results are in line with evidence that right hemispheric sympathetic outflow seems to be the more effective (Wittling, 1995): in dogs, as well as in humans and cats, there is evidence for a greater effectiveness of the sympathetic outflow from the right hemisphere to organs fundamental to the control of the 'fight or flight' behavioural response, such as the heart (Wittling, 1995; Wittling et al., 1998).

These findings have obvious and useful implications not only for dog welfare, e.g., choosing the best strategy in terms of sidedness for approaching dogs in potentially stressful situations, such as clinical examinations, or during training (e.g., guide dogs for the blind, dogs trained for animal-assisted therapy and activities) but also for a thorough understanding of social communication between different individuals observing, for example, sidedness bias during approach behaviour. The latter is, of course, an issue that deserves to be investigated on its own, for instance, by looking at which side semi-wild dogs prefer to approach each other and relating it to the social hierarchy in the pack.

Right hemispheric specialisation in processing threatening and alarming stimuli has also been reported in both the dogs' auditory and olfactory 'sensory domains' (Siniscalchi et al., 2008; Siniscalchi et al., 2011b).

Regarding the auditory sensory channel, during feeding behaviour, the head-turning response to acoustic stimuli played at the same time from two speakers placed symmetrically to the right and left side of the dog's head was recorded (Figure 13-6).

A striking left-orienting bias in the head-turning response was observed after thunderstorm playbacks (Figure 13-7): in the binaural auditory test used in this experiment, it is assumed that, when the subject turns towards the speaker leading with its left ear, the acoustic input is processed primarily by the right hemisphere, at least for the initial attention to the stimulus, and vice versa if it presents its right ear (Hauser et al., 1998). The direction of the head turn, which is an unconditioned response, is therefore considered to be an indicator of a contralateral hemispheric advantage in attention to the auditory stimulus (Scheumann & Zimmermann, 2008). Similarly to what has been found on visual stimuli, the activity of the right hemisphere enhanced the arousal state of subjects because dogs displayed longer latency to resume feeding from the bowl during left-ear-leading orienting turns.

In light of this evidence, we decided to investigate for the presence of lateralisation for the most relevant sensory domain in dogs, namely olfaction.

Previous research suggests the presence of a lateralised process in the analysis of olfactive stimuli in both vertebrate and invertebrate species (Rogers &

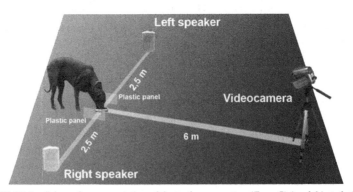

**FIGURE 13-6**   Schematic representation of the testing apparatus. *(From Siniscalchi et al., 2008.)*

Andrew, 2002; Royet & Plailly, 2004; McGreevy & Rogers, 2005; De Boyer Des Roches et al., 2008; Rogers & Vallortigara, 2008). In domestic chicks, for example, stronger head shaking was observed in response to a noxious odour under the right nostril (Burne & Rogers, 2002). In mammals, horses show a population bias to using the right nostril first in response to both stallion faeces (McGreevy & Rogers, 2005) and novel objects (De Boyer Des Roches et al., 2008).

To test for the presence of lateralisation in the dog olfactory system, we investigated whether dogs show nostril asymmetries in processing odorants that differ in terms of emotional valence during free sniffing behaviour under unrestrained conditions (Siniscalchi et al., 2011b).

Odorants used as test stimuli were presented on cotton swabs commonly used for canine vaginal cytology installed on a digital video camera located on a tripod in the centre of a large, isolated room (Figure 13-8). Since dog nostril use is very fast, sniffing behaviour was recorded continuously during inspection of odorants, and video recordings were subsequently analysed frame by frame.

During sniffing of clearly arousing odorants for dogs (adrenaline and veterinary sweat, the latter for its association with stressful activities required by routine clinical examinations), subjects showed a consistent right nostril bias throughout the experiment (Figure 13-9). Given that in mammals the olfactory system ascends mainly ipsilaterally, with most receptor information from each nostril projecting, via the olfactory bulb, to the primary olfactory cortex in the same hemisphere (Royet & Plailly, 2004), the use of the right nostril suggested involvement of the right hemisphere.

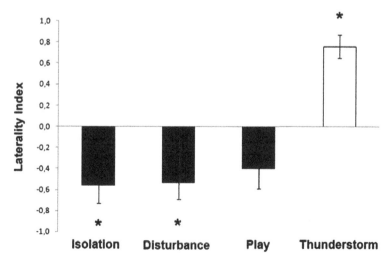

**FIGURE 13-7**   Laterality index (LI) for the head-orienting response of dogs to playbacks: a score of 1.0 represents exclusive head turning to the left side and −1.0 represents exclusive head turning to the right side; * = P < 0.01. *(From Siniscalchi et al., 2008.)*

In particular, right nostril bias in response to adrenaline stimulus is consistent with the idea that the sympathetic activation associated with enhanced tension or higher emotional responses is mainly under the control of the right hemisphere (Craig, 2005).

**FIGURE 13-8**  Schematic representation of the testing apparatus. *(From Siniscalchi et al., 2011b; with permission from Elsevier.)*

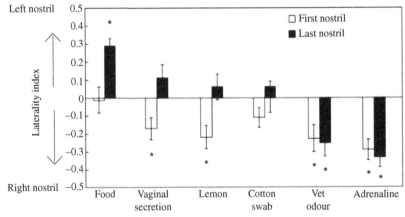

**FIGURE 13-9**  Laterality index (LI) of the first and last nostril used during inspection of different odorants (group means averaged across all trials with SEM are shown). Asterisks indicate significant biases (P < 0.05). *(From Siniscalchi et al., 2011b; with permission from Elsevier.)*

On the other hand, during sniffing of non-aversive stimuli such as food, lemon, vaginal secretions, and cotton swab, dogs showed just an initial use of the right nostril and then a shift towards use of the left nostril. This shift in nostril use (right → left) for non-aversive-arousal stimuli was evident for time spent sniffing, but less so for initial sniffing in which food did lead to any evident bias, possibly because of its reduced valence as a novel stimulus.

This pattern has been observed in a variety of animal species (e.g., birds: Vallortigara et al., 1999; fish: Sovrano, 2004), with the initial involvement of the right hemisphere in the analysis of novel stimuli followed by dominant control of behaviour by the left hemisphere when routine responses to stimuli emerge as a result of familiarisation/categorisation (reviewed in Vallortigara, 2000; Rogers & Andrew, 2002; MacNeilage et al., 2009; Vallortigara et al., 2011; Rogers et al., 2013).

For example, use of the left hemisphere in the routine task of finding food has also been reported during prey-catching responses in the toad (*Bufo marinus*) (Robins & Rogers, 2004). In this experiment, toads' predatory responses to an artificial prey stimulus (e.g., a model replica of an insect) were frequent when the prey target moved clockwise across the subject's visual midline into the right visual hemifield.

The specialisation of the left hemisphere in prey-catching behaviour has also been reported in fish and birds. In the zebra fish, for example, the right eye is used to fixate on a target that the fish intends to bite, but not when biting does not follow, even if the target is identical (Miklósi et al., 1998). In birds, black-winged stilts used their right monocular visual field during predatory pecking behaviour in a naturalistic setting (Ventolini et al., 2005).

We found something similar when studying detour behaviour in attack-trained dogs (Siniscalchi et al., 2013a) (see Figure 13-10).

The detour test is a technique largely employed to study visual lateralisation in several animal models (birds: Vallortigara et al., 1999; fish: Bisazza et al., 1998), since the decision to pass a barrier to the left or right reflected a preferential fixation by the contralateral eye (i.e., detour to the left → right-eye use; detour to the right → left-eye use). Over the years, researchers have demonstrated that eye use and the shift in eye use, which reflects different activation of hemispheres, depended on different target features, such as its degree of novelty and attractiveness (Vallortigara et al., 1999; Vallortigara, 2000). In our work, we found that left-turner dogs took less time to detour a barrier behind which a figurant (target) was located than right-turners (Figure 13-11).

Because in the experiment prey-catching behaviour is the crucial cue, driving subjects to reach the target, we hypothesised that the most logical explanation for shorter latencies to solve the task observed in left-turners (right visual hemifield) with respect to right-turners (left visual hemifield) was consistent with specialisation of the left hemisphere in the control of prey-catching behaviour.

Processing of familiar species-typical vocalisations is another specialisation of dogs' left hemisphere. Using the same head-turning paradigm described

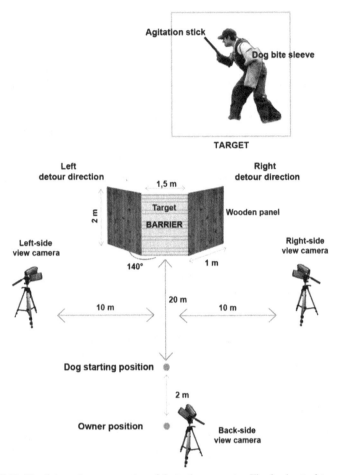

**FIGURE 13-10** Schematic representation of the testing apparatus. The dog has to detour the vertical barrier in order to reach the target (figurant). Times needed to perform the task and right/left detour directions were recorded. *(From Siniscalchi et al., 2013a; with permission from Taylor & Francis Ltd.)*

previously (refer to Figure 13-6), we found that dogs turned with the right ear leading (left hemisphere) in response to presentation of conspecific vocalisations that could be separated into three broad categories according to the work of Yin and McCowan (2004): (1) a disturbance situation in which a stranger knocked on the door of the owner's house, (2) an isolation situation in which the dog was in a room in the house isolated from its owner, and (3) a play situation in which either two dogs or a human and a dog played together.

Consistent with this, use of the left hemisphere to attend to the vocalisations of familiar conspecifics has also been observed in non-human primates (Poremba et al., 2004), horses (Basile et al., 2009), and sea lions (Böye et al., 2005).

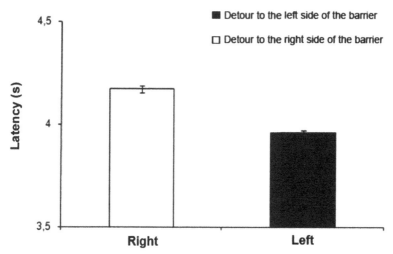

**FIGURE 13-11**    Latency to reach the target during left (black histogram) and right (white histogram) detours. Data presented are means calculated for right and left detours independently from dogs over trails. *(From Siniscalchi et al., 2013a; with permission from Taylor & Francis Ltd.)*

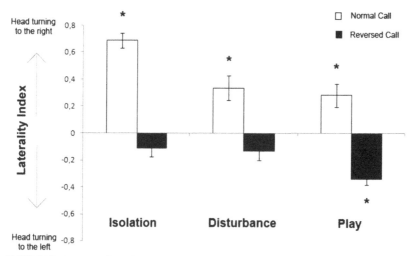

**FIGURE 13-12**    Laterality index (LI) for the head-orienting response of dogs to playbacks: a score of 1.0 represents exclusive head turning to the right side and −1.0 represents exclusive head turning to the left side. Normal calls, white histograms; reversed calls, black histograms. * = $P < 0.01$. *(From Siniscalchi et al., 2012a; with kind permission from Springer Science and Business Media.)*

Interestingly, there is evidence that this particular specialisation of the left hemisphere is dependent on temporal features in the calls, since reversing dog vocalisations caused a shift from a right-ear orienting bias (normal call versions) either to a left-ear orienting bias (play call) or to no asymmetry (disturbance and isolation calls) in their head-turning response (Siniscalchi et al., 2012a) (Figure 13-12).

Playbacks of time-reversed versions of the calls represented an experimental methodology largely employed to investigate the role of temporal cues in the structure of call signals in different animal models, since reversal signals had modified their temporal domain while preserving their spectral content, resulting in identical long-term frequency profiles as forward signals (Figure 13-13) (non-human primates: Ghazanfar et al., 2001; Le Prell and Moody, 2000; frog: Gerhardt, 1981; cat: Gehr et al., 2000).

## 13.4 PAW PREFERENCE AND ITS IMPLICATIONS FOR DOG COGNITION

Human handedness, a manifestation of cerebral dominance, is closely correlated with anatomical asymmetry and language lateralisation. Scientists proposed that both the direction and the different degree of handedness are associated with differences in behavioural reactivity (i.e., emotional state; Braccini & Caine, 2009) and cognitive ability (Rogers, 2010). This hypothesis is supported by the fact that in humans, individuals who develop psychotic disorders (e.g., schizophrenic symptoms) show lesser anatomical and functional asymmetries than the rest of the population (Klar, 1999; Angrilli et al., 2009). In addition, recent studies using a large population survey from 12 European countries have shown that left-handers are significantly more likely to have depressive symptoms than right-handers (Denny, 2009).

In dogs, the preferential use of one forepaw in a motor task (i.e., paw preference) has been measured in various activities such as removal of tape placed over the nose (Quaranta et al., 2004), removal of a blanket from over the head, and retrieval of food from a can (Wells, 2003) or from a Kong toy (Branson & Rogers, 2006). Although paw preference has several implications for dog welfare because clear evidence exists that dogs preferring the left and the right forepaw exhibit different patterns of immune response (Quaranta et al., 2006; Quaranta et al., 2008), relatively little research has been conducted on a potential association between reactivity (or cognitive bias) and paw preferences. It has been shown, however, that dogs with weaker paw preference when retrieving food from a Kong toy show greater reactivity to playbacks of thunderstorms and fireworks, thus supporting the hypothesis of greater reactivity in dogs with weaker hemispheric lateralisation (Branson & Rogers, 2006).

Consistent with this, recent work carried out in our laboratories showed that agility-trained dogs with weaker paw preference show greater distractibility and hence displayed greater latency when negotiating weave pole obstacles with their owner in their left visual field (Siniscalchi et al., 2013b). This demonstrates interference between analysis of the spatial information required to perform agility tasks and visual analysis of higher emotional valence stimuli (i.e., the owner), which are both under the control of the right hemisphere.

Weak lateralisation would mean that one hemisphere has a lessened ability to inhibit the other. Apart from having impaired ability to attend to two tasks

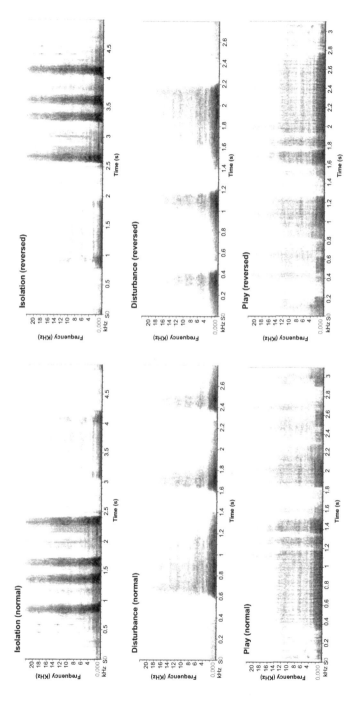

**FIGURE 13-13**    Acoustic spectrogram samples of the stimuli and their temporally reversed versions used in playback experiments. *(From Siniscalchi et al., 2012a; with kind permission from Springer Science and Business Media.)*

simultaneously (Rogers et al., 2004; Dharmaretnam & Rogers, 2005), such individuals would be relatively unable to inhibit responses of the right hemisphere after they have been elicited (as in the case of the detour experiment), or vice versa, depending on the prevailing stimuli and context (Rogers, 2010). This finding is in keeping with Batt et al. (2009), who found that dogs with greater strength of lateralisation in paw usage during the Kong Test exhibit more confident and relaxed behaviour when exposed to novel stimuli and unfamiliar environments.

In non-human primates, evidence for the relationship between handedness and emotional reactivity, which directly affects social cognition, comes from observations on chimpanzees and common marmosets. Generally, left-handedness in these species has been associated with 'reactive' personality traits, as demonstrated by the evidence that marmosets that preferentially use their left hand during a simple task (e.g., picking up food from the floor) are more fearful than right-handed marmosets after hearing the alarm calls of a predator (Braccini & Caine, 2009). Left-handed marmosets are also more likely to show elevated levels of cortisol (hormone mainly released in response to stressful events) than are right-handed marmosets (Rogers, 2009). On the other hand, right-handedness has been associated with 'proactive' personality traits because right-handed primates, compared to left-handed, are more likely to explore new objects and situations (chimpanzees: Hopkins and Bennett, 1994; common marmosets: Cameron & Rogers, 1999; Braccini & Caine, 2009). These results are also consistent with Rogers' (2010) hypothesis that the predominant use of the right or left hand indicates the individual's tendency to adopt a positive or negative cognitive bias, respectively. This would fit nicely with the Davidson valence hypothesis discussed earlier (see section 13.2), which states that negative emotions are associated with the right hemisphere and positive emotions are associated with the left hemisphere, suggesting that negative-positive cognitive biases would rely on processing, respectively, by the right and the left hemisphere. So far, this hypothesis has not been tested directly in dogs, but there is some indirect evidence to support it, as Tomkins et al. (2012) recently found that right-pawed dogs (possible proactive subjects) were more successful in completing a Guide Dog Training Programme. In dogs, a recent work demonstrated that subjects which exhibit high levels of separation-related behaviour (SRP) also appear to have a more negative 'cognitive bias' (i.e., 'reactive' subjects) (Mendl et al., 2010). Indeed, it would be interesting in the near future to study the direction and strength of pawedness in dogs with SRP and negative cognitive bias.

The differences between left- and right-preferent animals in the use of forelimbs may also be a direct consequence of structural asymmetries, since some differences in brain morphology between left- and right-handed non-human primates have been reported (Gorrie et al., 2008). Regarding a possible association between brain structural asymmetries and paw preference, although both *in vivo* (tomography CT brain scanning; Siniscalchi et al., 2011a) and postmortem techniques (Tan & Caliskan, 1987) revealed a right-biased hemispheric asymmetry with the right hemisphere greater than the left, this was not related to paw preference. It should be noted, however, that while in the first study, paw preference was not recorded, in the study by Tan and Caliskan, paw preference was measured using the preferential paw used to remove an adhesive plaster from the eyes: besides the obvious ethical considerations, in our opinion, this is not a suitable measurement of paw use because the necessary conditions for a hand-paw measure to be indicative of the dominant hemisphere in a given animal are that the task be simple (e.g., picking up food items from the floor) and that the subject be in a relaxed state (for details, see Rogers, 2009).

Consequently, the association between paw preference and brain structural lateralisation in dogs is a topic that still deserves further investigation.

## 13.5 CONCLUSIONS

Overall, data showed a different specialisation of the two sides of the dog brain, with the left-hemisphere dominance for routine behaviours and intraspecific communication, and a dominant role played by the right hemisphere in the analysis of arousal stimuli and novelty.

Consequently, in dogs, approach and withdrawal behaviours that are fundamental motivational dimensions are elicited, respectively, by the complementary activation of the left (approach) and right (withdrawal) brain hemispheres.

Accurate study of lateralised behaviour represents an interesting tool for estimating a dog's preference for a particular stimulus or individual: sidedness of the approach, left-right sensory bias (e.g., the constant use of one nostril during olfactory inspection of an object or a human being), and asymmetric tail wagging could reveal crucial cues on the emotional state of dogs even in the absence of clear behavioural signs.

Thus, the link between emotion, cognition, and lateralisation opens the door to new ways of measuring the affective states of dogs with direct implications for animal welfare and successful social communication.

### Future Directions

- Dog lateralisation (i.e., the different functional specialisation of the two sides of a dog's brain) is clearly apparent in a variety of behaviours (e.g., paw preference, tail wagging, nostril use).
- Minimally invasive techniques of measuring laterality (e.g., tail wagging, eye or nostril preferences to inspect a particular stimulus) could be applied in assessing dogs' emotional state.
- The crucial aspect relevant to dog welfare is that paw preferences may reflect individual differences in behavioural reactivity and cognitive style.
- The study of side biases in social displays in natural conditions (e.g., semi-wild dogs) will be extremely interesting in future research because the direction (left or right) of laterality expressed might reveal more about the displayed behaviour.

# REFERENCES

Angrilli, A., Spironelli, C., Elbert, T., Crow, T.J., Marano, G., Stegagno, L., 2009. Schizophrenia as failure of left hemispheric dominance for the phonological component of language. PLoS One 4 (2), e4507. http://dx.doi.org/10.1371/journal.pone.0004507.

Artelle, K.A., Dumoulin, L.K., Reimchen, T.E., 2011. Behavioral responses of dogs to asymmetrical tail wagging of a robotic dog replica. Laterality 16, 129–135.

Basile, M., Boivin, S., Boutin, A., Blois-Heulin, C., Hausberger, M., Lemasson, A., 2009. Socially dependent auditory laterality in domestic horses (Equus caballus). Anim. Cogn. 12, 611–619.

Batt, L.S., Batt, M.S., Baguley, J.A., McGreevy, P.D., 2009. The relationships between motor lateralization, salivary cortisol concentrations and behavior in dogs. J. Vet. Beh. Clin. Applic. Res. 4, 216–222.

Bisazza, A., Facchin, L., Pignatti, R., Vallortigara, G., 1998. Lateralisation of detour behaviour in poeciliid fishes: the effect of species, gender and sexual motivation. Behav. Brain Res. 91, 157–164.

Blakeslee, S. Published online April 24, 2007. If you want to know if Spot loves you so, it's in his tail. New York Times.

Böye, M., Güntürkün, O., Vauclair, J., 2005. Right ear advantage for conspecific calls in adults and subadults, but not infants, California sea lions (Zalophus californianus): hemispheric specialization for communication? Eur. J. Neurosci. 21, 1727–1732.

Braccini, S., Caine, N.G., 2009. Hand preference predicts reactions to novel foods and predators in marmosets (Callithrix geoffroyi). J. Comp. Psychol. 123, 18–25.

Bradshaw, J.L., Rogers, L.J., 1993. The evolution of lateral asymmetries, language, tool use, and intellect. Academic Press, New York. p. 463.

Branson, N.J., Rogers, L.J., 2006. Relationship between paw preference strength and noise phobia in Canis familiaris. J. Comp. Psychol. 120, 176–183.

Burne, T.H., Rogers, L.J., 2002. Chemosensory input and lateralization of brain function in the domestic chick. Behav. Brain Res. 133, 293–300.

Butler, S., Gilchrist, I.D., Burt, D.M., Perrett, D.I., Jones, E., Harvey, M., 2005. Are the perceptual biases found in chimeric face processing reflected in eye-movement patterns? Neuropsychologia 43, 52–59.

Buxton, D.F., Goodman, D.C., 1967. Motor function and the corticospinal tracts in the dog and raccoon. J. Comp. Neurol. 129, 341–360.

Cameron, R., Rogers, L.J., 1999. Hand preference of the common marmoset, problem solving and responses in a novel setting. J. Comp. Psychol. 113, 149–157.

Canli, T., Desmond, J.E., Zhao, Z., Glover, G., Gabrieli, J.D., 1998. Hemispheric asymmetry for emotional stimuli detected with fMRI. NeuroReport 9, 3233–3239.

Craig, A.D., 2005. Forebrain emotional asymmetry: a neuroanatomical basis? Trends Cognit. Sci. 912, 566–571.

Davidson, R.J., 1995. Cerebral asymmetry, emotion, and affective style. In: Davidson, R.J., Hugdahl, K. (Eds.), Brain asymmetry. MIT Press, Cambridge, MA, pp. 361–387.

De Boyer Des Roches, A., Richard-Yris, M.A., Henry, S., Ezzaouïa, M., Hausberger, M., 2008. Laterality and emotions: visual laterality in the domestic horse (Equus caballus) differs with objects' emotional value. Physiol. Behav. 94, 487–490.

Denny, K., 2009. Handedness and depression: evidence from a large population survey. Laterality 14, 246–255.

Dharmaretnam, M., Rogers, L.J., 2005. Hemispheric specialization and dual processing in strongly versus weakly lateralized chicks. Behav. Brain Res. 162, 62–70.

Fogle, B., 1992. The dog's mind. Pelham Editions, London. p. 203.

Frasnelli, E., Vallortigara, G., Rogers, L.J., 2012. Left-right asymmetries of behaviour and nervous system in invertebrates. Neurosci. Biobehav. Rev. 36, 1273–1291.

Gehr, D.D., Komiya, H., Eggermont, J.J., 2000. Neuronal responses in cat primary auditory cortex to natural and altered species-specific calls. Hear. Res. 150, 27–42.

Gerhardt, H.C., 1981. Mating call recognition in the barking treefrog (*Hyla gratiosa*): responses to synthetic mating calls and comparisons with the green treefrogs (*Hyla cinerea*). J. Comp. Physiol. A. Sensory, Neural, Behav. Physiol 144, 17–25.

Ghazanfar, A.A., Flombaum, J.I., Miller, C.T., Hauser, M.D., 2001. Units of perception in cotton-top tamarin (*Saguinus oedipus*) long calls. J. Comp. Physiol. A. Sensory, Neural, Behav. Physiol. 187, 27–35.

Gorrie, C.A., Waite, P.M.E., Rogers, L.J., 2008. Correlations between hand preference and cortical thickness in the secondary somatosensory (SII) cortex. Behav. Neurosci. 122, 1343–1351.

Guo, K., Meints, K., Hall, C., Hall, S., Mills, D., 2009. Left gaze bias in humans, rhesus monkeys and domestic dogs. Anim. Cogn. 12, 409–418.

Hanbury, D.B., Edens, K.D., Fontenot, M.B., Greer, T.F., McCoy, J.G., Watson, S.L., 2013. Handedness and lateralised tympanic membrane temperature in relation to approach-avoidance behaviour in Garnett's bushbaby (*Otolemur garnettii*). Laterality 18, 120–133.

Hauser, M.D., Agnetta, B., Perez, C., 1998. Orienting asymmetries in rhesus monkeys: the effect of time-domain changes on acoustic perception. Anim. Behav. 56, 41–47.

Hopkins, W.D., Bennett, A., 1994. Handedness and approach-avoidance behaviour in chimpanzees. J. Exp. Psychol. Anim. Behav. Proc. 20, 413–418.

Klar, A.J.S., 1999. Genetic models for handedness, brain lateralization, schizophrenia, and manic-depression. Schizophrenia Res. 39, 207–218.

Le Prell, C.G., Moody, D.B., 2000. Factors influencing the salience of temporal cues in the discrimination of synthetic Japanese monkey (*Macaca fuscata*) coo calls. J. Exp. Psychol. Anim. Behav. Proc. 26, 261–273.

Letzkus, P., Ribi, W.A., Wood, J.T., Zhu, H., Zhang, S.W., Srinivasan, M.V., 2006. Lateralization of olfactory learning in the honey bee *Apis mellifera*. Curr. Biol. 16, 1471–1476.

Lobue, V., DeLoache, J.S., 2008. Detecting the snake in the grass: attention to fear relevant stimuli by adults and young children. Psychological Sci. 19, 284–289.

MacNeilage, P.F. Published online October 23, 2007. Dog tails as tell-tales: the evolution of brain-hemisphere specialization. Sci. Am.

MacNeilage, P.F., Rogers, L.J., Vallortigara, G., 2009. Origins of the left and right brain. Sci. Am. 301, 60–67.

McGreevy, P.D., Rogers, L.J., 2005. Motor and sensory laterality in thoroughbred horses. Appl. Anim. Behav. Sci. 92, 337–352.

Mendl, M., Brooks, J., Basse, C., Burman, O., Paul, E., Blackwell, E., Casey, R., 2010. Dogs showing separation-related behaviour exhibit a 'pessimistic' cognitive bias. Curr. Biol. 12, R839–840.

Miklósi, A., Andrew, R.J., Savage, H., 1998. Behavioural lateralisation of the tetrapod type in the zebrafish (*Brachydanio rerio*). Physiol. Behav. 63, 127–135.

Miller, P.E., Murphy, C.J., 1995. Vision in dogs. J. Am. Vet. Med. Assoc. 207, 1623–1634.

Mitchell, R.L., Crow, T.J., 2005. Right hemisphere language functions and schizophrenia: the forgotten hemisphere. Brain 128, 963–978.

Poremba, A., Malloy, M., Saunders, R.C., Carson, R.E., Herscovitch, P., Mishkin, M., 2004. Species-specific calls evoke asymmetric activity in the monkey's temporal lobes. Nature 427, 448–451.

Quaranta, A., Siniscalchi, M., Albrizio, M., Volpe, S., Buonavoglia, C., Vallortigara, G., 2008. Influence of behavioural lateralisation on interleukin-2 and interleukin-6 gene expression in dogs before and after immunization with rabies vaccine. Behav. Brain Res. 186, 256–260.

Quaranta, A., Siniscalchi, M., Vallortigara, G., 2007. Asymmetric tail-wagging responses by dogs to different emotive stimuli. Curr. Biol. 17, R199–201.

Quaranta, A., Siniscalchi, M., Frate, A., Iacoviello, R., Buonavoglia, C., Vallortigara, G., 2006. Lateralised behaviour and immune response in dogs: Relations between paw preference and Interferon-g, Interleukin-10, and IgG antibodies production. Behav. Brain Res. 166, 236–240.

Quaranta, A., Siniscalchi, M., Frate, A., Vallortigara, G., 2004. Paw preference in dogs: relations between lateralised behaviour and immunity. Behav. Brain Res. 153, 521–525.

Racca, A., Guo, K., Meints, K., Mills, D.S., 2012. Reading faces: differential lateral gaze bias in processing canine and human facial expressions in dogs and 4-year-old children. PLoS One 7 (4), e36076. http://dx.doi.org/10.1371/journal.pone.0036076.

Robins, A., Rogers, L.J., 2004. Lateralised prey catching responses in the toad (Bufo marinus): analysis of complex visual stimuli. Anim. Behav. 68, 567–575.

Rogers, L.J., 2009. Hand and paw preferences in relation to the lateralized brain. Proc. R. Soc. London, Ser. B. 364, 943–954.

Rogers, L.J., 2010. Relevance of brain and behavioural lateralization to animal welfare. Appl. Anim. Behav. Sci. 127, 1–11.

Rogers, L.J., Andrew, R.J., 2002. Comparative vertebrate lateralization. Cambridge University Press, New York. p. 660.

Rogers, L.J., Vallortigara, G., 2008. From antenna to antenna: lateral shift of olfactory memory in honeybees. PLoS One 3, e2340. http://dx.doi.org/10.1371/journal.pone.0002340.

Rogers, L.J., Vallortigara, G., Andrew, R.J., 2013. Divided brains. The biology and behaviour of brain asymmetries. Cambridge University Press, New York. p. 229.

Rogers, L.J., Zucca, P., Vallortigara, G., 2004. Advantages of having a lateralized brain. Proc. R. Soc. London, Ser. B. 271, S420–S422.

Royet, J.P., Plailly, J., 2004. Lateralization of olfactory processes. Chem. Senses 29, 731–745.

Scheumann, M., Zimmermann, E., 2008. Sex-specific asymmetries in communication sound perception are not related to hand preference in an early primate. BMC Biol. 6, 3. http://dx.doi.org/10.1186/1741-7007-6-3.

Schneirla, T., 1959. An evolutionary and developmental theory of biphasic processes underlying approach and withdrawal. In: Jones, M. (Ed.), Nebraska symposium on motivation. University of Nebraska Press, Lincoln, NE, pp. 27–58.

Siniscalchi, M., Bertino, D., Quaranta, A., 2013b. Laterality and performance of agility trained dogs. Laterality. http://dx.doi.org/10.1080/1357650X.2013.794815. Jul 17. [Epub ahead of print] ISSN: 1357-650X.

Siniscalchi, M., Dimatteo, S., Pepe, A.M., Sasso, R., Quaranta, A., 2012b. Visual lateralization in wild striped dolphins (Stenella coeruleoalba) in response to stimuli with different degrees of familiarity. PLoS One 7 (1), e30001. http://dx.doi.org/10.1371/journal.pone.0030001.

Siniscalchi, M., Franchini, D., Pepe, A.M., Sasso, R., Dimatteo, S., Vallortigara, G., Quaranta, A., 2011a. Volumetric assessment of cerebral asymmetries in dogs. Laterality 16, 528–536.

Siniscalchi, M., Lusito, R., Sasso, R., Quaranta, A., 2012a. Are temporal features crucial acoustic cues in dog vocal recognition? Anim. Cogn. 15, 815–821.

Siniscalchi, M., Pergola, G., Quaranta, A., 2013a. Detour behaviour in attack-trained dogs: left-turners perform better than right-turners. Laterality 18, 282–293.

Siniscalchi, M., Quaranta, A., Rogers, L.J., 2008. Hemispheric specialization in dogs for processing different acoustic stimuli. PLoS One 3 (10), e3349. http://dx.doi.org/10.1371/journal.pone.0003349.

Siniscalchi, M., Sasso, R., Pepe, A.M., Dimatteo, S., Vallortigara, G., Quaranta, A., 2011b. Sniffing with right nostril: lateralization of response to odour stimuli by dogs. Anim. Behav. 82, 399–404.

Siniscalchi, M., Sasso, R., Pepe, A.M., Vallortigara, G., Quaranta, A., 2010. Dogs turn left to emotional stimuli. Behav. Brain Res. 208, 516–521.

Sovrano, V.A., 2004. Visual lateralization in response to familiar and unfamiliar stimuli in fish. Behav. Brain Res. 152, 385–391.

Tan, U., Caliskan, S., 1987. Allometry and asymmetry in the dog brain: the right hemisphere is heavier regardless of paw preference. Int. J. Neurosci. 35, 189–194.

Tomkins, L.M., Thomson, P.C., McGreevy, P.D., 2012. Associations between motor, sensory and structural lateralisation and guide dog success. Vet. J. 192, 359–367.

Toppino, T.C., Long, G.M., 2005. Top-down and bottom-up processes in the perception of reversible figures: towards a hybrid model. In: Ohta, N., MacLeod, C.M., Uttl, B. (Eds.), Dynamic Cognitive Processes. Springer, Tokyo, pp. 37–58.

Vallortigara, G., 2000. Comparative neuropsychology of the dual brain: a stroll through left and right animals' perceptual worlds. Brain Lang. 73, 189–219.

Vallortigara, G., Chiandetti, C., Sovrano, V.A., 2011. Brain asymmetry (animal). Wiley Interdiscip. Rev. Cogn. Sci. 2, 146–157. http://dx.doi.org/10.1002/wcs.100.

Vallortigara, G., Regolin, L., Pagni, P., 1999. Detour behaviour, imprinting, and visual lateralization in the domestic chick. Brain Res. Cogn. Brain Res. 7, 307–320.

Ventolini, N., Ferrero, E.A., Sponza, S., Chiesa, A.D., Zucca, P., Vallortigara, G., 2005. Laterality in the wild: preferential hemifield use during predatory and sexual behaviour in the black-winged stilt. Anim. Behav. 69, 1077–1084.

Wells, D.L., 2003. Lateralised behaviour in the domestic dog, Canis familiaris. Behav. Processes 61, 27–35.

Wittling, W., 1990. Psychophysiological correlates of human brain asymmetry: blood pressure changes during lateralized presentation of an emotionally laden film. Neuropsychologia 28, 457–470.

Wittling, W., 1995. Brain asymmetry in the control of autonomic-physiologic activity. In: Davidson, R.J., Hugdahl, K. (Eds.), Brain asymmetry. MIT Press, Cambridge, MA, pp. 305–357.

Wittling, W., Block, A., Schweiger, E., Genzel, S., 1998. Hemisphere asymmetry in sympathetic control of the human myocardium. Brain Cogn. 38, 17–35.

Yin, S., McCowan, B., 2004. Barking in domestic dogs: context specificity and individual identification. Anim. Behav. 68, 343–355.

# Index

# Color Plates

**FIGURE 2-1** Wolves and poodles living in a captive pack established and observed by Dorit Feddersen-Petersen's research group at the University of Kiel, Germany. Very often poodles took the leading positions in the dominance hierarchy, and the wolves readily submitted to them, as can be seen in this picture. Note the ear positions, averted gaze, and back posture of both wolves in this photo. *(Courtesy of Dorit Feddersen-Petersen.)*

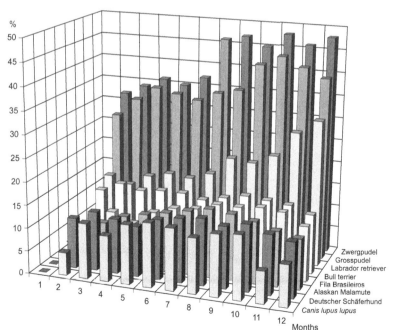

**FIGURE 2-5** Development of aggressive interactions (dyadic, ritualised, and non-ritualised agonistic interactions in proportion of all social interactions) during the first year of European wolves and different dog breeds, all living in groups (Feddersen-Petersen, 2004). *(Reprinted from Feddersen-Petersen, 2004.)*

**FIGURE 3-1** The 'Corridoio pack' on the move. Dogs were often observed moving in a single file. *(Photo by Simona Cafazzo.)*

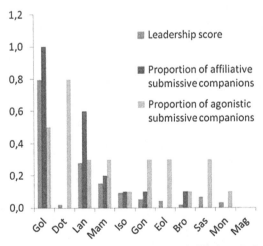

**FIGURE 3-3** Distribution of 'leadership score', 'proportion of affiliative submissive companions', and 'proportion of agonistic submissive companions', respectively, for the members of the Corridoio pack during 2007–2008. Spearman's correlation between 'leadership score' and 'affiliative/agonistic submissive companions': $rs = 0.75$, $P < 0.008$; $rs = 0.47$, $P = 0.14$, respectively. The latter failed to be statistically significant due to the behaviour of Dot. *(Modified from Bonanni et al., 2010a.)*

**FIGURE 4-1** Bentley, a 3-year-old golden retriever mix, gently bites Lela, a 5-month-old German shepherd, on her neck. Lela shows a play face. *(Copyright Barbara Smuts.)*

**FIGURE 4-2** Bentley and Lela rear up. *(Copyright Barbara Smuts.)*

**FIGURE 4-3** During a brief pause in play, Bentley and Lela show play faces. *(Copyright Barbara Smuts.)*

**FIGURE 4-4** Bentley 'defends' against his partner's mock bites. *(Copyright Barbara Smuts.)*

FIGURE 4-5 Role reversal: Lela, the smaller, younger dog, adopts the offensive role by biting Bentley's neck. She plays the offensive role less often than Bentley does (Figure 4-1). *(Copyright Barbara Smuts.)*

FIGURE 4-6 Bentley retracts his lips vertically, puckering his muzzle and revealing his teeth to make a fierce face. Lela's play face shows that she does not think this is a real threat. *(Copyright Barbara Smuts.)*

**FIGURE 4-7** Tex intervenes in dyadic play between Bentley and Lela by mock biting his favorite play partner. *(Copyright Barbara Smuts.)*

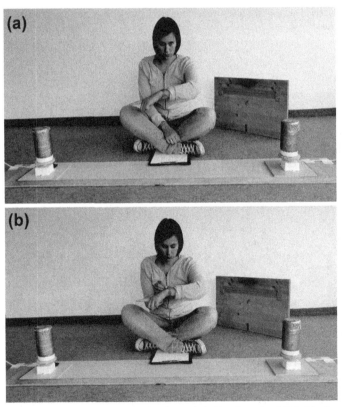

**FIGURES 10-2** Conditions (a) 'Intentional Point' and (b) Non-Intentional Point' in the study of Kaminski et al. (2012).

Printed in the United States
By Bookmasters